T0211602

The Elements of Big Data Value

Edward Curry • Andreas Metzger • Sonja Zillner •
Jean-Christophe Pazzaglia • Ana García Robles
Editors

The Elements of Big Data Value

Foundations of the Research and Innovation Ecosystem

Springer

Editors
Edward Curry
Insight Centre for Data Analytics
National University of Ireland
Galway, Ireland

Andreas Metzger
Paluno
Universität Duisburg-Essen
Essen, Germany

Sonja Zillner
Siemens AG, Munich, Germany

Jean-Christophe Pazzaglia
SAP, Mougins, France

Ana García Robles
Big Data Value Association
Bruxelles, Belgium

ISBN 978-3-030-68178-4   ISBN 978-3-030-68176-0   (eBook)
https://doi.org/10.1007/978-3-030-68176-0

This Springer imprint is published by the registered company Springer Nature Switzerland AG.
The registered company address is: Gewerbestrasse 11, 6330 Cham, Switzerland

Foreword

The global health crisis, growing concerns about the environment and mounting threats in the digital environment are changing our priorities. These threats and problems also come with opportunities and, very often, an important part of the solution to global problems lies in the digital transition, a better sharing of data and responsible, data-driven Artificial Intelligence (AI). Digital platforms have allowed us to keep society functioning in times of confinement. Data-driven AI helps to track infection chains, model disease-spreading patterns and assess the efficiency of alternative disease management options by means of simulation rather than by heavy, slow and expensive trial and error.

Although we have come a long way in terms of increasing the availability of data (especially for open data), there are still many obstacles to the sharing of personal, commercial and industrial data. *Common European data spaces* are a way to systematically eliminate obstacles to data sharing and enable a vibrant economy based on digitalisation and a safe and controlled flow of different kinds of data. Data spaces play a key role in making the world safer, more resilient towards threats and more friendly to the environment. For example, a data space in healthcare will allow an easy, yet safe and compliant, sharing of clinical and patient data to better track and combat diseases, as well as to develop better medicines and vaccines at a faster pace. An environmental data space will allow better models of climate, pollution and other environmental threats to be built. An energy data space will allow us to produce cleaner power efficiently, deliver it when and where it is needed, and reduce energy wastages.

The European Union is supporting the digital transition through its new 7-year framework programmes, Horizon Europe and Digital Europe. They will help create a greener society and economy, more resilience towards threats, and new opportunities for building businesses and prosperity. The Horizon Europe programme will support enabling technologies for secure data spaces, responsible AI and the green transition. The Digital Europe programme will support the actual building, operations and deployment of data spaces, gradually making large-scale, safe data sharing a reality.

Making data work for the economy and society is not only about technology. In order to progressively eliminate the legal, institutional and societal obstacles to data sharing, the European Commission recently proposed a data governance framework to allow the safe, fair and easy sharing of data – in compliance with all applicable legal and ethical requirements. The development of technology and the framework conditions need to be tightly coupled: one is not effective without the other. A broad involvement and constant interaction of businesses, academia, administrations and civil society is necessary to build a data economy that leads to prosperity, growth and jobs. Finally, it is of utmost importance that the whole value chain and computing continuum (cloud-fog-edge-IoT) is addressed when designing data-sharing infrastructures and facilities. This prerequisite is also clearly outlined in the European Strategy for Data, which was published by the European Commission on 19 February 2020.

To respond to these challenges, a structured and broad-based action is required. Until 2020 when it reached the end of its contractual term, the Big Data Value Public-Private Partnership (PPP) was a key instrument in supporting this response. This book and the upcoming PPP Monitoring Report 2019–20 will document an important milestone on the road to the data economy and will set the scene for the new Public-Private Partnership on AI, Data and Robotics, which is currently under preparation. The achievement of a thriving data economy – an ambitious goal set in 2014 when the first partnership was signed – is still a valid goal, and we are a big step closer to it. In the coming years, a much broader involvement of technology areas, research disciplines as well as sectors of business and society will be needed. As the Big Data Value PPP has in its past years of activity excelled in creating bridges to other relevant technology areas – high-performance computing, IoT, cybersecurity, Artificial Intelligence – the future looks particularly promising for the new endeavour, as many paths have already been opened.

DG Communications Networks,   Yvo Volman
Content and Technology,
European Commission,
Brussels, Belgium

Foreword

Artificial Intelligence (AI) is on everyone's lips. Many countries and companies have launched an AI action plan and have undertaken activities for the adoption of AI, from research to deployment. Almost everyone and every sector now realises the huge business potential of AI – a fact underscored by official forecasts, such as the IDC AI Worldwide Spending Guide.

As with any truly disruptive technology, AI also raises concerns. Some of them belong to the realm of science fiction; we are nowhere near having AI algorithms that could mimic "general intelligence". But even with the current state of the art, AI is a transformational technology that is bound to have a few unwanted side effects. Some of them are already well known, such as AI algorithms with a bias against certain individuals due to the way they have been trained, while others are yet to emerge. In his recent book *AI Superpowers* Kai-Fu Lee, former head of Google in China, rightfully acknowledges in his conclusion: "As both the creative and disruptive force of AI is felt across the world, we need to look to each other for support and inspiration".

For all these reasons we should ask ourselves how we will handle this technology – how can we get the most out of it, how can we mitigate risks? Having clear answers to these questions is crucial because the huge potential of AI can only be realised if society not only understands the potential of AI, but also trusts that those who design and implement AI algorithms are fully aware of the risks and know what they do. The difficult adoption of biotechnology in countries like Germany is a painful reminder that this trust is by no means a given and needs to be earned.

The development of AI in Europe thus depends on several critical success factors. One is the obvious need to focus AI-related efforts on domains such as manufacturing, infrastructure, mobility or healthcare, where Europe is already strong and can make a real difference – for Europe's competitiveness, but also in the fight against climate change and other societal challenges. The other is to strongly focus on responsible AI – the art of creating trustworthy AI solutions which are designed against transparent objectives in accordance with European values and implemented to reliably deliver on these objectives. This dual focus on industrial domain know-how and European values is key to making "AI made in Europe" a success story.

In this endeavour, speed is essential. AI can shift the balance of power from incumbents to newcomers almost overnight. In the race for industrial AI, Europe's strong domain know-how, embedded in world-class universities and research institutes, in a strong network of innovative small and medium-sized enterprises (SMEs), in world-leading suppliers of electrical and industrial equipment as well as industrial software, gives Europe a considerable head start. However, this head start is only temporary, and Europe is well-advised not to squander it. Fast-track programmes to exploit the opportunities offered by industrial AI are needed, the sooner the better. Europe also needs to get serious with the "better regulation" initiative and take bold steps to create a regulatory environment for AI-driven innovations to take root. Responsible AI is best developed and proven in practical projects, not in ethics councils. If needed, regulatory sandboxes, which have yet to be introduced at EU level, can be used to strike the right balance between innovative spirit and regulatory caution.

Last but not least, collaboration in ecosystems is indispensable in making Europe the pacemaker for industrial AI. Efforts by the European Public-Private Partnership on Big Data Value to establish a Data Innovation Ecosystem in Europe are exactly the right approach. Only through the sharing and joint exploitation of data, but without disregard for companies' obligation to return a profit to their shareholders, can we power a value-focused data-driven transformation of Europe's business and society. Most importantly, the Partnership acts as a hub for the European data community – researchers, entrepreneurs, businesses and citizens – to collaborate with one another across all the member states. Europe's wellbeing depends on a productive and effective data innovation ecosystem which positions Europe as a front runner in artificial intelligence.

Siemens,   Peter Körte
Berlin, Germany
March 2021

Foreword

Data is the defining characteristic of the twenty-first century, its importance such that it is often referred to as the "new oil". The ability to refine this resource, i.e. the ability to extract value from raw data through data analytics and artificial intelligence, is having a transformative effect on society, driving scientific breakthroughs and empowering citizens to create a smarter, better world.

Collaboration between researchers, industry and society to derive value from big data through data-driven innovations that enable better decision-making has been the driving force behind this transformation. Europe has been a leader in value-driven transformation through the Big Data Value PPP and the Big Data Value Association. This community has acted as the nucleus of the European data community to bring together businesses with leading researchers from across Europe to harness the value of data to benefit society, business, science and industry. As one of the largest research centres of its kind in Europe, the Insight SFI Research Centre for Data Analytics is proud to be at the heart of this community. In turn, we as a centre have significantly benefited from the openness of the European ecosystem and are committed to continue to invest in its collective endeavour to transform European society.

The book you are holding describes in detail the foundational "elements" needed to deliver value from big data. It clearly defines the enablers needed to grow data ecosystems, including technical research and innovation, business, skills, policy and societal elements. The book charts pathways to new value creation and new opportunities from big data. Decision-makers, policy advisors, researchers and practitioners at every level will benefit.

Insight
SFI Research Centre for Data Analytics,
Dublin, Ireland
March 2021

Noel O'Connor

Preface

Making use of technology to utilise and leverage resources has been a constant feature of human history. Advances in science moved humans from invention to reasoned invention, where a more sophisticated understanding of the elements led to an increased capacity to utilise their unique characteristics to drive the industrial and technological revolutions of the eighteenth, nineteenth and twentieth centuries. Scientists and inventors were the explorers who helped us to understand the world. Many scientists helped to develop the periodic system and the periodic table for classifying chemical elements by atomic mass. The first table had 63 elements, but the originators anticipated the discovery of more elements and left spaces in the table for them. Today the modern periodic table contains 118 elements and reflects the collective scientific endeavours of a community for over two centuries to understand the chemical and physical properties of the elements that make up the physical world and its natural ecosystems.

Today we live in the Information Age where our society, through reasoned invention, has created a new world beyond the physical one. This new world is a virtual world which contains a data ecosystem with information on every aspect of our society and the physical world. Today's researchers and inventors are investigating this virtual world to understand its elements and data ecosystems which drive the digital revolution of the twenty-first century. The virtual world keeps expanding as we continue the digital transformation of industry and society. The growth of data poses a continual challenge to devise new data management and processing capabilities to keep pace with the ever-increasing data resource. The ability to harness the value of this data is critical for society, business, science and industry. This challenge requires a collective effort from multiple different disciplines and society at large.

This book reports on such a collective effort undertaken by the European data community to understand the elements of data and to develop an increased capacity to exploit its unique characteristics to drive digital transformations through a process of sense-making and knowledge creation. The community had a firm conviction to focus on the value of data by analysing it for insights into decision-making and actions which can improve outcomes for individuals, organisations and society. The community identified the need to look holistically at data-driven innovation and consider the full spectrum of challenges from data to skills, legal, technical, application, business and social. The community gave rise to the Big Data Value Association as its home to pursue this mission.

The purpose of this book is to capture the initial discoveries of this community, providing the first set of *Elements of Big Data Value*. These elements provide readers of the book with insights on research and innovation roadmaps, technical architectures, business models, regulation, policy, skills and best practices which can support them in creating data-driven solutions, organisations and productive data ecosystems. The book is of interest to three primary audiences. First, researchers and students in the big data field and associated disciplines, e.g. computer science, information technology and information systems, among others. Second, industrial practitioners, who will find practical recommendations based on rigorous studies that contain insights and guidance in the area of big data across several technology and management areas. Third, the book will support policymakers and decision drivers at local, national and international level who aim to establish or nurture their data ecosystems.

This book arranges the elements into four groupings containing elements focusing on similar behaviours needed for big data value covering (1) ecosystem, (2) research and innovation, (3) business, policy and societal and (4) emerging elements.

Part I: Ecosystem Elements of Big Data Value focuses on establishing the big data value ecosystem using a holistic approach to make it healthy, vibrant and valuable to its stakeholders. The first chapter explores the opportunity to increase the competitiveness of European industries through a data ecosystem by tackling the fundamental elements of big data value. The second chapter discusses a stakeholder analysis concerning data ecosystems and stakeholder relationships within and between different industrial and societal case studies. A roadmap to drive adoption of data ecosystems is described in the third chapter, addressing a wide range of challenges from access to data and infrastructure, to technical barriers, skills, and policy and regulation. The fourth chapter details the impact of the Big Data Value Public-Private Partnership, which plays a central role in the implementation of the European data economy. The chapter provides an overview of the partnership and its objectives, together with an in-depth analysis of the impact of the PPP.

Part II: Research and Innovation Elements of Big Data Value details the key technical and capability challenges which must be addressed to deliver big data value. The fifth chapter details the technical priorities for big data value, covering key aspects such as real-time analytics, low latency and scalability in processing data, new and rich user interfaces, interacting with and linking data, information and content. The Big Data Value Reference Model is described in the sixth chapter, which has been developed with input from technical experts and stakeholders along the whole big data value chain. Data Protection and Data Technologies is the focus of the seventh chapter, where advances in privacy-preserving technologies are aimed at building privacy-by-design from the start into the back-end and front-end of digital services. The eighth chapter presents a best practice framework for Centres of Excellence for Big Data and AI. The ninth chapter describes the European Innovation Spaces which ensure that research on big data technologies and novel applications can be quickly tested, piloted and leveraged for the maximum benefit of all the stakeholders.

Part III: Business, Policy and Societal Elements of Big Data Value investigates the need to make more efficient use of big data and understand that data is an asset that has significant potential for the economy and society. The tenth chapter provides a collection of stories showing concrete examples of the value created thanks to big data value technologies. The eleventh chapter explores new data-driven business models as ways to generate value for companies along with the value chain and in different sectors. The Data-Driven Innovation (DDI) Framework is introduced in the twelfth chapter to support the process of identifying and scoping big data value. The thirteenth chapter covers the data skills challenge to ensure the availability of rightly skilled people who have an excellent grasp of the best practices and technologies for delivering big data value solutions. The critical topic of standards within the area of big data is the focus of the fourteenth chapter. The fifteenth chapter engages in the debate on data ownership and usage, data protection and privacy, security, liability, cybercrime and Intellectual Property Rights (IPR).

Part IV: Emerging Elements of Big Data Value explores the critical elements to maximising the future potential of big data value. The sixteenth chapter details the *European AI, Data and Robotics Framework* and its tremendous potential to benefit citizens, economy and society. The chapter also describes common European data spaces which can ensure that more data becomes available for use in the economy and society while keeping companies and individuals who generate the data in control.

With its origins tracing back over 200 years, the periodic table has been disputed, altered and improved as science has progressed, and new elements have been discovered. Today it is a vital tool for modern chemists and hangs on the wall of almost every classroom and lecture hall in the world. As society learns how to leverage and derive more value from data, we expect the elements of big data value to be challenged and to evolve as new elements are discovered. Just as the originators of the periodic table left room for new elements, The Periodic Table of the Elements of Big Data Value is open, and we invite you to be part of the evolution of this collective endeavour to explore, understand and extract value from the data resources of the Information Age.

Galway, Ireland   Edward Curry
March 2021

The original version of the front matter was revised: Affiliation for the author Davide Dalle Carbonare in the contributor section has been updated. The correction is available at https://doi.org/10.1007/978-3-030-68176-0_17

Acknowledgements

The editors and the chapter authors acknowledge the support, openness and collaborative atmosphere of the big data value community who contributed to this book in ways both big and small. Over the years, the community has produced a number of documents and white papers, including the Strategic Research and Innovation Agenda, which have formed the basis for several chapters in this book. We greatly acknowledge the collective effort of these contributors, including Antonio Alfaro, Jesus Angel García, Rosa Araujo, Sören Auer, Paolo Bellavista, Arne Berre, Freek Bomhof, Nozha Boujemaa, Stuart Campbell, Geraud Canet, Giuseppa Caruso, Alberto Crespo Garcia, Paul Czech, Stefano de Panfilis, Thomas Delavallade, Marija Despenic, Roberto Díaz Morales, Ivo Emanuilov, Ariel Farkash, Antoine Garnier, Wolfgang Gerteis, Aris Gkoulalas-Divanis, Nuria Gomez, Paolo Gonzales, Tatjana Gornosttaja, Thomas Hahn, Souleiman Hasan, Carlos Iglesias, Martin Kaltenböck, Bjarne Kjær Ersbøll, Yiannis Kompatasiaris, Paul Koster, Bas Kotterink, Antonio Kung, Oscar Lazaro, Yannick Legré, Giovanni Livraga, Yves Mabiala, Julie Marguerite, Ernestina Menasalves, Andreas Metzger, Elisa Molino, Thierry Nagellen, Dalit Naor, Angel Navia Vázquez, Axel Ngongo, Melek Önen, Ángel Palomares, Symeon Papadopoulos, Maria Perez, Juan-Carlos Perez-Cortes, Milan Petkovic, Roberta Piscitelli, Klaus-Dieter Platte, Pierre Pleven, Dumitru Roman, Titi Roman, Alexandra Rosén, Zoheir Sabeur, Nikos Sarris, Stefano Scamuzzo, Simon Scerri, Corinna Schulze, Bjørn Skjellaug, Cai Södergard, Francois Troussier, Colin Upstill, Josef Urban, Andrejs Vasiljevs, Meilof Veeningen, Tonny Velin, Akrivi Vivian Kiousi, Ray Walshe, Walter Waterfeld and Stefan Wrobel.

The editors thank Dhaval Salwala for his support in the preparation of the final manuscript. Thanks also go to Ralf Gerstner and all at Springer for their professionalism and assistance throughout the journey of this book. This book was made possible through funding from the European Union's Horizon 2020 research and innovation programme under grant agreement no. 732630 (BDVe).

We would like to thank our partners at the European Commission, in particular Commissioner Gabriel, Commissioner Kroes and the Director-General of DG CONNECT Roberto Viola who had the vision and conviction to develop the European data economy. Finally, we thank the current and past members of the European Commission's Unit for Data Policy and Innovation (Unit G.1) Yvo Volman, Márta Nagy-Rothengass, Kimmo Rossi, Beatrice Covassi, Stefano Bertolo, Francesco Barbato, Wolfgang Treinen, Federico Milani, Daniele Rizzi and Malte Beyer-Katzenberger. Together they have represented the public side of the big data partnership and were instrumental in its success.

Galway, Ireland   Edward Curry
Essen, Germany   Andreas Metzger
Munich, Germany   Sonja Zillner
Mougins, France   Jean-Christophe Pazzaglia
Bruxelles, Belgium   Ana García Robles
March 2021

Contents

Editors and Contributors

About the Editors

Edward Curry obtained his doctorate in Computer Science from NUI Galway in 2006. From 2006 to 2009 he worked as a postdoctoral researcher at the Digital Enterprise Research Institute (DERI). Currently, he holds a Research Lectureship at the Data Science Institute at NUI Galway, leads a research unit on Open Distributed Systems, and is a member of the Executive Management Team of the Institute. Edward has made substantial contributions to semantic technologies, incremental data management, event processing middleware, software engineering, as well as distributed systems and information systems. He combines strong theoretical results with high-impact practical applications. Edward is author/co-author of over 180 peer-reviewed scientific publications. The excellence and impact of his research have been acknowledged by numerous awards including best paper award and the NUIG President's Award for Societal Impact in 2017. The technology Edward develops with his team fuels many industrial applications, such as the energy, water and mobility management at Schneider Electric, Intel, DELL Technologies and Linate Airport. He is organiser and programme co-chair of renowned conferences and workshops, including CIKM 2020, AICS 2019, ECML 2018, IEEE BigData Congress and the European Big Data Value Forum. Edward is co-founder and elected Vice President of the Big Data Value Association, an industry-led European big data community, and has built consensus on a joint European big data research and innovation agenda and influenced European data innovation policy to deliver on the agenda.

Andreas Metzger received his Ph.D. in Computer Science (Dr.-Ing.) from the University of Kaiserslautern in 2004. He is a senior academic councilor at the University of Duisburg-Essen and heads the Adaptive Systems and Big Data Applications group at paluno, the Ruhr Institute for Software Technology. His background and research interests are software engineering and machine learning for adaptive systems. He has co-authored over 120 papers, articles and book chapters. His recent research on deep learning for proactive process adaptation received the Business Process Innovation Award at the International Conference on Business Process Management. He is co-organiser of over 15 international workshops and conference tracks, and programme committee member for numerous international conferences. Andreas was Technical Coordinator of the European lighthouse project TransformingTransport, which demonstrated in a realistic, measurable and replicable way the transformations that big data and machine learning can bring to the mobility and logistics sector. In addition, he was a member of the Big Data Expert Group of PICASSO, an EU-US collaboration action on ICT topics. Andreas serves as steering committee vice chair of NESSI, the European Technology Platform dedicated to Software, Services and Data, and as deputy secretary general of the Big Data Value Association.

Sonja Zillner studied mathematics and psychology at the Albert-Ludwigs-University Freiburg, Germany, and received her PhD in computer science specialising in the topic of Semantics at Technical University in Vienna. Since 2005 she has been working at Siemens AG, Corporate Technology as a key expert focusing on the definition, acquisition and management of global innovation and research projects in the domain of semantics and artificial intelligence. Since 2020 she has been Lead of the Core Company Technology Module "Trustworthy AI" at Siemens Corporate Technology. Previously, from 2016 to 2019 she was invited to consult the Siemens Advisory Board in strategic decisions regarding artificial intelligence. She is chief editor of the Strategic Research Innovation and Deployment Agenda of the new Partnership in AI, Data and Robotics, leading editor of the Strategic Research and Innovation Agenda of the Big Data Value Association (BDVA), and member of the editing team of the strategic agenda of the European On-Demand Platform AI4EU. Between 2012 and 2018 she was a professor at Steinbeis University in Berlin, between 2017 and 2018 she was a guest professor at the Technical University of Berlin and since 2016 she has been a lecturer at Technical University of Munich. She is author of more than 80 publications and more than 25 patents in the area of semantics, artificial intelligence and data-driven innovation.

Jean-Christophe Pazzaglia studied informatics and received his engineering degree from *Ecole Superieure en Sciences Informatiques* (now Polytech) of the University of Nice (1992). He completed a Ph.D. on the usage of behavioural reflection in the CNRS laboratory I3S (1997). He graduated from the Essentials of Management programme of the University of St Gallen (2009). Jean-Christophe is a Design Thinking coach. He initially worked on AI – multi-agent systems, neural networks and reflexive languages – and later embraced the field of ICT Security and Privacy. After 8 years working abroad, he returned to the South of France and since 2006 he has been working for SAP. Former director of the SAP Research Center Sophia Antipolis, he was the principal investigator for SAP of several European and

French research projects. Today, he is Chief Support Architect Higher Education & Research and is supporting SAP involvement in the BDVA, managing the Big Data Value ecosystem project while also leading the pilot AI4Citizen in the AI4EU project. In a complementary role, within SAP University Alliance, he is giving lectures on SAP Technologies and Design Thinking workshops. He also enjoys teaching Scratch to children (Europe/Africa Code week) and co-developed the OpenSAP lecture on Scratch for teenagers within the SAP Corporate Social Responsibility initiative.

Ana García Robles is Secretary General of the Big Data Value Association (BDVA) and holds a Master's Degree in Telecommunications Engineering and an International Executive MBA. Ana has a strong ICT industrial background in the telecommunications sector, with over 10 years' experience in the design, implementation and configuration of large-scale telecom networks and services, and in the research and techno-economical assessment of new technologies and solutions for large-scale implementation. Ana has specialised in innovation management and ecosystems and has extensive experience at both local/regional and international level in open innovation ecosystems, Living Labs, and socio-economic impacts of technology, with over 5 years' experience managing international associations and projects in this area. Ana has participated in multiple research and innovation collaborative projects and programmes in the areas of smart cities and urban innovation, open and big data, IoT, open platforms, digital social innovation, e-health, digital cultural heritage, ICT for education, ICT for food and intelligent mobility. She is a speaker at conferences, an inventor, and a contributor to various research papers and publications in the field of smart cities and innovation ecosystems.

Contributors

Daniel Alonso ITI, Valencia, Spain

Sören Auer Leibniz Universität Hannover, Hannover, Germany

Martina Barbero Big Data Value Association, Bruxelles, Belgium

Arne J. Berre SINTEF Digital, Oslo, Norway

Alessandra Boggio-Marzet Universidad Politécnica de Madrid, Madrid, Spain

Södergård Caj VTT, Espoo, Finland

Davide Dalle Carbonare Engineering Ingegneria Informatica, Rome, Italy

Gabriella Cattaneo IDC, Milan, Italy

Edward Curry Insight SFI Research Centre for Data Analytics, NUI, Galway, Ireland

Nuria De Lama Atos, Madrid, Spain

Marija Despenic ABN AMRO Bank, Amsterdam, the Netherlands

Wolfgang Gerteis SAP, Walldorf, Germany

Jon Ander Gomez Universitat Politècnica de València, València, Spain

Thomas Hahn Siemens AG, Erlangen, Germany

Souleiman Hasan Insight SFI Research Centre for Data Analytics, NUI Galway, Galway, Ireland

Marissa Hoekstra Strategy, Analysis & Policy Department, TNO, The Hague, The Netherlands

Jim Kenneally Intel, Leixlip, Ireland

Laure Le Bars SAP, Paris, France

Zoltan Mann paluno, University of Duisburg-Essen, Essen, Germany

Dirk Mayer Software AG, Saarbrücken, Germany

Ernestina Menasalvas Universidad Politécnica de Madrid, Madrid, Spain

Andreas Metzger paluno, University of Duisburg-Essen, Essen, Germany

Andrés Monzón Universidad Politécnica de Madrid, Madrid, Spain

Ana Moreno Universidad Politécnica de Madrid, Madrid, Spain

Adegboyega Ojo Insight SFI Research Centre for Data Analytics, NUI Galway, Galway, Ireland

Edo Osagie Insight SFI Research Centre for Data Analytics, NUI Galway, Galway, Ireland

Niki Pavlopoulou Insight SFI Research Centre for Data Analytics, NUI, Galway, Ireland

Jean-Christophe Pazzaglia SAP, Mougins, France

Milan Petkovic Philips and Eindhoven University of Technology, Eindhoven, The Netherlands

Ana García Robles Big Data Value Association, Bruxelles, Belgium

Dumitru Roman SINTEF Digital, Oslo, Norway

Aristide Rothweiler paluno, University of Duisburg-Essen, Essen, Germany

Dhaval Salwala Insight SFI Research Centre for Data Analytics, NUI Galway, Galway, Ireland

Simon Scerri Fraunhofer IAIS, Sankt Augustin, Germany

Robert Seidl Nokia Bell Labs, Munich, Germany

Nik Swoboda Universidad Politécnica de Madrid, Madrid, Spain

Tjerk Timan Strategy, Analysis & Policy Department, TNO, The Hague, The Netherlands

Marie Claire Tonna Digital Catapult, London, UK

Umair Ul Hassan Insight SFI Research Centre for Data Analytics, NUI Galway, Galway, Ireland

Charlotte van Oirsouw Tilburg University, Tilburg, The Netherlands

Ray Walshe ADAPT SFI Centre for Digital Content, Dublin City University, Dublin, Ireland

Walter Waterfeld Saarbrücken, Germany

Sonja Zillner Siemens AG, Munich, Germany

Part I
Ecosystem Elements of Big Data Value

The European Big Data Value Ecosystem

Edward Curry, Andreas Metzger, Sonja Zillner, Jean-Christophe Pazzaglia, Ana García Robles, Thomas Hahn, Laure Le Bars, Milan Petkovic, and Nuria De Lama

Abstract The adoption of big data technology within industrial sectors facilitates organizations to gain competitive advantage. The impacts of big data go beyond the commercial world, creating significant societal impact, from improving healthcare systems to the energy-efficient operation of cities and transportation infrastructure, to increasing the transparency and efficiency of public administration. In order to exploit the potential of big data to create value for society, citizens and businesses, Europe needs to embrace new technology, applications, use cases and business models within and across various sectors and domains. In the early part of the 2010s, a clear strategy centring around the notion of the European Big Data Value Ecosystem started to take form with the aim of increasing the competitiveness of European industries through a data ecosystem which tackles the fundamental elements of big data value, including the ecosystem, research and innovation, business,

E. Curry (✉)
Insight SFI Research Centre for Data Analytics, NUI, Galway, Ireland
e-mail: edward.curry@nuigalway.ie

A. Metzger
paluno, University of Duisburg-Essen, Duisburg, Germany

S. Zillner
Siemens AG, Munich, Germany

J.-C. Pazzaglia
SAP, Mougins, France

A. García Robles
Big Data Value Association, Bruxelles, Belgium

T. Hahn
Siemens AG, Erlangen, Germany

L. Le Bars
SAP, Paris, France

M. Petkovic
Philips and Eindhoven University of Technology, Eindhoven, the Netherlands

N. De Lama
Atos, Madrid, Spain

© The Author(s) 2021
E. Curry et al. (eds.), *The Elements of Big Data Value*,
https://doi.org/10.1007/978-3-030-68176-0_1

3

policy and regulation, and the emerging elements of data-driven AI and common European data spaces. This chapter describes the big data value ecosystem and its strategic importance. It details the challenges of creating this ecosystem and outlines the vision and strategy of the Big Data Value Public-Private Partnership and the Big Data Value Association, which together formed the core of the ecosystem, to make Europe the world leader in the creation of big data value. Finally, it details the elements of big data value which were addressed to realise this vision.

Keywords Data ecosystem · Big Data Value · Data innovation

1 Introduction

For many businesses and governments in different parts of the world, the ability to effectively manage information and extract knowledge is now seen as a critical competitive advantage, and many organisations are building their core business on their ability to collect and analyse information, to extract business knowledge and insight (Cavanillas et al. 2016a). The capability to meaningfully process and analyse large volumes of data (big data) constitutes an essential resource for driving value creation, fostering new products, processes and markets and enabling the creation of new knowledge (OECD 2014). The adoption of big data technology within indus-trial sectors facilitates organisations in gaining competitive advantage. The impacts of big data go beyond the commercial world, creating significant societal impact, from improving healthcare systems to the energy-efficient operation of cities and transportation infrastructure, to increasing the transparency and efficiency of public administration.

Europe must exploit the potential of big data to create value for society, citizens and businesses. Europe needs to embrace new technology, applications, use cases and business models within and across various sectors and domains (Cavanillas et al. 2016b). A clear strategy was needed to increase the competitiveness of European industries through a data ecosystem which tackled the fundamental elements of big data value, including the ecosystem, research and innovation, business, policy and regulation, and the emerging elements of data-driven AI and common European data spaces. This chapter describes the notion of big data value and its strategic impor-tance. It details the challenges of creating a European Big Data Value Ecosystem, and outlines the vision and strategy of the Big Data Value Public-Private Partnership (BDV PPP) to make Europe competitive in data technologies and the extraction of value from data. Finally, it details the elements of big data value which were addressed to realise this vision.

In what follows, Sect. 2 aims to define the notion of big data value. Section 3 elaborates on the strategic importance of big data value for Europe. Section 4 summarises the process that was followed in developing a European big data value ecosystem. Section 5 drills down into the different elements of this ecosystem, along which the remaining chapters of this book are structured.

2 What Is Big Data Value?

In recent years, the term "big data" has been used by various major players to label data with different attributes (Hey et al. 2009; Davenport et al. 2012). Several definitions of big data have been proposed over the last decade (see Table 1).

Big data brings together a set of data management challenges for working with data under new scales of size and complexity. Many of these challenges are not new. What is new are the challenges raised by the specific characteristics of big data related to the 3 Vs:

- **Volume (amount of data):** dealing with large scales of data within data processing (e.g. Global Supply Chains, Global Financial Analysis, Large Hadron Collider).
- **Velocity (speed of data):** dealing with streams of high-frequency incoming real-time data (e.g. Sensors, Pervasive Environments, Electronic Trading, Internet of Things).
- **Variety (range of data types/sources):** dealing with data using differing syntactic formats (e.g. Spreadsheets, XML, DBMS), schemas and meanings (e.g. Enterprise Data Integration).

The 3 Vs of big data challenge the fundamentals of existing technical approaches and require new forms of data processing to enable enhanced decision-making, insight discovery, and process optimization. As the big data field has matured, other Vs have been added, such as Veracity (documenting quality and uncertainty) and Value (Rayport and Sviokla 1995; Biehn 2013). The definition of Value within

Table 1 Definitions of big data (Curry 2016)

Big data definition	Source
"Big data is high volume, high velocity, and/or high variety information assets that require new forms of processing to enable enhanced decision making, insight discovery and process optimization."	Laney (2001), Manyika et al. (2011)
"When the size of the data itself becomes part of the problem and traditional techniques for working with data run out of steam."	Loukides (2010)
Big data is "data whose size forces us to look beyond the tried-and-true methods that are prevalent at that time."	Jacobs (2009)
"Big data is a field that treats ways to analyse, systematically extract information from, or otherwise deal with data sets that are too large or complex to be dealt with by traditional data-processing application software."	Wikipedia (2020)
"Big Data is a term encompassing the use of techniques to capture, process, analyse and visualize potentially large datasets in a reasonable timeframe not accessible to standard IT technologies. By extension, the platform, tools and software used for this purpose are collectively called 'Big Data technologies.'"	NESSI (2012)
"Big data can mean big volume, big velocity, or big variety."	Stonebraker (2012)

Table 2 Definitions of big data value

Big data value definition	Source
"Top-performing organizations use analytics five times more than lower performers...a widespread belief that analytics offers value."	Lavalle et al. (2011)
"The value of big data isn't the data. It's the narrative."	Hammond (2013))
"Companies need a strategic plan for collecting and organizing data, one that aligns with the business strategy of how they will use that data to create value."	Wegener and Velu (2013)
"We define prescriptive, needle-moving actions and behaviors and start to tap into the fifth V from Big Data: value."	Biehn (2013)
"Data value chain recognizes the relationship between stages, from raw data to decision making, and how these stages are interdependent."	Miller and Mork (2013)

the context of big data also varies. Table 2 lists a few of those definitions, which clearly show a pattern of common understanding that the Value dimension of big data resets upon successful decision-making through analytics. The value of big data can be described in the context of the dynamics of knowledge-based organizations, where the processes of decision-making and organizational action are dependent on the process of sense-making and knowledge creation (Choo 1996).

3 Strategic Importance of Big Data Value

Economic and social activities have long relied on data. But the increased volume, velocity, variety and social and economic value of data signals a paradigm shift towards a data-driven socio-economic model. The significance of data is continuing to grow in importance as it is used to make critical decisions in our everyday lives, from the course of treatment for a critical illness to safely driving a car. The exploitation of big data in various sectors has already had a significant socio-economic impact. According to International Data Corporation (IDC),[1] the global investment in AI and Big Data is projected to reach 86.6 billion euro worldwide in 2023, whereas the European share of industrial investments for this market is estimated at 18.8 billion euro. Since 2017 "Developing the European Data Economy" (Economy 2017) has been one of the new pillars of the extended European Digital Single Market strategy designed to keep up with emerging trends and challenges. It focuses on defining and implementing the framework conditions for a European Data Economy, ensuring a fair, open and secure digital environment. The main focus was on ensuring the effective and reliable cross-border flow of non-personal data, and access to and reuse of such data, as well as looking at the challenges to the safety and liabilities posed by the Internet of Things (IoT).

---

[1]For this analysis of the AI and Data sector we are using data from the Worldwide Semiannual Artificial Intelligence Systems Spending Guide 2018.

Large companies and SMEs in Europe see the real potential of big data value in causing disruptive change in markets and business models. Companies intending to build and rely on data-driven solutions appear to have begun to fruitfully address challenges that extend well beyond technology usage. The successful adoption of big data requires changes in business orientation and strategy, processes, procedures and organisational set-up. European enterprises are creating new knowledge and are starting to hire new experts, enhancing a new ecosystem.

In 2020 the EC renewed its Data strategy (*Communication: A European strategy for data* 2020) and identified Data as an essential resource for economic growth, competitiveness, innovation, job creation and societal progress. A critical driver for the emerging AI business opportunities is the significant growth of data volume and the rates at which data is generated. By 2025, there will be more than 175 zettabytes of data),[2] reflecting a fivefold growth of data from 2018 to 2025. At the same time, we see a shift of data to the Edge. In 2020, 80% of processing and analysis takes place within data centres, and the move is on to process more data at the Edge of the network in smart connected devices and machines. This creates new opportunities for Europe to lead this form of data processing and for European actors to maintain and control the processing of their data. As EU Commissioner Thierry Breton stated, *"**My goal is to prepare ourselves so the data produced by Europeans will be used for Europeans, and with our European values**."*

Data enables AI innovation, and AI makes data actionable. Data flows link together the emerging value chains disrupted by new AI services and tools, where new skills, business models and infrastructures are needed. The data governance models and issues such as data access, data sovereignty and data protection are an essential factor in the development of sustainable AI-driven value chains respecting all stakeholder interests, particularly SMEs, who are currently lagging in AI adoption.

AI innovation can generate value not only for business but also for society and individuals. There is increasing attention to AI's potential for social good, for example contributing to achieving the UN's sustainable development goals and the environmental goals of the EU Green Deal, and fighting against COVID-19 (Coronavirus disease) and other pandemics (Vaishya et al. 2020). Enterprises are developing sustainability programmes in the context of their CSR strategies, leveraging data and AI to reduce their environmental footprint, cutting costs and contributing to social welfare at the same time. Business and social value can be pursued at the same time, encouraging the reuse and sharing of data collected and processed for AI innovation (sharing private data for the public good, Business to Government (B2G) and not only Business to Business (B2B)). Expertise is needed to increase awareness about the potential value for society and people, as well as the business of data-driven innovation combined with AI, and to use this assessment to prioritise public funding.

---

[2]Vernon Turner, John F. Gantz, David Reinsel and Stephen Minton, *The digital universe of opportunities: rich data and the increasing value of the Internet of Things*, Report from IDC for EMC April 2014.

For the European Data Economy to develop further and meet expectations, large volumes of cross-sectoral, unbiased, high-quality and trustworthy data need to be made available. There are, however, important business, organisational and legal constraints that can hinder this scenario, such as the lack of motivation to share data due to ownership concerns, loss of control, lack of trust, the lack of foresight in not understanding the value of data or its sharing potential, the lack of data valuation standards in marketplaces, the legal blocks to the free flow of data and the uncertainty around data policies. The exploration of ethical, secure and trustworthy legal, regulatory and governance frameworks is needed. European values, e.g. democracy, privacy safeguards and equal opportunities, can become the trademark of European Data Economy technologies, products and practices. Rather than be seen as restrictive, legislation enforcing these values should be considered as a unique competitive advantage in the global data marketplace.

4   Developing a European Big Data Value Ecosystem

A Data Ecosystem is a socio-technical system enabling value to be extracted from data value chains supported by interacting organizations and individuals. Within an ecosystem, data value chains can be oriented to business and societal purposes. The ecosystem can create the conditions for a marketplace competition between participants or enable collaboration among diverse, interconnected participants that depend on each other for their mutual benefit. Data Ecosystems can be formed in different ways around an organisation or community technology platforms, or within or across sectors (Curry 2016).

Creating a European data ecosystem would "bring together data owners, data analytics companies, skilled data professionals, cloud service providers, companies from the user industries, venture capitalists, entrepreneurs, research institutes and universities" (DG Connect 2013). However, in the early 2010s, there was no coherent data ecosystem at the European level (DG Connect 2013), and Europe was lagging behind in the adoption of big data. To drive innovation and competitiveness, Europe needed to foster the development and broad adoption of data technologies, value-adding use cases and sustainable business models. There were significant challenges to overcome.

4.1   Challenges

To understand the difficulties that existed in establishing a European data ecosystem, it is useful to look at the multiple challenges (Cavanillas et al. 2016a) that needed to be overcome:

- *Low rates of big data adoption:* The European industry was lagging in the adoption of big data solutions. Many businesses and NGOs were uncertain of how to apply the technology within their operations, what the return on investment would be and how to deal with non-technical issues such as data privacy.
- *A disconnection between data owners and data innovators:* Many data owners (often large organisations) possessed large datasets, but they could not fully utilize big data's innovation potential. Data entrepreneurs and innovators (often SMEs and researchers) had vital insights on how to extract the value but lacked access to the data to prove their innovation. This mismatch created an impasse which needed to be overcome if innovation was to flourish.
- *Lack of technical and non-technical big data skills:* A key challenge for Europe was the provision of appropriately skilled people who had an excellent grasp of the best practices and technologies for delivering big data solutions. There was a shortage of data scientists and engineers who had expertise in analytics, statistics, machine learning, data mining and data management. Strong domain knowledge of how to apply big data know-how within organisations to create value was and still is a critical but rare skill.
- *Next-generation technologies:* US organizations had mainly driven the first generation of big data technology. It was essential to develop European leadership in the next generation of big data technology. Leadership in this space was critical for job creation and prosperity by creating a European-wide competency in technology and applications.

A thriving data ecosystem would need to overcome these challenges and bring together the ecosystem stakeholders to create new business opportunities, more access to knowledge and benefits for society. For Europe to seize this opportunity, action was needed.

4.2 A Call for Action

Big data offers tremendous untapped potential value for many sectors, however, there was no coherent data ecosystem in Europe. As Commissioner Kroes explained, "The fragmentation concerns sectors, languages, as well as differences in laws and policy practices between EU countries" (European Commission 2013; Neelie 2013). To develop its data ecosystem, Europe needed strong players along the big data value chain, in areas ranging from data generation and acquisition, through data processing and analysis, to curation, usage, service creation and provisioning. Each link in the value chain needed to be strong so that a vibrant big data value ecosystem could evolve.

The cross-fertilisation of a broad range of organisations (business, research and society) and data was seen as the critical enabler for advancing the data economy in Europe. Stakeholders from all along the Data Value Chain needed to be brought together to create a basis for cooperation to tackle the complex and multidisciplinary challenges to create an optimal business environment for big data that would accelerate adoption within Europe. During the ICT 2013 Conference, Commissioner Kroes called for a European public-private partnership on big data to create a coherent European data ecosystem that stimulates research and innovation around data, as well as the uptake of cross-sector, cross-lingual and cross-border data services and products.

4.3   The Big Data Value PPP (BDV PPP)

Europe needed to aim high and mobilise stakeholders throughout society, industry, academia and research to enable the creation of a European big data value economy. It needed to support and boost agile business actors; deliver products, services and technology; and provide highly skilled data engineers, scientists and practitioners along the entire big data value chain. The goal was an innovation ecosystem in which value creation from big data flourishes.

To achieve these **goals the European contractual Public-Private Partnership on Big Data Value (BDV PPP)** was signed on 13 October 2014. This marked the commitment of the European Commission, industry and partners from academia to build a data-driven economy across Europe, mastering the generation of value from big data and creating a significant competitive advantage for European industry, thus boosting economic growth and jobs.

The BDV PPP commenced in 2015 and was operationalised with the launch of the Leadership in Enabling and Industrial Technologies (LEIT) work programme of Horizon 2020. The BDV PPP activities addressed the development of technology and applications, business model discovery, ecosystem validation, skills profiling, regulatory and IPR environments, and many social aspects.

With an initial indicative budget from the European Union of €534M for the period 2016–2020 and €201M allocated in total by the end of 2018, the BDV PPP has already mobilised €1570M in private investments since the launch of the PPP (€467M for 2018). Forty-two projects were running at the beginning of 2019 and the BDV PPP in only 2 years developed 132 innovations of exploitable value (106 delivered in 2018, 35% of which are significant innovations), including technologies, platforms, services, products, methods, systems, components and/or modules, frameworks/architectures, processes, tools/toolkits, spin-offs, datasets, ontologies, patents and knowledge. Ninety-three percent of the innovations delivered in 2018 had economic impact and 48% had societal impact. By 2020, the BDV PPP had projects covering a spectrum of data-driven innovations in sectors including advanced manufacturing, transport and logistics, health, and bioeconomy. These projects have advanced the state of the art in key enabling technologies for big data value and in non-technological areas such as providing solutions, platforms, tools, frameworks, best practices and invaluable general innovations, setting up firm foundations for a data-driven economy and future European competitiveness in data and AI.

The BDV PPP has supported the emergence of a comprehensive data innovation ecosystem for achieving and sustaining European leadership in big data and delivering the maximum economic and societal benefits to Europe – its businesses and citizens. In 2018 alone, the BDV PPP organised 323 events (including European Big Data Value Forum, BDV PPP Summit, seminars and conferences) outreaching over 630,000 participants, and taking into account mass media. The number of people outreached and engaged in dissemination activities has been estimated at 7.8 million by the Monitoring Report 2018 (*Big Data Value PPP Monitoring Report 2018* 2019). According to the European Data Market Study,[3] there has been a significant expansion of the European Data Economy in recent years:

- The *number of Data Companies* increased to 290,000 in 2019, compared to 283,300 in 2018.
- The *revenues of Data Companies* in the European Union reached €83.5B in 2019 compared to €77B in the previous year, with a growth rate of 8%.
- The *baseline for Data Professionals* in the European Union in 2013 was 5.77 million. The number of data professionals increased to a total of 7.6 million by 2019 in the EU28, corresponding to 1.836 million jobs created for data professionals since 2013.

4.4 Big Data Value Association

The Big Data Value Association (BDVA) is an industry-driven international non-profit organisation which has grown over the years to over 220 members all over Europe, with a well-balanced composition of large, small and medium-sized industries as well as research and user organisations. BDVA has over 25 working groups organised in Task Forces and subgroups, tackling all the technical and non-technical challenges of big data value.

BDVA served as a private counterpart to the European Commission to implement the Big Data Value PPP programme. BDVA and the Big Data Value PPP pursued a common shared vision of positioning Europe as the world leader in the creation of big data value. BDVA is also a private member of the EuroHPC Joint Undertaking and one of the leading promoters and driving forces of the AI, Data and Robotics Partnership planned for the next framework programme Multiannual Financial Framework (MFF) 2021–2027.

The mission of BDVA was "*to develop the Innovation Ecosystem that will enable the data-driven digital transformation in Europe delivering maximum economic and societal benefit, and, to achieve and to sustain Europe's leadership on Big Data Value creation and Artificial Intelligence.*" BDVA enabled existing regional multi-partner cooperation, to collaborate at European level through the provision of tools and know-how to support the co-creation, development and experimentation of pan-European data-driven applications and services, and know-how exchange. To achieve its mission, in 2017 BDVA defined four strategic priorities (Zillner et al. 2017):

- **Develop Data Innovation Recommendations**: Providing guidelines and recommendations on data innovation to the industry, researchers, markets and policy-makers
- **Develop Ecosystem**: Developing and strengthening the European big data value ecosystem
- **Guiding Standards**: Driving big data standardisation and interoperability priorities and influencing standardisation bodies and industrial alliances
- **Know-How and Skills**: Improving the adoption of big data through the exchange of knowledge, skills and best practices

BDVA developed a joint Strategic Research & Innovation Agenda (SRIA) on Big Data Value (Zillner et al. 2017). It was initially fed by a collection of technical papers and roadmaps (Cavanillas et al. 2016a) and extended with a public consultation that included hundreds of additional stakeholders representing both the supply and the demand side. The BDV SRIA defined the overall goals, main technical and non-technical priorities, and a research and innovation roadmap for the BDV PPP. The SRIA set out the strategic importance of big data, described the Data Value Chain and the central role of Ecosystems, detailed a vision for big data value in Europe in 2020, analysed the associated strengths, weaknesses, opportunities and threats, and set out the objectives and goals to be accomplished by the BDV PPP within the European research and innovation landscape of Horizon 2020 and at national and regional level.

5 The Elements of Big Data Value

To foster, strengthen and support the development and wide adoption of big data value technologies within an increasingly complex landscape requires an interdisciplinary approach that addresses the multiple elements of big data value. This book captures the early discoveries of the big data value community as an initial set of *Elements of Big Data Value*. This book arranges these elements into a classification system which is inspired by the periodic table for classifying chemical elements by atomic mass. Within our periodic table we have four groupings (see Fig. 1) containing elements focusing on similar behaviours needed for big data value covering (1) ecosystem, (2) research and innovation, (3) business, policy and societal elements, and (4) emerging elements. As we learn more about how to leverage and derive more value from data, we expect the elements of big data value to be challenged and to evolve as new elements are discovered. Just as the originators of the periodic table left room for new elements, The Periodic Table of the Elements of Big Data Value is open to future contributions.

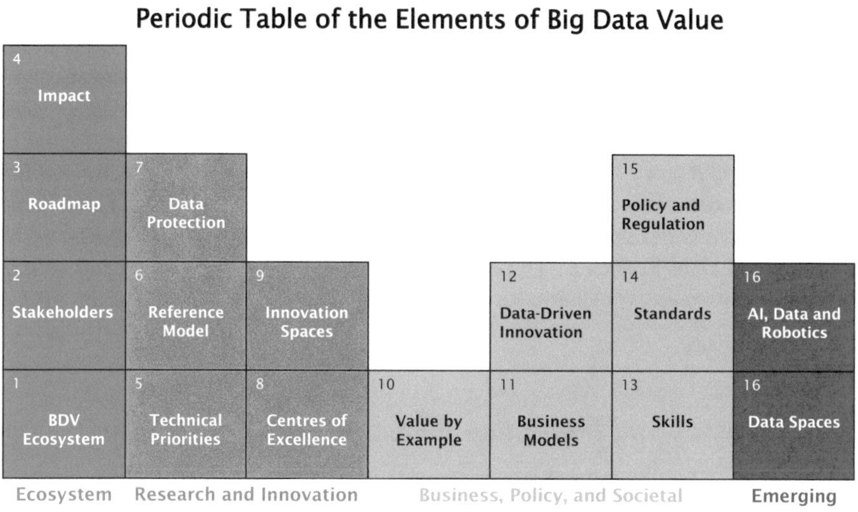

Periodic Table of the Elements of Big Data Value

4 Impact							
3 Roadmap	**7** Data Protection					**15** Policy and Regulation	
2 Stakeholders	**6** Reference Model	**9** Innovation Spaces			**12** Data-Driven Innovation	**14** Standards	**16** AI, Data and Robotics
1 BDV Ecosystem	**5** Technical Priorities	**8** Centres of Excellence	**10** Value by Example	**11** Business Models	**13** Skills		**16** Data Spaces
Ecosystem	Research and Innovation			Business, Policy, and Societal			Emerging

Fig. 1 The elements of big data value

5.1 Ecosystem Elements of Big Data Value

The establishment of the big data value ecosystem and promoting its accelerated adoption required a holistic approach to make it strong, vibrant and valuable to its stakeholders. The main elements that needed to be tackled to create and sustain a robust data ecosystem are as follows:

- **BDV Ecosystem:** This chapter explores the opportunity to increase the competitiveness of European industries through a data ecosystem by tackling the fundamental elements of big data value, including the ecosystem, research and innovation, business, policy and regulation, and the emerging elements of data-driven AI and common European data spaces.
- **Stakeholders:** Chapter "Stakeholder Analysis of Data Ecosystems" discusses a stakeholder analysis concerning data ecosystems and stakeholder relationships within and between different industrial and societal case studies. The stakeholder analysis helps determine how to incentivise stakeholders to participate in the activities of the data ecosystem. Each case study within the analysis focuses on big data practices across a range of industrial sectors to gain an understanding of the economic, legal, social, ethical and political externalities. A horizontal analysis is conducted to identify how positive externalities can be amplified and negative externalities diminished.
- **Roadmap:** A roadmap to drive adoption of data value ecosystems is described in Chap. "A Roadmap to Drive Adoption of Data Ecosystems". Creating a productive ecosystem for big data and driving accelerated adoption requires an interdisciplinary approach addressing a wide range of challenges from access to data and infrastructure, to technical barriers, skills, and policy and regulation. Overcoming

these challenges requires collective action from all stakeholders working together in an effective, holistic and coherent manner. To this end, the Big Data Value Public-Private Partnership was established to develop the European data ecosystem and enable data-driven digital transformation, delivering maximum economic and societal benefit.

- **Impact:** Chapter "Achievements and Impact of the Big Data Value Public-Private Partnership: The Story so Far" details the impact of the Big Data Value Public-Private Partnership, which plays a central role in the implementation of the European Data Economy. The chapter provides an overview of the partnership and its objectives, together with an in-depth analysis of the impact of the PPP.

5.2 Research and Innovation Elements of Big Data Value

New technical concepts will emerge for data collection, processing, storing, analysing, handling, visualisation and, most importantly, usage, and new data-driven innovations will be created using them. The key research and innovation elements of big data value are as follows:

- **Technical Priorities:** Chapter "Technical Research Priorities for Big Data" details the technical priorities for big data value covering key aspects such as real-time analytics, low latency and scalability in processing data, new and rich user interfaces, interacting with and linking data, information and content, all of which have to be developed to open up new opportunities and to sustain or develop competitive advantages. As well as having agreed approaches, the interoperability of datasets and data-driven solutions is essential to ensure broad adoption within and across sectors.
- **Reference Model:** Chapter "A Reference Model for Big Data Technologies" describes the Big Data Value Reference Model, which has been developed with input from technical experts and stakeholders along the whole big data value chain. The BDV Reference model serves as a common reference framework to locate data technologies on the overall IT stack. It addresses the main concerns and aspects to be considered for big data value systems.
- **Data Protection:** Data Protection and Data Technologies are the focus of Chap. "Data Protection in the Era of Artificial Intelligence: Trends, Existing Solutions and Recommendations for Privacy-Preserving Technologies", where advances in privacy-preserving technologies are aimed at building privacy-by-design from the start into the back-end and front-end of digital services. They make sure that data-related risks are mitigated both at design time and run time, and they ensure that data architectures are safe and secure. The chapter discusses recent trends in the development of tools and technologies that facilitate secure and trustworthy data analytics.
- **Centres of Excellence:** Chapter "A Best Practice Framework for Centres of Excellence in Big Data and Artificial Intelligence" presents a best practice

framework for Centres of Excellence for Big Data and AI. Within universities, academic departments and schools, it often works towards the establishment of a special-purpose organizational unit within a national system of research and education that provides leadership in research, innovation and training for Big Data and AI technologies. Centres of Excellence can serve as a common practice for the accumulation and creation of knowledge that addresses the scientific challenges of Big Data and AI, opens new avenues of innovation in collaboration with industry, engages in the policy debates, and informs the public about the externalities of technological advances.

- **Innovation Spaces:** Within the European data ecosystem, cross-organisational and cross-sectorial experimentation and innovation environments play a central role. Chapter "Data Innovation Spaces" describes the European Innovation Spaces, which are the main elements to ensure that research on big data value technologies and novel applications can be quickly tested, piloted and exploited to the maximum benefit of all the stakeholders.

5.3   Business, Policy and Societal Elements of Big Data Value

Big data is an economic and societal asset that has significant potential for the economy and society. New sustainable economic models within a policy environment that respects data owners and individuals are needed to the deliver value from big data. Critical elements of big data value for business and policy are as follows:

- **Value Creation:** Chapter "Big Data Value Creation by Example" provides a collection of stories showing concrete examples of the value created thanks to big data value technologies. These novel solutions have been developed and validated by stakeholders in the big data value ecosystems and provide proof points of how data can drive innovation across industries to transform business practices and society. Meanwhile, start-ups are working at the confluence of emerging data sources (e.g. IoT, DNA, high-definition images, satellite data) and new or revisited processing paradigms (e.g. Edge computing, blockchain, machine learning) to tackle new use cases and provide disruptive solutions.
- **Business Models:** Chapter "Business Models and Ecosystem for Big Data" explores new data-driven business models as ways to generate value for companies along the value chain, regardless of sector or domain: optimising and improving the core business; selling data services; and, perhaps most importantly, creating entirely new business models and business development. Identifying sustainable business models and ecosystems in and across sectors and platforms will be an important challenge. In particular, many SMEs that are now involved in highly specific or niche roles will need support to help them align and adapt to new value chain opportunities.

- **Data-Driven Innovation:** Chapter "Innovation in Times of Big Data and AI: Introducing the Data-Driven Innovation (DDI) Framework" introduces the Data-Driven Innovation (DDI) Framework to support the process of identifying and scoping big data value. The framework guides start-ups, entrepreneurs and established companies alike in scoping promising data business opportunities by analysing the dynamics of both supply and demand.
- **Skills:** Chapter "Recognition of Formal and Non-formal Training in Data Science" covers the data skills challenge to ensure the availability of appropriately skilled people who have an excellent grasp of the best practices and technologies for delivering big data value solutions. Promoting the "transparency and recognition of skills and qualifications" is particularly relevant to the task of recognizing both formal and informal data science training, and consequently the challenge will be to provide a framework in order to validate these skills.
- **Standards:** Chapter "The Road to Big Data Standardisation" covers the critical topic of standards within the area of big data where the use of standardised services and products is needed to effectively drive the adoption of common data solutions and services around the world. This chapter provides an overview of the key standardisation activities within the European Union and the current status and future trends of big data standardisation.
- **Policy and Regulation**: Chapter "The Role of Data Regulation in Shaping AI: An Overview of Challenges and Recommendations for SMEs" engages in the debate on data ownership and usage, data protection and privacy, security, liability, cybercrime and Intellectual Property Rights (IPR). A necessary first step is to frame this policy and regulatory debate about the non-technical aspects of big data value creation as part of the data-driven economy. These issues need to be resolved to remove the barriers to adoption. Favourable European regulatory environments are required to facilitate the development of a genuine pan-European big data market. For an accelerated adoption of big data, it is critical to increase awareness of the benefits and the value that big data offers, and to understand the obstacles to building solutions and putting them into practice.

5.4   Emerging Elements of Big Data Value

Artificial Intelligence (AI) has tremendous potential to benefit citizens, economy and society. From a big data value perspective, AI techniques can extract new value from data to enable data-driven systems that in turn enable machines and people with digital capabilities, such as perception, reasoning, learning and even autonomous decision-making. Data ecosystems are an essential driver for data-driven AI to exploit the continued growth of data. Developing both of these elements together is critical to maximising the future potential of big data value:

- **Artificial Intelligence, Data and Robotics:** To maximise the potential of AI, a solid foundation is needed for successfully deploying AI solutions. To this end, Chap. "Data Economy 2.0: From Big Data Value to AI Value and a European Data Space" details the *European AI, Data and Robotics Framework* (Zillner et al. 2020), which represents the legal and societal fabric that underpins the impact of AI on stakeholders and users of the products and services that businesses will provide. The *AI, Data and Robotics Innovation Ecosystem Enablers* represent essential ingredients for significant innovation and deployment to take place within this framework. Finally, *Cross Sectorial AI, Data and Robotics Technology Enablers* are needed to provide the core technical competencies that are essential for the development of data-driven AI systems.
- **Data Spaces:** As part of the continued development of the European Big Data Value Ecosystem, Chap. "Data Economy 2.0: From Big Data Value to AI Value and a European Data Space" describes common European data spaces which will be established to ensure that more data becomes available for use in the economy and society while keeping companies and individuals who generate data in control. These data spaces (in both a technical Curry 2020] and regulatory [European Commission 2018] sense) will be critical to fuelling data-driven AI innovations.

6 Summary

Exploiting big data offers enormous potential to create value for European society, citizens and businesses. Europe needs to embrace new technology, applications, use cases and business models within and across various sectors and domains. In this chapter, we presented the European strategy followed by the European big data value ecosystem to increase the competitiveness of European industries by addressing fundamental elements of big data value. These elements will enable data-driven digital transformation in Europe, delivering maximum economic and societal benefit, and achieving and sustaining Europe's leadership in the fields of big data value creation and Artificial Intelligence.

References

Biehn, N. (2013). *The missing V's in Big Data: Viability and value*.

Big Data Value cPPP Monitoring Report 2018. (2019). Retrieved from https://www.bdva.eu/MonitoringReport2018

Cavanillas, J. M., Curry, E., & Wahlster, W. (Eds.). (2016a). *New horizons for a data-driven economy: A roadmap for usage and exploitation of big data in Europe*. New York: Springer. https://doi.org/10.1007/978-3-319-21569-3

Cavanillas, J. M., Curry, E., & Wahlster, W. (2016b). The big data value opportunity. In J. M. Cavanillas, E. Curry, & W. Wahlster (Eds.), *New horizons for a data-driven economy: A roadmap for usage and exploitation of big data in Europe* (pp. 3–11). New York: Springer. https://doi.org/10.1007/978-3-319-21569-3_1

Choo, C. W. (1996). The knowing organization: How organizations use information to construct meaning, create knowledge and make decisions. *International Journal of Information Management, 16*(5), 329–340. https://doi.org/10.1016/0268-4012(96)00020-5

Communication: A European strategy for data. (2020). Retrieved from https://ec.europa.eu/info/sites/info/files/communication-european-strategy-data-19feb2020_en.pdf

Curry, E. (2016). The Big Data value chain: Definitions, concepts, and theoretical approaches. In J. M. Cavanillas, E. Curry, & W. Wahlster (Eds.), *New horizons for a data-driven economy: A roadmap for usage and exploitation of big data in Europe*. https://doi.org/10.1007/978-3-319-21569-3_3

Curry, E. (2020). Real-time linked dataspaces: A data platform for intelligent systems within Internet of things-based smart environments. In *Real-time linked dataspaces* (pp. 3–14). https://doi.org/10.1007/978-3-030-29665-0_1

Davenport, T. H., Barth, P., & Bean, R. (2012). How 'big data' is different. *MIT Sloan Management Review, 54*, 22–24.

DG Connect. (2013). *A European strategy on the data value chain*. Retrieved from http://ec.europa.eu/information_society/newsroom/cf/dae/document.cfm?doc_id=3488

Economy, E. D. (2017). *Public consultation on building the European data economy.*

European Commission. (2013). *Digital agenda for Europe, Session Reports, ICT for industrial leadership: Innovating by exploiting big and open data and digital content.*

European Commission. (2018). *Communication from the commission to the European parliament, the council, the European Economic and social committee and the committee of the regions "Towards a common European data space" COM/2018/232 final.*

Hammond, K. J. (2013). The value of big data isn't the data. *Harvard Business Review*. Retrieved from https://hbr.org/2013/05/the-value-of-big-data-isnt-the

Hey, T., Tansley, S., & Tolle, K. M. (Eds.). (2009). *The fourth paradigm: Data-intensive scientific discovery*. Redmond, WA: Microsoft Research.

Jacobs, A. (2009). The pathologies of big data. *Communications of the ACM, 52*(8), 36. https://doi.org/10.1145/1536616.1536632

Laney, D. (2001). 3D data management: Controlling data volume, velocity, and variety. *Application Delivery Strategies, 2*.

Lavalle, S., Lesser, E., Shockley, R., Hopkins, M. S., & Kruschwitz, N. (2011). Big data, analytics and the path from insights to value. *MIT Sloan Management Review, 52*(2), 21–32.

Loukides, M. (2010, June). What is data science? *O'Reilly Radar*. Retrieved from http://radar.oreilly.com/2010/06/what-is-data-science.html

Manyika, J., Chui, M., Brown, B., Bughin, J., Dobbs, R., Roxburgh, C., & Byers, A. H. (2011). *Big data: The next frontier for innovation, competition, and productivity*. Retrieved from McKinsey Global Institute website http://scholar.google.com/scholar.bib?q=info:kkCtazs1Q6wJ:scholar.google.com/&output=citation&hl=en&as_sdt=0,47&ct=citation&cd=0

Miller, H. G., & Mork, P. (2013). From data to decisions: A value chain for big data. *IT Professional*. https://doi.org/10.1109/MITP.2013.11

Neelie K. (2013). *Big data for Europe - ICT 2013 Event – Session on Innovating by exploiting big and open data and digital content*. Vilnius.

NESSI. (2012, December). Big Data: A new world of opportunities. *NESSI White Paper*.

OECD. (2014). *Data-driven innovation for growth and well-being.*

Rayport, J. F., & Sviokla, J. J. (1995). Exploiting the virtual value chain. *Harvard Business Review, 73*, 75–85. https://doi.org/10.1016/S0267-3649(00)88914-1

Stonebraker, M. (2012). What does 'big data' mean. *Communications of the ACM, BLOG@ ACM.*

Vaishya, R., Javaid, M., Khan, I. H., & Haleem, A. (2020). Artificial Intelligence (AI) applications for COVID-19 pandemic. *Diabetes and Metabolic Syndrome: Clinical Research and Reviews*. https://doi.org/10.1016/j.dsx.2020.04.012

Wegener, R., & Velu, S. (2013). The value of Big Data: How analytics differentiates winners. *Bain Brief*.

Wikipedia. (2020). *Big data. Wikipedia article*. Retrieved from http://en.wikipedia.org/wiki/Big_data

Zillner, S., Curry, E., Metzger, A., Auer, S., & Seidl, R. (Eds.). (2017). *European big data value strategic research & innovation agenda*. Retrieved from Big Data Value Association website www.bdva.eu

Zillner, S., Bisset, D., Milano, M., Curry, E., Hahn, T., Lafrenz, R., et al. (2020). *Strategic research, innovation and deployment agenda – AI, data and robotics partnership. Third Release (3rd)*. Brussels: BDVA, euRobotics, ELLIS, EurAI and CLAIRE.

Stakeholder Analysis of Data Ecosystems

Umair ul Hassan and Edward Curry

Abstract Stakeholder analysis and management have received significant attention in management literature primarily due to the role played by key stakeholders in the success or failure of projects and programmes. Consequently, it becomes important to collect and analyse information on relevant stakeholders to develop an understanding of their interest and influence. This chapter provides an analysis of stakeholders within the European data ecosystem. The analysis identifies the needs and drivers of stakeholders concerning big data in Europe; furthermore, it examines stakeholder relationships within and between different sectors. For this purpose, a two-stage methodology was followed for stakeholder analysis, which included sector-specific case studies and a cross-case analysis of stakeholders. The results of the analysis provide a basis for understanding the role of actors as stakeholders who make consequential decisions about data technologies and the rationale behind the incentives targeted at stakeholder engagement for active participation in a data ecosystem.

Keywords Data ecosystem · Stakeholder analysis · Case study · Data value chain

1 Introduction

This chapter discusses the stakeholder analysis performed within the scope of the "Big data roadmap and cross-disciplinarY community for addressing socieTal Externalities" (BYTE[1]) project, between 2014 and 2017. The BYTE project analysed stakeholders in relation to data ecosystems as well as their relationships within and between different sectors. This analysis enabled the project to determine how to incentivise stakeholders to participate in its activities. The BYTE project was

---

[1] https://cordis.europa.eu/project/id/619551

U. ul Hassan (✉) · E. Curry
Insight SFI Research Centre for Data Analytics, NUI Galway, Galway, Ireland
e-mail: umair.ulhassan@nuigalway.ie

© The Author(s) 2021
E. Curry et al. (eds.), *The Elements of Big Data Value*,
https://doi.org/10.1007/978-3-030-68176-0_2

21

aimed at assisting European science and industry in capturing the positive externalities and diminishing the negative externalities associated with big data to gain a
more significant market share. BYTE accomplished its goals by leveraging an
international advisory board and an additional network of contacts to conduct a
series of case studies. Each case study focused on big data practices across an
industrial sector to gain an understanding of the economic, legal, social, ethical,
and political externalities. A horizontal analysis was conducted to identify how
positive externalities can be amplified and negative externalities diminished.

The rest of this chapter is organised as follows. Section 2 underlines the need for
stakeholder analysis, and Sect. 3 defines a stakeholder in the context of the BYTE
project. Sections 4 and 5 detail the methodology and dimensions of stakeholder
analysis. Section 6 introduces the sector-wise case studies and the results of the
cross-case analysis. Section 7 summarises the chapter.

2 Stakeholder Analysis

According to Grimble et al., "stakeholder analysis can be defined as an approach for
understanding a system by identifying the key actors or stakeholders in the system
and assessing their respective interest in that system" (Grimble et al. 1995). To map
the relevant stakeholders within the European data ecosystem, the BYTE project
started with industry contacts, academic experts, and civil society representatives
active with big data, statistics, computer science, economics, open access, social
science, and legal and ethical experts (Curry 2016). As the project progressed,
industry and public sector representatives from the case study sectors, policymakers,
institutional representatives, standards organisations, funding bodies, and any other
relevant stakeholders were all engaged.

Grimble and Wellard have emphasised the importance of stakeholder analysis in
understanding the complexity and compatibility problems between objectives and
stakeholders (Grimble and Wellard 1997). Two questions must be answered before
any stakeholder analysis: "Who is a stakeholder?" and "Why is their role needed?"
To answer the first question, stakeholders are identified based on many factors,
including their interest in and influence on a system, their knowledge about the
system, and their networks internal and external to the system. With respect to the
second question, it is also important to note that the roles played by stakeholders are
dynamic rather than static over time. Depending on circumstances, the same people
or groups can take on different roles at different times; furthermore, stakeholder roles
may also be blended. It is also possible for stakeholders to move between roles, and
specific actions can be targeted to "move" stakeholders from one role to another.

3 Who Is a Stakeholder?

Stakeholder theory has become the mainstream of management literature across different disciplines since Freeman's seminal work on *Strategic Management: A Stakeholder Approach* (Freeman 1984). Within this work, the primary purpose of stakeholder theory was to assist managers in identifying stakeholders and strategically manage them. Freeman defines stakeholders as "any group or individual who can affect or is affected by the achievement of the organisation's objectives". Since this early work, stakeholder theory has been applied in many contexts and disciplines outside of management. Weryer describes it as a "slippery creature", "used by different people to mean widely different things" (Weyer 1996). Miles has established that stakeholder is an essentially contested concept, and therefore requiring a universal definition is unfeasible (Miles 2012). Nonetheless, it is essential to define stakeholder and provide the basis for necessary stakeholder analysis. The following definition of stakeholder was agreed and adopted after considering existing definitions in the literature and taking into account the objectives of the BYTE project:

> A stakeholder is any group or individual who can affect or is affected by the information ecosystem in a positive or negative manner.

This definition served as the starting point to identify the stakeholders within each of the case studies. Subsequently, the same definition was used for analysis while following the methodology detailed in the next section.

4 Methodology

Both normative and instrumental approaches have been applied in different disciplines for stakeholder analysis. For instance, Reed et al. provide a comprehensive overview of the wide variety of techniques and approaches for stakeholder analysis (Reed et al. 2009). As illustrated in Fig. 1, they have categorised the methods used for: (i) identifying stakeholders, (ii) differentiating between and categorising stakeholders, and (iii) investigating relationships between stakeholders.

The stakeholder analysis within BYTE took place in two phases. The first phase focused on sector-specific case studies that built a logical chain of evidence to support the stakeholder analysis (Miles 2012; Yin 2013). The second phase involved a cross-case examination in identifying if generalities or commonalities existed across case studies.

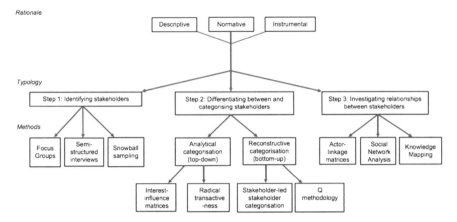

Fig. 1 Schematic representation of rationale, typology, and methods for stakeholder analysis (Reed et al. 2009). (Reprinted from Journal of Environmental Management, 90/5, Mark S. Reed, Anil Graves, Norman Dand, Helena Posthumus, Klaus Hubacek, Joe Morris, Christina Prell, Claire H. Quinn, Lindsay C. Stringer, Who's in and why? A typology of stakeholder analysis methods for natural resource management, 1933–1949., Copyright (2009), with permission from Elsevier.)

4.1   Phase 1: Case Studies

The first phase of stakeholder analysis includes eight steps, as follows:

1. **Identify the focus of the case study:** At the start of the case study, it is important to establish a clear focus. This defines the context of the case study and makes it possible to determine those who are affected or can affect decisions relating to the issues under investigation.

2. **Identify the boundary of analysis of the case study:** With a clear focus of the case study defined, the next step is to establish a clear system boundary for the stakeholder analysis. The system boundary is used to limit the scope of the analysis to ensure that it tackles the identified focus. Steps 1 and 2 may follow a participatory approach that involves the stakeholders directly in the identification of foci and boundaries. This necessitates an iterative feedback loop. It should be noted that stakeholder participation in the analysis may not be necessary if the project team have sufficient knowledge of the case study.

3. **Identification of stakeholders and their stake (i.e. interviews, case studies, workshops):** The project team, in collaboration with the case study liaisons, prepares a list of possible stakeholders for the case study. The stakeholders are listed according to the role that best describes their involvement in the case study (policymakers, data scientist, data engineer, managers, end-users, consultants, and consumers). The initial list can be as exhaustive as possible to ensure the inclusion of all relevant stakeholders in the case study. Once completed, the initial list is circulated to the key actors in the case study for feedback. The feedback is then used to add missing members or delete others who are not relevant. To

provide a systematic tool for the identification of stakeholders in the complex context of case studies, Pouloudi has suggested a set of principles of stakeholder behaviour that guide stakeholder identification and analysis (Pouloudi 1999).

4. **Differentiate between and categorise stakeholders**: Several methods are available for categorising stakeholders and understanding their inter-relationships (Step 5). The project team classifies the list of stakeholders based on their interest in each case study. The initial classification is qualitative as it is based on the subjective judgement of the project team. To further validate the categories, feedback can be sought from the key actors in the case study. The categories should be dynamic as stakeholders' interests and influence can change over time, depending on the dynamicity of the environment. Stakeholders can also be placed in multiple categories.

5. **Investigate relationships between stakeholders:** Once stakeholders have been identified and classified, the next step is to understand what relationships exist between the stakeholders. Understanding the interplay between stakeholders can reveal common motivations, alliances, and conflicts that exist within and across cases. It can also help us understand the motivations of stakeholders, which can help to support their incentivisation.

6. **Identify stakeholder incentivisation and communication plan**: Before any initiatives are designed to engage stakeholders, it is crucial to identify the most relevant stakeholders within the case and ensure their participation in the analysis process. The engagement of stakeholders is time consuming and not a trivial issue. Many potential stakeholders within a case study may lack interest, whereas some may have strong (and specific) interests that could dominate the agenda. Careful consideration of stakeholder interests may persuade less interested stakeholders to join the process. For instance, civil society organisations can prove challenging to engage.

7. **Feedback into Step 1 or Step 2:** At the end of the process, the feedback received is input back into the process to improve the quality of the analysis. Typically, stakeholder analysis will take place over several iterations of the process. As the analysis is refined, it is important to consider issues such as the legitimacy, representation, and credibility of the analysis. Where additional stakeholders have been identified in the process, they should be included in the next iteration.

8. **Engage stakeholders for validation:** The last step of the process is the validation of the stakeholder analysis with a selected group of stakeholders. The validation takes the form of interviews with key actors within the case study and stakeholder engagement workshops. In these workshops, the project team directly works with stakeholders to elicit required input for validation and consensus, where possible, on the stakeholder analysis. At the beginning of the interview/workshop, the purpose of the stakeholder analysis is detailed to ensure that relevant stakeholders actively participate. The feedback is then used to update the stakeholder analysis, as necessary.

4.2   Phase 2: Cross-Case Analysis

As part of the second phase, cross-case analysis is used to examine themes, similarities, and differences across several cases. It provides further insight into issues concerning the case and reveals the potential for generalising the case study results. Cross-case analysis can also be used to delineate the combination of factors that may contribute to the outcomes of the individual case. It can be used to determine an explanation as to why one case is different from or the same as others. Multiple cases are examined to build a logical chain of evidence to support the stakeholder analysis (Miles 2012; Yin 2013). The cross-case analysis consists of the following steps:

  (i)   Within-case stakeholder analysis
 (ii)   Analysis of consistencies identified across the cases in the various relationships, along with reasons why these relationships exist
(iii)   Formulating systematic cross-case observations

5   Sectoral Case Studies

A key fallacy associated with big data is that the processing of large data sets will lead directly to either benefit or harm. However, economic experts have noted that data only becomes information once it guides strategy, motivates action, and leads to observable changes in behaviour. More information does provide strategic options with which to deal with strategic, environmental, or technical challenges. But these options require the correct environment to obtain a competitive advantage. Likewise, the capability to exploit information for harm does not guarantee that societal harm will occur. Expected harm can be minimised by ensuring the correct institutional or legal framework for addressing negative externalities of big data.

Through the Digital Agenda for Europe, European policymakers have expressed that they expect big data to result in positive competitive advantages across various sectors of the economy. At a high level, these sectors include transport, healthcare, environment, smart city, energy, crisis management, and culture. The BYTE project threaded case studies in these sectors through the course of the project, as listed in Table 1. These case studies involved organisations actively using big data for their operational and strategic purposes. The case studies enabled BYTE to understand strategies, actions, and changes in behaviour associated with big data, with the aim of identifying their resultant positive and negative externalities (Cuquet et al. 2017). Furthermore, they enabled BYTE to better predict the type of regulatory environment that would allow European actors to take advantage of potential positive externalities and diminish negative externalities.

Table 1 List of stakeholders considered as part of the case studies in the BYTE project

Case study sector	Stakeholder	Secondary sector
Crisis	RICC	Computer science
	International Government Organization (OCHA)	Humanitarian organisation
	International Humanitarian Organization (ICRC)	Humanitarian organisation
Culture	National cultural heritage institutions, including libraries, museums, galleries, etc.	Cultural
	National data aggregator	Cultural
	Pan-European cultural heritage data	Cultural
	Policymakers and legal professionals	Government
	Citizens	Citizens
	Educational institutions	Public sector
	Open data advocates	Society organisation
Energy	Statoil	Oil & gas operator
	ConocoPhillips	Oil & gas operator
	Lundin	Oil & gas operator
	Eni Norge	Oil & gas operator
	SUPPLIER	Oil & gas supplier
	Norwegian Petroleum Directorate	Oil & gas regulator in Norway
Environment	EC	Public sector (EU)
	EEA	Public sector (EU)
	EPA	Public sector (USA)
	EuroGeoSurveys	Public sector (EU)
	EUSatCen	Public sector (EU)
	IEEE	Professional association
	NASA	Space (USA)
	SANSA	Space (South Africa)
	UNEP	Public sector
Healthcare	Public sector health research initiative	Healthcare, medical research
	Geneticists	Healthcare, medical research,
	Clinicians	Healthcare (private and public)
	Data scientists	Healthcare, medical research
	Pharmaceutical companies	Commercial
	Translational medicine specialists	Healthcare (private and public sector)
	Public health research initiative	Healthcare, translational medicine specialist
	NHS Regional genetics laboratory	Public sector healthcare laboratory
	Charity organisations	Civil society organisations

(continued)

Table 1 (continued)

Case study sector	Stakeholder	Secondary sector
	Privacy and data protection policymakers and lawyers	Public and private sector
	Citizens	Society at large
	Patients and immediate family members	Public sector
Transport	Established ship owner	Transport
	New ship owner	Transport
	European yard	Manufacturing
	Navigation equipment supplier	Manufacturing
	Machinery subsystem supplier	Manufacturing
	Shipping association	Transport
	Maritime consulting company	Transport
	Classification society	Transport/legal
	Natl. Coastal Authority	Legal
Smartcity	European city	Public sector
	Technology provider	Start-up, energy
	Technology provider	Non-profit, mobility
	Technology provider & research	Multinational, smart city

6 Cross-Case Analysis

This section specifies the dimensions used in the cross-case analysis of stakeholders. The relevance of the dimensions may vary between stakeholders and use cases. Based on the case studies described earlier, this section compares the stakeholders of the BYTE project. This cross-case analysis aims to identify the commonalities of stakeholders and highlight the differences (Lammerant et al. 2015). The analysis informed the activities of the BYTE project, including big data community formation and long-term stakeholder engagement.

6.1 Technology Adoption Stage

The diffusion of innovations is a theory that seeks to explain how, why, and at what rate new ideas and technology spread through cultures. The seminal work on this theory was undertaken by Everett Rogers (Rogers 1962). He describes diffusion as the process by which an innovation is communicated through specific channels over time among the members of a social system. Adoption implies accepting something created by another or foreign to one's nature. For a technology to be adopted by many users, it needs to be successfully diffused. Rogers describes the five adopters as follows:

- **Innovators** are the first 2.5% of individuals to adopt an innovation. They are adventurous, comfortable with a high degree of complexity and uncertainty, and typically have access to substantial financial resources
- **Early Adopters** are the next 13.5% to adopt innovation. They are well integrated into their social system and have great potential for opinion leadership. Other potential adopters look to early adopters for information and advice. Thus early adopters make excellent "missionaries" for new products or processes
- **Early Majority** are the next 34%. They adopt innovations slightly before the average member of a social system. They are typically not opinion leaders, but they frequently interact with their peers
- **Late Majority** are the next 34%. They approach innovation with a sceptical air and may not adopt the innovation until they feel pressure from their peers. They may have scarce resources
- **Laggards** are the last 16%. They base their decisions primarily on experience and possess almost no opinion leadership. They are highly sceptical of innovations and innovators and must feel confident that an innovation will not fail before adopting it.

In terms of technology adoption, the BYTE case studies highlight some specifics of and similarities between the stakeholders. As shown in Fig. 2, the stakeholders in these case studies follow the Rogers curve, i.e. 6% innovators, 21% early adopters, 33% early majority, 23% late majority, and 17% laggards. Some sectors are more advanced in their adoption of data technologies. For instance, the stakeholders in smart cities and crisis management case studies are either early adopters or early majority. This underlines their natural dependence on data-driven decision-making and operations. Only the stakeholders in the environment case study included

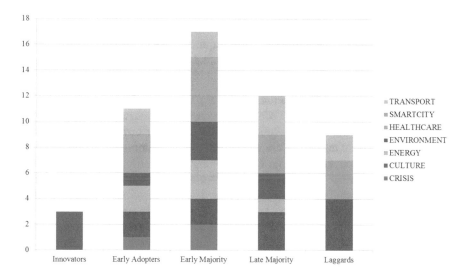

Fig. 2 Stakeholders against the technology adoption stages

innovators that encompassed space agencies and technology standards organisations. The majority stakeholders in the transport, healthcare, and culture sectors fall in the late stages of technology adoption. Therefore, some stakeholder engagement activities can be tailored towards these sectors to encourage participation in the big data community and amplification of positive externalities. Late adoption might be due to higher regulatory standards or lower levels of technology readiness.

6.2 Data Value Chain

Value chains have been used as a decision support tool to model the chain of activities that an organisation performs to deliver a valuable product or service to the market. A value chain categorises the generic value-adding activities of an organisation, allowing them to be understood and optimised. A value chain is made up of a series of subsystems, each with inputs, transformation processes, and outputs. As an analytical tool, the value chain can be applied to the information systems to understand the value-creation of data technologies. The *Data Value Chain* models the high-level activities that comprise an information system. A typical data value chain comprises the following activities:

1. **Data Acquisition** is the process of gathering, filtering, and cleaning data before it is put in a data warehouse or any other storage solution on which data analysis can be carried out.
2. **Data Analysis** is concerned with making acquired raw data easy to use in decision-making as well as for domain-specific purposes.
3. **Data Curation** is the active management of data over its life cycle to ensure that it meets the necessary data quality requirements for its effective usage.
4. **Data Storage** is concerned with storing and managing data in a scalable way, satisfying the needs of applications that require access to the data.
5. **Data Usage** covers the business goals that require access to data and its analysis, and the tools needed to integrate analysis in business decision-making.

Figure 3 shows the distribution of the BYTE stakeholders in the activities associated with the Data Value Chain. Among the stakeholders analysed, 56% explicitly consider the data acquisition activities, 56% perform some form of data analysis, 44% curate data, 40% are concerned with data storage solutions, and the majority of 88% actively use data for decision-making and operations. The crisis management sector has a primary focus on data usage, with minimal consideration for data acquisition and data analysis activities. The cultural sector is mainly focused on data acquisition, curation, and usage. Designing incentives that target the specific activities of the value chain can help engage with the relevant stakeholders. The sharing of best practices from stakeholders may also serve as an incentive for engagement with the big data community. Significantly, the stakeholders can share their expertise on one type of activity on the Data Value Chain with others.

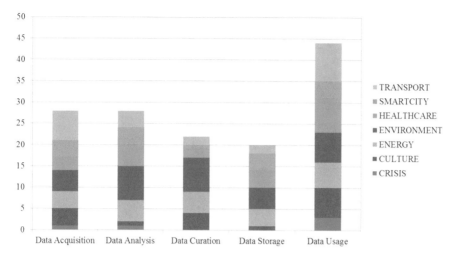

Fig. 3 Distribution of stakeholders in terms of activities on the Data Value Chain

6.3 Strategic Impact of IT

The strategic impact grid is an analytical tool proposed by Nolan and McFarlan that is used by managers to evaluate their firm's current and future information system's needs (Nolan and McFarlan 2005). The grid defines the use of information systems resources going forward, by enabling managers to:

- Identify the current need for reliable information systems by focusing on current day-to-day operations and the functionalities of the existing information systems
- Identify future needs for new information system functionalities by focusing on the strategic role that new IT capabilities play in the organisation

Based on this analysis, the grid helps managers to identify if they need to take a defensive or offensive approach in their information systems (IS) strategy. As depicted in Fig. 4, the grid classifies the approaches into four roles:

- **Support Role:** Information systems constitute a tool to support and enable operations. IS are not mission-critical for current business operations. New systems offer little strategic differentiation to significantly benefit the organisation.
- **Factory Role:** IS infrastructure is critical to the operation of the firm. Service outages can endanger the firm's well-being and future viability. However, limited potential exists for new systems and functionalities to make a substantial contribution to the firm.
- **Turnaround Role:** The firm's current IS are not mission-critical for current business operations. However, new IS functionalities will be critical for the business's future viability and success. The firm needs to engage in a transformation of its IT.

Defensive IT
Offensive IT
Factory
Strategic
Support
Turnaround
Low to High need for reliable IT
Low to High need for new IT

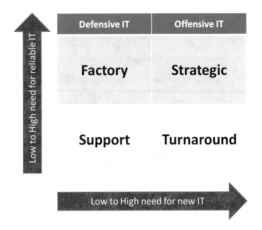

Fig. 4 Strategic Impact Grid

ROLE OF IT
Factory
Strategic
Support
Turnaround
ENVIRONMENT
3 4
1 1
CULTURE
4 3
5 3
SMARTCITY
0 2
0 2
CRISIS
0 1
2 0
HEALTH
4 1
4 4
ENERGY
1 4
0 1
TRANSPORT
2 3
2 2

Fig. 5 Distribution of stakeholders on the Strategic Impact Grid

- **Strategic Role:** IS are critical to the firm's current business operations. New IS functionalities will be critical for the future viability and prosperity of the business. Such firms have a very offensive IT posture and are proactive concerning IT investments.

Figure 5 shows the distribution of the BYTE stakeholders on the Strategic Impact Grid. Among the stakeholders analysed, 18 stakeholders were identified as having a strategic role in IT. This highlights the need to balance engagement activities to encourage participation from stakeholders in the community in other roles, which may not consider big data to be critical to their decision-making and operations management.

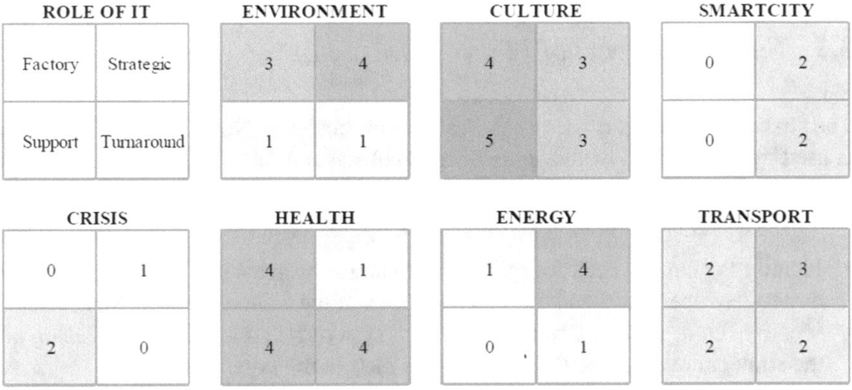

We also analysed the IT intensity of each case study as defined in a big data report published by McKinsey Global Institute (MGI) (Manyika et al. 2011). IT intensity indicates the ease of technology adoption and utilisation for a section. The report ranked the sectors according to their IT intensity and then divided them into five quantiles (first, second, third, fourth, fifth). The more IT assets a sector has on average, the easier it is to overcome barriers to data technologies. Each case study was mapped to the sectors indicated in the MGI report. The following list provides a summary of the analysis:

- **Environment:** The environment case study is mapped to the "Natural Resources" sector in the MGI report, which lies in the third quantile of IT intensity. The stakeholders in the environment case study are divided into distinct groups. The first group is focused on operations support and maintaining existing infrastructure, hence remaining in the factory role. The second group employs IT for strategic decisions and implements groundbreaking technologies, hence achieving the strategic role.
- **Crisis Management:** The crisis management case study is mapped to the "Health & Social Care" sector in the MGI report, which lies in the fifth quantile of IT intensity. Crisis management stakeholders require more reliable IT processes due to the mission criticality of their operations.
- **Smart City:** The smart city case study is mapped to the "Utilities" sector in the MGI report, which lies in the second quantile of IT intensity. Stakeholders in the smart city case study indicated the need for offensive IT strategies. This is understandable due to the data-dependent nature of the businesses and services that enable the concept of the smart city.
- **Culture:** The smart city case study is mapped to the "Arts, Entertainment, and Recreation" sector in the MGI report, which lies in the second quantile of IT intensity. The stakeholders of the culture case study are interested in both reliable IT and innovative IT.
- **Energy:** The energy case study is mapped to the "Natural Resources" sector in the MGI report, which lies in the third quantile of IT intensity. For the stakeholders in the energy case study, the role of IT is primarily strategic for both business operations and competitive advantage.
- **Health:** The health case study is mapped to the "Healthcare and Social Assistance" sector in the MGI report, which lies in the fifth quantile of IT intensity. The stakeholders in the heath case study are more oriented towards reliable IT, which is a prerequisite of the health sector. However, there are stakeholders that are dependent on new tools for drug discovery and improved healthcare.
- **Transport:** The transport case study is mapped to the "Transportation and warehousing" sector in the MGI report, which lies in the first quantile of IT intensity. In the transport case study, we observe an even distribution of the role of IT on the Strategic Impact Grid. This indicates a balance between maintaining operations through big data and using big data to gain a competitive advantage.

6.4 Stakeholder Characteristics

In addition to the dimensions introduced above, the stakeholder analysis captures a few additional attributes that are used to profile stakeholders. This section details these specific attributes and how they are represented for the purpose of analysis to establish the roles and communication needs of stakeholders. These attributes are as follows:

- **Knowledge:** Level of information and understanding possessed by the representative about the case study. This information is obtained by asking the representative a set of questions. Knowledge attribute could be expressed as a five-scale value: Very High, High, Average, Low, and Very Low.
- **Position:** Attitude and perspective of the representative towards the exercise, in terms of the degree of opposition or support expressed by the stakeholder representative. This attribute can be represented using a five-scale value: Supporter, Moderate Supporter, Neutral, Moderate Opponent, and Opponent.
- **Interest:** Level of interest shown by the representative in the case study, represented as a five-scale value: Very High, High, Average, Low, and Very Low.

In addition to the organisation-level analysis of stakeholder dimensions, the case studies also involved interviewing stakeholder individuals (or organisation representatives). The following figures show the distribution of stakeholders in terms of their knowledge, position, and interest (Figs. 6, 7, and 8).

Most stakeholders belong to the data providers and data users categories. This underlines the focus on the usage and exploitation of big data by the case studies. In general, the case study stakeholders rated high in terms of knowledge and interest, which could be attributable to the fact that each case study had an active big data solution. It also shows that the stakeholders across different sectors are actively involved in big data with an interest in facilitating the positive impacts of big data externalities. We coded the Likert scale for knowledge (1 to 5 scale), interest (1 to 5 scale), and position (-2 to $+2$ scale) levels indicated by the stakeholder individuals. Figure 9 shows the average characteristics of stakeholders to cross the case studies.

	Very low	Low	Average	High	Very high
Crisis				2	5
Culture					12
Energy					7
Environment		1	1	2	2
Healthcare					10
Smart City				1	3
Transport					10

Fig. 6 Knowledge level of stakeholder individuals in BYTE case studies

	Opponent	Moderate Opponent	Neutral	Moderate Supporter	Supporter
Crisis				2	6
Culture	1				13
Energy				5	2
Environment				3	3
Healthcare					8
Smart City				2	2
Transport					

Fig. 7 Position of stakeholder individuals in support of the BYTE case studies

	Very low	Low	Average	High	Very high
Crisis				2	5
Culture				1	12
Energy			3	3	2
Environment			3		3
Healthcare					9
Smart City				1	3
Transport		3	1	1	5

Fig. 8 Interest of stakeholder individuals in BYTE case studies

	Knowledge	Interest	Position
Crisis	4.71	4.71	1.71
Culture	5	4.92	2
Energy	5	4	1.28
Environment	3.83	4	1.5
Healthcare	5	5	2
Smartcity	4.75	4.75	1.5
Transport	5	3.8	

Fig. 9 Average levels of knowledge, support position, and interest of stakeholders

6.5   Stakeholder Influence

Identification of stakeholder influence is an important step to classify stakeholders. By understanding a stakeholder's influence, we can better understand their relationships within the case study. Influence can be understood in terms of the amount of power a stakeholder has over the system. Influence can be both formal and informal. Formal influence is primarily based on rules or rights as laid down in legislation or formal agreements (i.e. law and rights to enforce the law, or usage rights). Informal influences are based on other factors such as interest groups or non-governmental organisations that can mobilise media, use resources, or lobby to put pressure on the ecosystem.

Table 2 Influence of different data stakeholders based on case studies

Stakeholder	Type	Influence
Environment case study		
EC (European Commission)	Governmental organisation	High
EEC (European Economic Committee)	Governmental organisation	High
EPA (Environmental Protection Agency)	Governmental organisation	High
EuroGeoSurveys	Governmental organisation	Medium
EUStatCen	Governmental organisation	Medium
IEEE	Not-for-profit organisation	High
UNEP	International organisation	Medium
Crisis management case study		
RICC	Research institute	Medium
ICRC	International organisation	Medium
OCHA	International organisation	Medium
Cultural case study		
National cultural heritage institution	Governmental organisation	Low
National data aggregator	Governmental organisation	Medium
Pan-European cultural heritage organisation	International organisation	Medium
National policy office	Governmental organisation	High
Citizens	Citizens	Low
Educational institutions	Educational institution	Medium
Open data advocates	Non-governmental organisation	Medium
Private sector cultural data consultancy	Small & medium Enterprise	Medium
Energy case study		
StatOil	Large corporation	Medium
ConcoPhillips	Large corporation	Medium
Lundin	Large corporation	Medium
EniNorge	Large corporation	Medium
Supplier	Large corporation	Low
Norwegian Petroleum Directorate	Governmental organisation	Medium
Health case study		
Public sector health research initiative	Research institute	High
Geneticists	Skilled professionals	Medium
Clinicians	Skilled professionals	Medium
Data scientists	Skilled professionals	High
Pharmaceutical companies	Large corporation	Medium
Translational medicine specialists	Experts	Medium
Public health research initiative	Research institute	High
NHS regional genetics laboratory	Government organisation	Medium
Charity organisations	Charity organisations	Low
Privacy and data protection policymakers	Policymakers	High
Privacy and data protection policy lawyers	Skilled professionals	Medium
Citizens	Citizens	Low
Patients and immediate family members	Citizens	Low

(continued)

Table 2 (continued)

Stakeholder	Type	Influence
Transport case study		
Established ship owner	Large corporation	Medium
New ship owner	Large corporation	Low
European yard	Governmental organisation	High
Navigation equipment supplier	Large corporation	Medium
Machinery subsystem supplier	Large corporation	Medium
Shipping association	Not-for-profit corporation	Low
Maritime consulting company	Small & medium enterprise	Medium
Classification society	Not-for-profit corporation	Medium
National Coastal Authority	Governmental organisation	Low
Utilities/smart cities case study		
European City	Governmental organisation	Medium
Technology provider – SME	Small & medium enterprise	Medium
Technology provider	Not-for-profit corporation	Medium
Technology provider & research	Large corporation	High

This section provides a cross-case analysis of the power or influence of the stakeholders in the data ecosystem. This cross-case analysis was performed using a questionnaire, interviews, and workshops conducted as part of the BYTE project. We provide an analysis of stakeholders in terms of their influence on the data ecosystem and its externalities (Table 2). This analysis is performed at the group level of stakeholders. The objective of the analysis is to classify stakeholder groups and organisations according to their capability to affect or influence the data ecosystem. In general, civil society organisations and citizens have low to medium influence on data ecosystems, which is a cause for concern. This is also true for stakeholders in the cultural sector. To address this, better incentives and a better engagement approach are required for these stakeholders to meaningfully contribute to the big data community.

7 Summary

This chapter analysed the stakeholders in European big data with the help of sectoral case studies. It also examined the stakeholder relationships within and between different categories. Although preliminary, the results of the analysis indicate that, in general, the innovation in data technologies is driven by sector-specific demands. Environment, energy, and smart city sectors show maturity in data technologies. Transport, healthcare, crisis management, and culture sectors require more engagement with the big data community for better adoption of useful technologies and influencing the European policy to address their needs.

Acknowledgements We thank the participants of the BYTE project focus groups and workshops for their insightful contributions. This work was funded by the European Union's Seventh Framework Programme FP7/2007-2013/CSA under grant agreement n° 619551. This publication has emanated from research supported in part by a research grant from Science Foundation Ireland (SFI) under Grant Number SFI/12/RC/2289_P2, co-funded by the European Regional Development Fund.

References

Cuquet, M., Vega-Gorgojo, G., Lammerant, H., et al. (2017). *Societal impacts of big data: Challenges and opportunities in Europe*. arXiv:170403361

Curry, E. (2016). The Big data value chain: Definitions, concepts, and theoretical approaches. In J. M. Cavanillas, E. Curry, & W. Wahlster (Eds.), *New horizons for a data-driven economy: A roadmap for usage and exploitation of big data in Europe*. New York: Springer.

Freeman, R. E. (1984). *Strategic management: A stakeholder approach.*

Grimble, R., & Wellard, K. (1997). Stakeholder methodologies in natural resource management: A review of principles, contexts, experiences and opportunities. *Agricultural Systems, 55*, 173–193.

Grimble, R., Chan, M.-K., Aglionby, J., & Quan, J. (1995). Trees and trade-offs: A stakeholder approach to natural resource management. *Gatekeeper Series, 52*, 18.

Lammerant, H., De Hert, P., Lasierra Beamonte, N., et al. (2015). *Horizontal analysis of positive and negative societal externalities.*

Manyika, J., Chui, M., Brown, B., et al (2011). *Big data: The next frontier for innovation, competition, and productivity*. McKinsey Global Institute.

Miles, S. (2012). Stakeholder: Essentially contested or just confused? *Journal of Business Ethics*. https://doi.org/10.1007/s10551-011-1090-8

Nolan, R., & McFarlan, F. W. (2005). Information technology and the board of directors. *Harvard Business Review, 96–106*(157), 83.

Pouloudi, A. (1999). Aspects of the stakeholder concept and their implications for information systems development. *Proceedings of the 32nd Annual Hawaii International Conference on Systems Sciences*. https://doi.org/10.1109/HICSS.1999.772776

Reed, M. S., Graves, A., Dandy, N., et al. (2009). Who's in and why? A typology of stakeholder analysis methods for natural resource management. *Journal of Environmental Management, 90*, 1933–1949. https://doi.org/10.1016/j.jenvman.2009.01.001

Rogers, E. M. (1962). *Diffusion of innovations*. The Free Press.

Weyer, M. V. (1996). In an ideal world. *Management Today*, 34–38.

Yin RK (2013) Case Study Research: Design and Methods. SAGE Publications 26:93–96. https://doi.org/10.1017/CBO9781107415324.004

A Roadmap to Drive Adoption of Data Ecosystems

Sonja Zillner, Laure Le Bars, Nuria de Lama, Simon Scerri, Ana García Robles, Marie Claire Tonna, Jim Kenneally, Dirk Mayer, Thomas Hahn, Södergård Caj, Robert Seidl, Davide Dalle Carbonare, and Edward Curry

The original version of this chapter was revised with an update in the affiliation of the author Davide Dalle Carbonare. A correction to this chapter can be found at https://doi.org/10.1007/978-3-030-68176-0_17

S. Zillner (✉)
Siemens AG, Munich, Germany
e-mail: sonja.zillner@siemens.com

L. Le Bars
SAP, Paris, France

N. de Lama
Atos, Madrid, Spain

S. Scerri
Fraunhofer IAIS, Sankt Augustin, Germany

A. García Robles
Big Data Value Association, Bruxelles, Belgium

M. C. Tonna
Digital Catapult, London, UK

J. Kenneally
Intel, Leixlip, Ireland

D. Mayer
Software AG, Saarbrücken, Germany

T. Hahn
Siemens AG, Erlangen, Germany

S. Caj
VTT, Espoo, Finland

R. Seidl
Nokia Bell Labs, Munich, Germany

D. D. Carbonare
Engineering Ingegneria Informatica, Rome, Italy

E. Curry
Insight SFI Research Centre for Data Analytics, NUI Galway, Galway, Ireland

Abstract To support the adoption of big data value, it is essential to foster, strengthen, and support the development of big data value technologies, successful use cases and data-driven business models. At the same time, it is necessary to deal with many different aspects of an increasingly complex data ecosystem. Creating a productive ecosystem for big data and driving accelerated adoption requires an interdisciplinary approach addressing a wide range of challenges from access to data and infrastructure, to technical barriers, skills, and policy and regulation. In order to overcome the adoption challenges, collective action from all stakeholders in an effective, holistic and coherent manner is required. To this end, the Big Data Value Public-Private Partnership (BDV PPP) was established to develop the European data ecosystem and enable data-driven digital transformation, delivering maximum economic and societal benefit, and achieving and sustaining Europe's leadership in the fields of big data value creation and Artificial Intelligence. This chapter describes the different steps that have been taken to address the big data value adoption challenges: first, the *establishment of the BDV PPP* to mobilise and create coherence with all stakeholders in the European data ecosystem; second, the introduction of *five strategic mechanisms* to encourage cooperation and coordination in the data ecosystem; third, a *three-phase roadmap* to guide the development of a healthy European data ecosystem; and fourth, a systematic and strategic approach towards actively *engaging the key communities* in the European Data Value Ecosystem.

Keywords Big data value · Public Private Partnership · European data ecosystem · Adoption of big data

1 Introduction

To support the adoption of big data value, it is essential to foster, strengthen and support the development of big data value technologies, successful use cases and data-driven business models. At the same time, it is necessary to deal with many different aspects of an increasingly complex data ecosystem. Creating a productive ecosystem for big data and driving accelerated adoption was possible by relying on an interdisciplinary approach addressing a wide range of central challenges from access to data and infrastructure, to technical barriers, skills, and policy and regulation. Given the broad range of challenges and opportunities with big data value, new instruments, an aligned implementation roadmap and a strategic approach towards cooperation were needed. In this chapter, we set out such a strategy, the formulation of which is the result of an inclusive discussion process involving a large number of relevant European Big Data Value (BDV) stakeholders. The result is an interdisciplinary approach that integrates expertise from the different fields necessary to tackle both the strategic and specific objectives. To this end, the Big Data Value Public-Private Partnership was established to develop the European data ecosystem and enable data-driven digital transformation, delivering maximum economic and societal benefit, and achieving and sustaining Europe's leadership in the fields of big data value creation and Artificial Intelligence.

This chapter starts by detailing the adoption challenges of big data value and all the different steps that were taken to overcome the adoption challenges: first, the *establishment of the Big Data Value Public-Private Partnership (BDV PPP)* to mobilise and create coherence with all stakeholders in the European data ecosystem; second, the introduction of *five strategic mechanisms* to encourage cooperation and coordination in the data ecosystem; third, a *three-phase roadmap* to guide the development of a healthy European data ecosystem; and fourth, a systematic and strategic approach towards actively *engaging the key communities* in the European Data Value Ecosystem.

2   Challenges for the Adoption of Big Data Value

To support the adoption of big data value, it was important to foster, strengthen and support the development of big data value technologies, successful use cases and data-driven business models. At the same time, it was necessary to deal with many different aspects of an increasingly complex data ecosystem. Building on the analysis provided in the literature (Cavanillas et al. 2016; Zillner et al. 2017, 2020), the main challenges that needed to be tackled to create and sustain a robust big data ecosystem have been as follows:

- **Access to Data and Infrastructures:** Availability of data sources and access to data infrastructures is paramount. There is a broad range of data types and data sources: structured and unstructured data, multi-lingual data sources, data generated from machines and sensors, data-at-rest and data-in-motion. Value is created by acquiring data, combining data from different sources and providing access to it with low latency, while ensuring data integrity and preserving privacy. Pre-processing, validating, augmenting data, and ensuring data integrity and accuracy add value. Both academics and innovators (SMEs and start-ups in particular) need proper access to world-class innovation infrastructures, including to data and infrastructure resources such as High Performance Computing (HPC) and test environments.
- **Higher Complexity of Data-driven Applications in Industry and Public Domain:** Novel applications and solutions must be developed and validated in ecosystems to deliver value creation from the data ecosystem. However, implementing data value and data-driven AI in industrial and public environments relies on incorporating the domain knowledge of underlying processes. Handling these challenges requires combining domain-specific process knowledge with knowledge on data-driven approaches.
- **Lack of Skills and Know-How:** To leverage the potential of big data value, a key challenge is to ensure the availability of highly and appropriately skilled people who have an excellent grasp of the best practices and technologies for delivering big data value within applications and solutions. Data experts need to be connected to other experts with strong domain knowledge and the ability to apply this know-how within organisations for value creation. Many European

organisations lack the skills to manage or deploy data-driven solutions with global competition for talent under way.

- **Policy and Regulation Uncertainty:** The increased importance of data will intensify the debate on data ownership and usage, data protection and privacy, security, liability, cybercrime, Intellectual Property Rights (IPR), and the impact of insolvencies on data rights. These issues have to be resolved to remove the adoption barriers. In the area of data-driven AI, policy and regulation are still unclear in areas including liability, right to explain and data access. Many organisations have concerns about compliance.
- **Technical Barriers:** There is considerable complexity and cost in creating systems with the ability to collect, process, and analyse large quantities of data to make robust and trustworthy decisions and implement autonomy. Key aspects such as real-time analytics, low latency and scalability in processing data, new and rich user interfaces, and interacting with and linking data, information and content all have to be advanced to open up new opportunities and to sustain or develop competitive advantages. Interoperability of data sets and data-driven solutions, as well as agreed approaches, is essential for a wide adoption within and across sectors.
- **Digitalisation of Business:** Businesses have to increase their digitalisation effort to maintain their competitive advantage within a Digital Single Market. A more efficient use of big data, and understanding data as an economic asset, carries great potential for the economy and society. The setup of big data value ecosystems and the development of appropriate business models on top of a strong big data value chain must be supported to generate the desired impact on the economy and employment.
- **Societal Trust in Data:** Big data will provide solutions for major societal challenges in Europe, such as improved efficiency in healthcare information processing and reduced CO_2 emissions through climate impact analysis. However, there are many misconceptions and much misinformation about data-driven systems in societal debates, and the technology seems not to be fully accepted by society in all application areas. It is critical for accelerated adoption of big data to increase awareness of the benefits and the value that big data can create for business, the public sector, the citizen, and the environment.
- **EU Private Investment Environment:** Still lagging behind other parts of the world within its investments in digitalisation, Europe needs to create a competitive, forward-looking private investments ecosystem to boost innovation in data and data-driven AI in a fast and focused way.

Creating a productive ecosystem for big data and driving accelerated adoption requires an interdisciplinary approach addressing all of the challenges above in collective action from all stakeholders working together in an effective, holistic and coherent manner.

3 Big Data Value Public-Private Partnership

Europe must aim high and mobilise stakeholders in society, industry, academia and research to enable a European big data value economy, supporting and boosting agile business actors, delivering products, services and technology, and providing highly skilled data engineers, scientists and practitioners along the entire big data value chain. This will result in an innovation ecosystem in which value creation from big data flourishes.

To achieve these goals, the **European contractual Public-Private Partnership** on **Big Data Value (BDV PPP)** was signed on 13 October 2014. This signature marks the commitment by the European Commission, industry and academia partners to build a data-driven economy across Europe, mastering the generation of value from big data and creating a significant competitive advantage for European industry, boosting economic growth and employment. The **Big Data Value Association (BDVA)** is the private counterpart to the EU Commission in implementing the BDV PPP programme. BDVA has a well-balanced composition of large, small and medium-sized industries and enterprises as well as research organisations to support the development and deployment of the PPP work programme and to achieve the Key Performance Indicators (KPI) committed in the PPP contract. The BDV PPP commenced in 2015 and was operationalised with the launch of the LEIT work programme 2016/2017. The BDV PPP activities address technology and applications development, business model discovery, ecosystem validation, skills profiling, regulatory and IPR environment, and social aspects. The BDV PPP did lead to a comprehensive innovation ecosystem fostering and sustaining European leadership on big data and delivering maximum economic and societal benefit to Europe – its business and its citizens (see Chap. "Achievements and Impact of the Big Data Value Public-Private Partnership: The Story so Far" for more details).

3.1 The Big Data Value Ecosystem

A data ecosystem is a socio-technical system enabling value to be extracted from data value chains supported by interacting organisations and individuals (Curry 2016). Within an ecosystem, data value chains are oriented to business and societal purposes. The ecosystem can create the conditions for marketplace competition between participants or can enable collaboration among diverse, interconnected participants that depend on each other for their mutual benefit.

The clear goal of the BDV PPP was to develop a European data ecosystem that enables data-driven digital transformation in Europe, delivers maximum economic and societal benefit, and fosters and sustains Europe's leadership in the fields of big data value creation and Artificial Intelligence. The ecosystem is established on a set of principles to ensure openness, inclusion and incubation (see Table 1).

Table 1 The principles of the big data value ecosystem

Focus	BDV ecosystem principle
Openness	The ecosystem should embrace and contribute to openness in terms of data, technology standards, interoperability, best practice, education and innovation.
Connectedness	The ecosystem should prioritise connectedness and synergies between actors, industrial sectors, and languages, and across borders.
Cross-sectorial	The ecosystem should involve all relevant sectors of society and scientific disciplines.
Sustainable	The ecosystem should strive for self-sustainability in structure and function, in order to maintain long-term ecosystem services.
Co-evolution	The ecosystem should support competition, cooperation and co-evolution between actors.
Incubation	The ecosystem should incubate start-ups and entrepreneurs.

4 Five Mechanism to Drive Adoption

In order to implement the research and innovation strategy, and to align technical issues with aspects of cooperation and coordination, five major types of mechanisms were identified:

- **Innovation Spaces (i-Spaces):** Cross-organisational and cross-sectorial environments that allow challenges to be addressed in an interdisciplinary way and serve as a hub for other research and innovation activities
- **Lighthouse projects:** To raise awareness of the opportunities offered by big data and the value of data-driven applications for different sectors, acting as incubators for data-driven ecosystems
- **Technical projects:** To tackle specific big data issues, addressing targeted aspects of the technical priorities
- **Data platforms:** To support the sharing and trading of industrial and personal data (free flow of data) as a key enabler of the data economy
- **Cooperation and coordination projects:** To foster international cooperation for efficient information exchange and coordination of activities within the ecosystem

4.1 European Innovation Spaces (i-Spaces)

Extensive consultation with many stakeholders from areas related to big data value (BDV) had confirmed that in addition to technology and applications, several key issues required consideration. First, infrastructural, economic, social and legal issues have to be addressed. Second, the private and public sectors need to be made aware of the benefits that BDV can provide, thereby motivating them to be innovative and to adopt BDV solutions.

To address all of these aspects, European cross-organisational and cross-sectorial environments, which rely and build upon existing national and European initiatives, play a central role in a European big data ecosystem. These so-called European Innovation Spaces (or i-Spaces for short) are the main elements to ensure that research on BDV technologies and novel BDV applications can be quickly tested, piloted and thus exploited in a context with the maximum involvement of all the stakeholders of BDV ecosystems. As such, i-Spaces enable stakeholders to develop new businesses facilitated by advanced BDV technologies, applications and business models. They contribute to the building of communities, providing a catalyst for community engagement and acting as incubators and accelerators of data-driven innovation.

In this sense, i-Spaces are hubs for uniting technical and non-technical activities, for instance, by bringing technology and application development together and by fostering skills, competence and best practices. To this end, i-Spaces offer both state-of-the-art and emerging technologies and tools from industry, as well as open-source software initiatives; they also provide access to data assets. In this way, i-Spaces foster community building and an interdisciplinary approach to solving BDV challenges along the core dimensions of technology, applications, legal, social and business issues, data assets, and skills.

The creation of i-Spaces is driven by the needs of large and small companies alike to ensure that they can easily access the economic opportunities offered by BDV and develop working prototypes to test the viability of actual business deployments. This does not necessarily require moving data assets across borders; rather, data analytic tools and computation activities are brought to the data. In this way, valuable data assets are made available in environments that simultaneously support the legitimate ownership, privacy and security policies of corporate data owners and their customers, while facilitating ease of experimentation for researchers, entrepreneurs and small and large IT providers.

Concerning the discovery of value creation, i-Spaces support various models: at one end, corporate entities with valuable data assets can specify business-relevant data challenges for researchers or software developers to tackle; at the other end, entrepreneurs and companies with business ideas to be evaluated can solicit the addition and integration of desired data assets from corporate or public sources. i-Spaces also contribute to filling the skills gap Europe is facing in providing (controlled) access to real use cases and data assets for education and skills improvement initiatives.

i-Spaces themselves are data-driven, both at the planning and the reporting stage. At the planning stage, they prioritise the inclusion of data assets that, in conjunction with existing assets, present the greatest promise for European economic development (while taking full account of the international competitive landscape); at the reporting stage, they provide methodologically sound quantitative evidence on important issues such as increases in performance for core technologies or reductions in costs for business processes. These reports have been an important basis to foster learning and continuous improvement for the next cycle of technology and applications.

The particular value addition of i-Spaces in the European context is that they federate, complement and leverage activities of similar national incubators and environments, existing PPPs, and other national or European initiatives. With the aim of not duplicating existing efforts, complementary activities considered for inclusion have to stand the test of expected economic development: new data assets and technologies are considered for inclusion to the extent that they can be expected to open new economic opportunities when added to and interfaced with the assets maintained by regional or national data incubators or existing PPPs.

Over recent years, the successive inclusion of data assets into i-Spaces, in turn, has driven and prioritised the agenda for addressing data integration or data processing technologies. One example is the existence of data assets with homogenous qualities (e.g. geospatial factors, time series, graphs and imagery), which called for optimising the performance of existing core technology (e.g. querying, indexing, feature extraction, predictive analytics and visualisation). This required methodologically sound benchmarking practices to be carried out in appropriate facilities. Similarly, business applications exploiting BDV technologies have been evaluated for usability and fitness for purpose, thereby leading to the continuous improvement of these applications.

Due to the richness of data that i-Spaces offer, as well as the access they afford to a large variety of integrated software tools and expert community interactions, the data environments provide the perfect setting for the effective training of data scientists and domain practitioners. They encourage a broader group of interested parties to engage in data activities. These activities are designed to complement the educational offerings of established European institutions.

4.2 Lighthouse Projects

Lighthouse projects[1] are projects with a high degree of innovation that run large-scale data-driven demonstrations whose main objectives are to create high-level impact and to promote visibility and awareness, leading to faster uptake of big data value applications and solutions.

They form the major mechanism to demonstrate big data value ecosystems and sustainable data marketplaces, and thus promote increased competitiveness of established sectors as well as the creation of new sectors in Europe. Furthermore, they propose replicable solutions by using existing technologies or very near-to-market technologies that show evidence of data value and could be integrated in an innovative way.

Lighthouse projects lead to explicit business growth and job creation, which is measured by the clear indicators and success factors that had been defined by all projects in both a qualitative and quantitative manner beforehand.

---

[1]Sometimes also labelled as large-scale demonstrations or pilots.

Increased competitiveness is not only a result of the application of advanced technologies; it also stems from a combination of changes that expand the technological level, as well as political and legal decisions, among others. Thus, Lighthouse projects were expected to involve a combination of decisions centred on data, including the use of advanced big data-related technologies, but also other dimensions. Their main purpose has been to render results visible to a widespread and high-level audience to accelerate change, thus allowing the explicit impact of big data to be made in a specific sector, and a particular economic or societal ecosystem.

Lighthouse projects are defined through a set of well-specified goals that materialise through large-scale demonstrations deploying existing and near-to-market technologies. Projects may include a limited set of research activities if that is needed to achieve their goals, but it is expected that the major focus will be on data integration and solution deployment.

Lighthouse projects are different from Proof of Concepts (which are more related to technology or process) or pilots (which are usually an intermediate step on the way to full production): they need to pave the way for a faster market roll-out of technologies (big data with Cloud and HPC or the IoT), they need to be conducted on a large scale, and they need to use their successes to rapidly transform the way an organisation thinks or the way processes are run.

Sectors or environments that were included were not pre-determined but had been in line with the goal mentioned above of creating a high-level impact.

The first call for Lighthouse projects made by the BDV PPP resulted in two actions in the domains of bioeconomy (including agriculture, fisheries and forestry) and transport and logistics. The second call resulted in two actions for health and smart manufacturing.

Lighthouse projects operate primarily in a single domain, where a meaningful (as evidenced by total market share) group of EU industries from the same sector can jointly provide a safe environment in which they make available a proportion of their data (or data streams) and demonstrate, on a large scale, the impact of big data technologies. Lighthouse projects used data sources other than those of the specific sector addressed, thereby contributing to breaking silos. In all cases, projects did enable access to appropriately large, complex and realistic datasets.

Projects needed to show sustainable impact beyond the specific large-scale demonstrators running through the project duration. Whenever possible, this was addressed by projects through solutions that could be replicated by other companies in the sector or by other application domains.

All Lighthouse projects were requested to involve all relevant stakeholders to reach their goals. This again did lead to the development of complete data ecosystems of the addressed domain or sector. Whenever this was appropriate, Lighthouse projects did rely on the infrastructure and ecosystems facilitated by one or more i-Spaces.

Some of the indicators that were used to assess the impact of Lighthouse projects have been the number and size of datasets processed (integrated), the number of data sources made available for use and analysis by third parties, and the number of services provided for integrating data across sectors. Market indicators are obviously of utmost importance.

Key elements for the implementation of Lighthouse projects include at least the following areas.

The Use of Existing or Close-to-Market Technologies Lighthouses have not been expected to develop entirely new solutions; instead, they have been requested to make use of existing or close-to-market technologies and services by adding and/or adapting current relevant technologies, as well as accelerating the roll-out of big data value solutions using the Cloud and the IoT or HPC. Solutions should provide answers for real needs and requirements, showing an explicit knowledge of the demand side. Even though projects were asked to concentrate on solving concrete problems which again might easily lead to specific deployment challenges, the replicability of concepts was always a high priority to ensure impact beyond the particular deployments of the project. Lighthouse projects have been requested to address frameworks and tools from a holistic perspective, considering, for example, not only analytics but also the complete data value chain (data generation, the extension of data storing and analysis).

Interoperability and Openness All projects did take advantage of both closed and open data; during the project, they could determine if open source or proprietary solutions were the most suitable to address their challenges. However, it was always requested that projects promote the interoperability of solutions to avoid locking in customers.

The involvement of smaller actors (e.g. through opportunities for start-ups and entrepreneurs) who can compete in the same ecosystem in a fair way was always a must. For instance, open Application Programming Interfaces (APIs) had been identified as an important way forward (e.g. third-party innovation through data sharing). In addition, projects have been requested to focus on re-usability and ways to reduce possible barriers or gaps resulting from big data methods impacting end-users (break the 'big data for data analysts only' paradigm).

Performance All projects have been requested to contribute to common data collection systems and to have a measurement methodology in place. Performance monitoring was accomplished over at least two-thirds of the duration of the project.

The Setting Up of Ecosystems Lighthouse projects have a transformational power, that is, they had never been restricted to any type of narrow-minded experiments with limited impact. All projects demonstrated that they could improve (sometimes changing associated processes) the competitiveness of the selected industrial sector in a relevant way. To achieve this, the active involvement of different stakeholders is mandatory. For that reason, the supporting role of the ecosystem that enabled such changes is an important factor to keep in mind: All Lighthouse projects had been connected to communities of stakeholders from the design phase. Ecosystems evolved, extended or connected with existing networks of stakeholders and hubs, whenever this was possible.

As is well known, the European industry is characterised by a considerable number of small and medium-sized enterprises. Therefore, the adequate consideration of SME integration in the projects was always a central requirement to create a healthy environment.

Even though all projects had been requested to primarily focus on one particular sector, the use of data from different sources and industrial fields had always been encouraged, with priority given to avoiding the 'silo' effect.

Long-Term Commitment and Sustainability The budgets assigned to the projects have been envisioned as seeds for more widely implemented plans. All funded activities had been integrated into more ambitious strategies that allowed for the involvement of additional stakeholders and further funding (preferably private but also possibly a combination of public and private).

After the launch of the four initial Lighthouse projects, all learnings related to the concept of Lighthouse projects could be consolidated. As a result, a more advanced concept had been proposed including more concrete requirements for the upcoming large-scale pilots, in some cases further specifying aspects that had already been worked out. The following list served as guidance without the claim of completeness:

- It is important to reuse technologies and frameworks by combining and adapting relevant existing technologies (big data with the Cloud, HPC or IoT) that are already in the market or close to it (i.e. those with a high technology readiness level) to avoid the development of new platforms where a reasonable basis already exists (e.g. as part of the Open Source community). In addition, projects are especially encouraged to build on the technologies created by the ongoing projects of the Big Data PPP that fit their requirements (e.g. in the area of privacy-preserving technologies).
- Particular attention should be paid to interoperability. This applies to all layers of the solution, including data (here, some of the results of the projects funded under the Big Data PPP with a focus on data integration could be particularly useful), and to relevant efforts within the HPC, Cloud and IoT communities.
- It is expected that projects will combine the use of open and closed data. While it is understandable that some closed data will remain as such, we also expect these projects to contribute to the increasing availability of datasets that could be used by other stakeholders, such as SMEs and start-ups. This could happen under different regimes (not necessarily for free). Projects should declare how they will contribute to this objective by quantifying and qualifying datasets (when possible) and by including potential contributions to the ongoing data incubators/accelerators and Innovation Spaces.
- Lighthouse projects have to contribute to the horizontal activities of the Big Data PPP as a way of helping in the assessment of the PPP implementation and increasing its potential impact. Some of the targeted activities include contributing to the standardisation of activities, the measurement of KPIs, and coordination with the PPP branding, or active participation in training and educational activities proposed by the PPP.

4.3 Technical Projects

Technical projects focus on addressing one issue or a few specific aspects identified as part of the BDV technical priorities. In this way, technical projects provide the technology foundation for Lighthouse projects and i-Spaces. Technical projects may be implemented as Research and Innovation Actions (RIA) or Innovation Actions (IA), depending on the amount of research work required to address the respective technical priorities.

To identify the most important technical priorities to be addressed within these projects, the stakeholders within the data ecosystem had been engaged within a structured methodology to produce a set of consolidated cross-sectorial technical research requirements. The result of this process was the identification of five key technical research priorities (data management, data processing architectures, deep analytics, data protection and pseudonymisation, advanced visualisation and user experience) together with 28 sub-level challenges to delivering big data value (Zillner et al. 2017). Based on this analysis, the overall, strategic technical goal could be summarised as follows:

> Deliver big data technology empowered by deep analytics for data-at-rest and data-in-motion, while providing data protection guarantees and optimised user experience, through sound engineering principles and tools for data-intensive systems.

Further details on the technical priorities and how they were defined are provided in Chap. "Technical Research Priorities for Big Data". The Big Data Value Reference Model, which structures the technical priorities identified during the requirements analysis, is detailed in Chap. "A Reference Model for Big Data Technologies".

4.4 Platforms for Data Sharing

Platform approaches have proved successful in many areas of technology (Gawer and Cusumano 2014), from supporting transactions among buyers and sellers in marketplaces (e.g. Amazon), to innovation platforms which provide a foundation on top of which to develop complementary products or services (e.g. Windows), to integrated platforms which are a combined transaction and innovation platform (e.g. Android and the Play Store).

The idea of large-scale "data" platforms has been touted as a possible next step to support data ecosystems (Curry and Sheth 2018). An ecosystem data platform would have to support continuous, coordinated data flows, seamlessly moving data among intelligent systems. The design of infrastructure to support data sharing and reuse is still an active area of research (Curry and Ojo 2020).

Data sharing and trading are seen as important ecosystem enablers in the data economy, although closed and personal data present particular challenges for the free flow of data. The following two conceptual solutions – Industrial Data Platforms (IDP) and Personal Data Platforms (PDP) – introduce new approaches to addressing this particular need to regulate closed proprietary and personal data.

4.4.1 Industrial Data Platforms (IDP)

IDPs have increasingly been touted as potential catalysts for advancing the European Data Economy as a solution for emerging data markets, focusing on the need to offer secure and trusted data sharing to interested parties, primarily from the private sector (industrial implementations). The IDP conceptual solution is oriented towards proprietary (or closed) data, and its realisation should guarantee a trusted, secure environment within which participants can safely, and within a clear legal framework, monetise and exchange their data assets. A functional realisation of a continent-wide IDP promises to significantly reduce the existing barriers to a free flow of data within an advanced European Data Economy. The establishment of a trusted data-sharing environment will have a substantial impact on the data economy by incentivising the marketing and sharing of proprietary data assets (currently widely considered by the private sector as out of bounds) through guarantees for fair and safe financial compensations set out in black and white legal terms and obligations for both data owners and users. The 'opening up' of previously guarded private data can thus vastly increase its value by several orders of magnitude, boosting the data economy and enabling cross-sectorial applications that were previously unattainable or only possible following one-off bilateral agreements between parties over specific data assets.

The IDP conceptual solution complements the drive to establish BDVA i-Spaces by offering existing infrastructure and functional technical solutions that can better regulate data sharing within the innovation spaces. This includes better support for the secure sharing of proprietary or 'closed' data within the trusted i-Space environment. Moreover, i-Spaces offer a perfect testbed for validating existing implementations of conceptual solutions such as the IDP.

The identified possibilities for action can be categorised into two branches:

- **Standardisation**: Addressing the lack of an existing standard platform (technical solution) that limits stakeholders from participating in the European Digital Single Market, and the availability of clear governance models (reference models, guidelines and best practices) regulating the secure and trusted exchange of proprietary data.
- **Implementation**: Establishing, developing or aligning existing IDP implementations to provide a functional European-wide infrastructure within which industrial participants can safely, and within a clear legal framework, monetise and exchange data assets.

Standardisation activities outlined by the Strategic Research and Innovation Agenda (SRIA) (Zillner et al. 2017) and in Chap. "Recognition of Formal and Non-formal Training in Data Science" have taken into account the need to accommodate activities related to the evolving IDP solutions. The opportunity to drive forward emerging standards also covers the harmonisation of reference archi- tectures and governance models put forward by the community. Notable advanced contributions in this direction include the highly relevant white paper and the reference architecture[2] provided by the Industrial Data Space (IDS) Association. The Layered Databus, introduced by the Industrial Internet Consortium,[3] is another emerging standard advocating the need for data-centric information-sharing tech- nology that enables data market players to exchange data within a virtual and global data space.

The implementation of IDPs needs to be approached on a European level, and existing and planned EU-wide, national and regional platform development activi- ties could contribute to these efforts. The industries behind existing IDP implementations, including the IDS reference architecture and other examples such as the MindSphere Open Industrial Cloud Platform,[4] can be approached to move towards a functional European Industrial Data Platform. The technical prior- ities outlined by the SRIA (Zillner et al. 2017), particularly the Data Management priority, need to address data management across a data ecosystem comprising both open and closed data. The broadening of the scope of data management is also reflected in the latest BDVA reference model, which includes an allusion to the establishment of a digital platform whereby marketplaces regulate the exchange of proprietary data.

4.4.2 Personal Data Platforms (PDP)

So far, consumers have trusted companies, including Google, Amazon, Facebook, Apple and Microsoft, to aggregate and use their personal data in return for free services. While EU legislation, through directives such as the Data Protection Directive (1995) and the ePrivacy Directive (1998), has ensured that personal data can only be processed lawfully and for legitimate use, the limited user control offered by such companies and their abuse of a lack of transparency have undermined consumers' trust. In particular consumers experience everyday leakage of their data, traded by large aggregators in the marketing networks for value only returned to consumers in the form of often unwanted digital advertisements. This has recently led to a growth in the number of consumers adopting adblockers to protect

---

[2]Reference Architecture Model for the Industrial Data Space, April 2017, https://www.fraunhofer. de/content/dam/zv/de/Forschungsfelder/industrial-data-space/Industrial-Data-Space_Reference-Architecture-Model-2017.pdf

[3]The Industrial Internet of Things, Volume G1: Reference Architecture, January 2017, https:// www.iiconsortium.org/IIC_PUB_G1_V1.80_2017-01-31.pdf

[4]*MindSphere: The cloud-based, open IoT operating system for digital transformation*, Siemens, 2017, https://www.plm.automation.siemens.com/media/global/en/Siemens_MindSphere_ Whitepaper_tcm27-9395.pdf

their digital life,[5] while at the same time they are becoming more conscious of and suspicious about their personal data trail.

In order to address this growing distrust, the concept of Personal Data Platforms (PDP) has emerged as a possible solution that could allow data subjects and data owners to remain in control of their data and its subsequent use.[6] PDPs leverage 'the concept of user-controlled cloud-based technologies for storage and use of personal data ("personal data spaces")'.[7] However, so far consumers have only been able to store and control access to a limited set of personal data, mainly by connecting their social media profiles to a variety of emerging Personal Information Management Systems (PIMS). More successful (but limited in number) uses of PDPs have involved the support of large organisations in agreeing to their customers accumulating data in their own self-controlled spaces. The expectation here is the reduction of their liability in securing such data and the opportunity to access and combine them with other data that individuals will import and accumulate from other aggregators. However, a degree of friction and the lack of a successful business model are still hindering the potential of the PDP approach.

A new driver behind such a self-managed personal data economy has recently started to appear. As a result of consumers' growing distrust, measures such as the General Data Protection Regulation (GDPR), which has been in force since May 2018, have emerged. The GDPR constitutes the single pan-European law on data protection, and, among other provisions and backed by the risk of incurring high fines, it will force all companies dealing with European consumers to (1) increase transparency and (2) provide users with granular control for data access and sharing and will (3) guarantee consumers a set of fundamental individual digital rights (including the right to rectification, erasure, data portability and to restrict processing). In particular, by representing a threat to the multi-billion euro advertising business, we expect individuals' data portability right, as enshrined in the GDPR, to be the driver for large data aggregators to explore new business models for personal data access. As a result, this will create new opportunities for PDPs to emerge. The rise of PDPs and the creation of more decentralised personal datasets will also open up new opportunities for SMEs that might benefit from and

---

[5]Used by 615 million devices at the end of 2016, http://uk.businessinsider.com/pagefair-2017-ad-blocking-report-2017-1?r=US&IR=T

[6]See a Commission paper on 'Personal information management services – current state of service offers and challenges' analysing feedback from public consultation: https://ec.europa.eu/digital-single-market/en/news/emerging-offer-personal-information-management-services-current-state-service-offers-and

[7]A Personal Data Space is a concept, framework and architectural implementation that enables individuals to gather, store, update, correct, analyse and/or share personal data. This is also a marked deviation from the existing environment where distributed data is stored throughout organisations and companies internally, with limited to no access or control from the user that the information concerns. This is a move away from the B2B (business to business) and B2C (business to consumer) models, with a move towards Me2B – when individuals start collecting and using data for their own purposes and sharing data with other parties (including companies) under their control (https://www.ctrl-shift.co.uk/news/2016/09/19/shifting-from-b2c-to-me2b/).

investigate new secondary uses of such data, by gaining access to them from user-controlled personal data stores – a privilege so far available only to large data aggregators. However, further debate is required to reach an understanding on the best business models (for demand and supply) to develop a marketplace for personal data donors, and on what mechanisms are required to demonstrate transparency and distribute rewards to personal data donors. Furthermore, the challenges organisations face in accessing expensive data storage, and the difficulties in sharing data with commercial and international partners due to the existence of data platforms which are considered to be unsafe, need to be taken into account. Last but not least, questions around data portability and interoperability also have to be addressed.

4.5 Cooperation and Coordination Projects

Cooperation and coordination projects aimed to work on detailed activities that ensured coordination and coherence in the PPP implementation and provided support to activities. The portfolio of support activities comprised support actions that addressed complementary, non-technical issues alongside the European Innovation Spaces, Lighthouse projects, data platforms, and research and innovation activities. In addition to the activities addressed, the governance of the data ecosystem, cooperation and coordination activities focused on the following.

Skills Development The educational support for data strategists and data engineers needs to meet industry requirements. The next generation of data professionals needs this wider view to deliver the data-driven organisation of the future. Skill development requirements need to be identified that can be addressed by collaborating with higher education institutes, education providers and industry to support the establishment of:

- New educational programmes based on interdisciplinary curricula with a clear focus on high-impact application domains
- Professional courses to educate and re-skill/up-skill the current workforce with the specialised skillsets needed to be data-intensive engineers, data scientists and data-intensive business experts
- Foundational modules in data science, statistical techniques and data management within related disciplines such as law and the humanities
- A network between scientists (academia) and industry that leverages Innovation Spaces to foster the exchange of ideas and challenges
- Datasets and infrastructure resources provided by industry that enhance the industrial relevance of courses.

Business Models and Ecosystems The big data value ecosystem will comprise many new stakeholders and will require a valid and sustainable business model. Dedicated activities for investigating and evaluating business models will be connected to the innovation spaces where suppliers and users will meet. These activities include:

- Delivering means for the systematic analysis of data-driven business opportunities
- Establishing a mapping of technology providers and their value contribution
- Identifying mechanisms by which data value is determined and value is established
- Providing a platform for data entrepreneurs and financial actors, including venture capitalists, to identify appropriate levels of value chain understanding
- Describing and validating business models that can be successful and sustainable in the future data-driven economy

Policy and Regulation The stakeholders of the data ecosystem need to contribute to the policy and regulatory debate about non-technical aspects of the future big data value creation as part of the data-driven economy. Dedicated activities addressed the aspects of data governance and usage, data protection and privacy, security, liability, cybercrime, and Intellectual Property Rights (IPR). These activities enabled the exchange between stakeholders from industry, end-users, citizens and society to develop input to ongoing policy debates where appropriate. Of equal importance was the identification of concrete legal problems for actors in the Value Chain, particularly SMEs that have limited legal resources. The established body of knowledge on legal issues was of high value for the wider community.

Social Perceptions and Societal Implications Societal challenges cover a wide range of topics including trust, privacy, ethics, transparency, inclusion efficacy, manageability and acceptability in big data innovations. There needs to be a common understanding in the technical community leading to an operational and validated method that applies to data-driven innovations development. At the same time, it is critical to develop a better understanding of inclusion and collective awareness aspects of big data innovations that enable a clear profile of the social benefits provided by big data value technology. By addressing the listed topics, the PPP ensured that citizens' views and perceptions were taken into account so that technology and applications were developed with a chance to be widely accepted.

5   Roadmap for Adoption of Big Data Value

The roadmap ensured and guided the development of the ecosystem in distinct phases, each with a primary theme. The three phases, as depicted in Fig. 1, are as follows:

- **Phase I**: Establish the ecosystem (governance, i-Spaces, education, enablers) and demonstrate the value of existing technology in high-impact sectors (Lighthouses, technical projects)

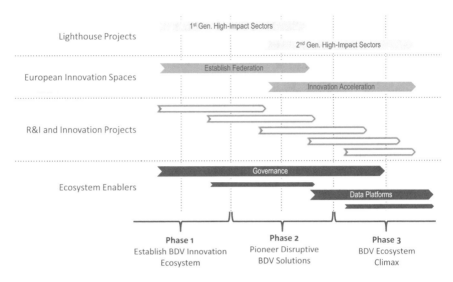

Fig. 1 Three-phase timeline of the adoption of Big Data Value PPP

- **Phase II**: Pioneer disruptive new forms of big data value solutions (Lighthouses and technical projects) in high-impact domains of importance for EU industry, addressing emerging challenges of the data economy
- **Phase III**: Develop long-term ecosystem enablers to maximise sustainability for economic and societal benefit, including the establishment of data platforms

Phase I: Establish an Innovation Ecosystem

The first phase of the roadmap focused on laying the foundations necessary to establish a sustainable European data innovation ecosystem. The key activities of Phase I included:

- Establishing a European network of i-Spaces for cross-sectorial and cross-lingual data integration, experimentation and incubation
- Demonstrating big data value solutions via large-scale pilot projects in domains of strategic importance for EU industry, using existing technologies or very near-to-market technologies.
- Tackling the main technology challenges of the data economy by improving the technology, methods, standards and processes for big data value
- Advancing state-of-the-art in privacy-preserving big data technologies and exploring the societal and ethical implications
- Establishing key ecosystem enablers, including support and coordination structures for industry skills and benchmarking.

Phase II: Disruptive Big Data Value

Building on the foundations established in Phase I, the second phase had a primary focus on Research and Innovation (R&I) activities to deliver the next generation of big data value solutions. The key activities of Phase II included:

- Supporting the emergence of the data economy with a particular focus on accelerating the progress of SMEs, start-ups and entrepreneurs, as well as best practices and standardisation
- Pioneering disruptive new forms of big data value solutions with the Cloud, HPC or IoT technologies via large-scale pilot projects in emerging domains of importance for EU industry using advanced platforms, tools and testbeds
- Tackling the next generation of big data research and innovation challenges for extreme-scale analytics
- Addressing ecosystem roadblocks and inhibitors to the take-up of big data value platforms for data ecosystem viability, including platforms for personal and industrial data
- Providing (continuing) support, facilitating networking and cooperation among ecosystem actors and projects, and promoting community building among BDV, Cloud, HPC and IoT activities.

Phase III: Long-Term Ecosystem Enablers
While the sustainability of the ecosystem has been considered from the start of the PPP, the third phase had a specific focus on activities that could ensure long-term self-sustainability. The key activities of Phase III included:

- Sowing the seeds for long-term ecosystems enablers to ensure self-sustainability
- Creating innovation projects within a federation of i-Spaces (European Digital Innovation Hubs for Big Data) to validate and incubate innovative big data value solutions and business models
- Ensuring continued support for technology outputs of PPP (Lighthouse projects, R&I, CSA), including non-technical aspects (training and Open Source Community, Technology Foundation)
- Establishing a Foundation for European Innovation Spaces with a charter to continue collaborative innovation activity, in line with the concept of European Digital Innovation Hub for Big Data
- Liaising with private funding (including Venture Capital) to accelerate entry into the market and socio-economic impacts, including the provision of ancillary services to develop investment-ready proposals and support scaling for BDV PPP start-ups and SMEs to reach the market
- Tackling the necessary strategy and planning for the BDV ecosystem until 2030, including the identification of new stakeholders, emerging usage domains, technology, business and policy roadmapping activity.

6   European Data Value Ecosystem Development

Developing the European Data Value Ecosystem is at the core of the mission and strategic priorities of the Big Data Value Association and the Big Data Value PPP. The European Data Value Ecosystem brings together communities (all the different stakeholders who are involved, affected or stand to benefit), technology, solutions and data platforms, experimentation, incubation and know-how resources, and the business models and framework conditions for the data economy. In this section, we refer to the 'community' and stakeholder aspect of the European big data value ecosystem (see Fig. 2).

A dimension to emphasise in the European Data Value Ecosystem is its twofold nature of vertical versus horizontal in respect to the different sector or application domains (transport health, energy, etc.). While specific data value ecosystems are needed per sector (concerning targeted markets, stakeholders, regulations, type of users, data types, challenges, etc.), one of the main values identified for the Big Data Value Association and the PPP is its horizontal nature, allowing cross-sector value creation, considering both the reuse of value from one sector to another, and the creation of innovations based on cross-sector solutions and consequently new value chains.

Establishing collaborations with other European, international and local organisations is crucial for the development of the ecosystem, to generate synergies between communities and to impact research and innovation, standards, regulations, markets and society.

Collaborations, in particular with other PPPs, European and international standardisation bodies, industrial technology platforms, data-driven research and innovation initiatives, user organisations and policymakers, had been identified and developed at national, European and international level since the launch of the PPP and the creation of the Association, influencing the level of maturity of these collaborations.

A key part of ensuring the sustainability of the BDV ecosystem was to develop collaborations with complementary ecosystems with an impact on technology integration and the digitisation of industry challenges. These collaborations, detailed in Fig. 2, include the ETP4HPC (European Technology Platform for HPC) (for HPC),

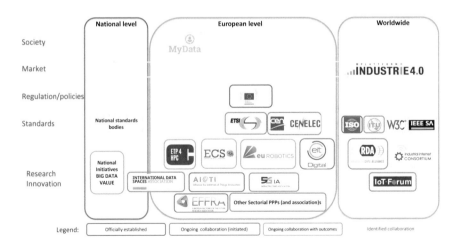

Fig. 2 Map of collaboration for BDV ecosystem

ECSO (for cybersecurity), AIOTI (for IoT), 5G (through 5G PPP), the European Open Science Cloud (EOSC) (for the Cloud) and the European Factories of the Future Research Association (EFFRA) (for factories of the future).

7 Summary

Creating a productive ecosystem for big data and driving accelerated adoption requires an interdisciplinary approach addressing a wide range of challenges from access to data and infrastructure, to technical barriers, skills, and policy and regulation. To overcome these challenges, collective action is needed from all stakeholders working together in an effective, holistic and coherent manner. To this end, the Big Data Value Public-Private Partnership was established to develop the European data ecosystem and enable data-driven digital transformation, delivering maximum economic and societal benefit, and achieving and sustaining Europe's leadership in the fields of big data value creation and Artificial Intelligence. The BDV PPP follows a phased roadmap with the use of five strategic mechanisms to drive the adoption of big data value and to encourage cooperation and coordination in the data ecosystem. The PPP proactively engaged with the key communities, which helped to enhance the development of the European Data Value Ecosystem.

Acknowledgements We greatly acknowledge the collective effort of the SRIA teams: Carlos A. Iglesias, Antonio Alfaro, Jesus Angel, Sören Auer, Paolo Bellavista, Arne Berre, Freek Bomhof, Stuart Campbell, Geraud Canet, Giuseppa Caruso, Paul Czech, Stefano de Panfilis, Thomas Delavallade, Marija Despenic, Wolfgang Gerteis, Aris Gkoulalas-Divanis, Nuria Gomez, Paolo Gonzales, Thomas Hahn, Souleiman Hasan, Bjarne Kjær Ersbøll, Bas Kotterink, Yannick Legré, Yves Mabiala, Julie Marguerite, Ernestina Menasalves, Andreas Metzger, Elisa Molino, Thierry Nagellen, Dalit Naor, Maria Perez, Milan Petkovic, Roberta Piscitelli, Klaus-Dieter Platte, Pierre Pleven, Dumitru Roman, Titi Roman, Alexandra Rosén, Nikos Sarris, Stefano Scamuzzo, Simon Scerri, Corinna Schulze, Bjørn Skjellaug, Francois Troussier, Colin Upstill, Josef Urban, Meilof Veeningen, Tonny Velin, Ray Walshe, Walter Waterfeld and Stefan Wrobel.

References

Cavanillas, J. M., Curry, E., & Wahlster, W. (Eds.). (2016). *New horizons for a data-driven economy: A roadmap for usage and exploitation of big data in Europe*. New York: Springer. https://doi.org/10.1007/978-3-319-21569-3

Curry, E. (2016). The big data value chain: Definitions, concepts, and theoretical approaches. In J. M. Cavanillas, E. Curry, & W. Wahlster (Eds.), *New horizons for a data-driven economy: A roadmap for usage and exploitation of big data in Europe*. New York: Springer. https://doi.org/10.1007/978-3-319-21569-3_3

Curry, E., & Ojo, A. (2020). Enabling knowledge flows in an intelligent systems data ecosystem. In *Real-time linked dataspaces* (pp. 15–43). https://doi.org/10.1007/978-3-030-29665-0_2

Curry, E., & Sheth, A. (2018). Next-generation smart environments: From system of systems to data ecosystems. *IEEE Intelligent Systems, 33*(3), 69–76. https://doi.org/10.1109/MIS.2018.033001418

Gawer, A., & Cusumano, M. A. (2014). Industry platforms and ecosystem innovation. *Journal of Product Innovation Management, 31*(3), 417–433. https://doi.org/10.1111/jpim.12105

Zillner, Sonja, Curry, E., Metzger, A., Auer, S., & Seidl, R. (Eds.). (2017). *European big data value strategic research & innovation agenda*. Retrieved from Big Data Value Associate website www.bdva.eu

Zillner, S., Bisset, D., Milano, M., Curry, E., Hahn, T., Lafrenz, R., et al. (2020). *Strategic research, innovation and deployment agenda – AI, data and robotics partnership. Third Release (3rd)*. Brussels: BDVA, euRobotics, ELLIS, EurAI and CLAIRE.

Achievements and Impact of the Big Data Value Public-Private Partnership: The Story so Far

Ana García Robles, Sonja Zillner, Wolfgang Gerteis, Gabriella Cattaneo, Andreas Metzger, Daniel Alonso, Martina Barbero, Ernestina Menasalvas, and Edward Curry

Abstract The European contractual Public-Private Partnership on Big Data Value (BDV PPP) has played a central role in the implementation of the revised Digital Single Market strategy, contributing to multiple pillars, including "Digitising European Industry", "Digital Skills", "Building the European Data Economy" and "Developing a European Data Infrastructure". The BDV PPP and the Big Data Value Association have also played a pivotal role in the European Artificial Intelligence and Data Strategies launched by the European Commission in 2018. This chapter provides an overview and an in-depth analysis of the impact of the PPP by mid-2019, with a focus on the achievements and the overall impact since the launch of the PPP.

Keywords Public-private partnership · Data impact · Data PPP · Big data value

A. García Robles (✉) · M. Barbero
Big Data Value Association, Bruxelles, Belgium
e-mail: ana.garcia@core.bdva.eu

S. Zillner
Siemens AG, Munich, Germany

W. Gerteis
SAP, Walldorf, Germany

G. Cattaneo
IDC, Milan, Italy

A. Metzger
paluno, University of Duisburg-Essen, Duisburg, Germany

D. Alonso
ITI, Valencia, Spain

E. Menasalvas
Universidad Politécnica de Madrid, Madrid, Spain

E. Curry
Insight SFI Research Centre for Data Analytics, NUI Galway, Galway, Ireland

1 Introduction

The European contractual Public-Private Partnership on Big Data Value (BDV PPP) was signed on 13 October 2014. It marked the commitment of the European Commission, industry and partners from research to build a data-driven economy across Europe, mastering the generation of value from Big Data and creating a significant competitive advantage for European industry, thus boosting economic growth and employment. The BDV PPP started in 2015 and was operationalised with the launch of the Leadership in Enabling and Industrial Technologies (LEIT) work programme 2016/2017 of Horizon 2020 (H2020) with the first PPP projects (Call 1) starting in January 2017. With 57 projects,[1] an allocated investment of public funding of €301 million by the end of 2019 and around 300 organisations as part of the private association[2] (Big Data Value Association, BDVA) over the years, the Big Data Value PPP has played a central role in the implementation of the revised Digital Single Market (DSM) strategy, contributing to multiple pillars including "Digitising European Industry", "Digital Skills", "Building the European Data Economy" and "Developing a European Data Infrastructure". The BDV PPP and the BDVA have also played an important role in the European AI and Data Strategies launched by the European Commission in 2018. This chapter provides an overview and an in-depth analysis of the impact of the PPP by mid-2019, with a focus on the achievements and the overall impact since the launch of the PPP.

This chapter details the achievements and the impact of the Big Data Value PPP. After explaining the key elements of the Big Data Value PPP in Sect. 2, and presenting a summary of the achievements and impact created during 2018 discussed in Chap. "A Roadmap to Drive Adoption of Data Ecosystems", an in-depth analysis of the overall progress towards the mains goals of the partnership by mid-2019[3] is given in Sect. 4. Finally, the Sect. 5 concludes with a summary and perspectives on the future.

2 The Big Data Value PPP

The vision, overall goals, main technical and non-technical priorities and a research and innovation roadmap for the European Public-Private Partnership (PPP) on Big Data Value are defined in the Big Data Value Strategic Research and Innovation Agenda (BDV SRIA) (Zillner et al. 2017).

---

[1]Considering projects selected for funding by end of December 2019.

[2]Includes all BDVA members, including active and terminated/resigned (source: BDVA).

[3]The BDVA is responsible for providing a full monitoring report on its activities. Since 2019 and in accordance with the European Commission, the full monitoring report of the Partnership will only be submitted every 2 years. The most recent version was delivered in 2019 covering the period from beginning 2018 to beginning 2019 (https://bdva.eu/MonitoringReport2018).

The BDV PPP SRIA defined the roadmap and methodology by describing three different phases:

- Phase I: Establish an Innovation Ecosystem (H2020 WP 2016–2017 calls)
- Phase II: Disruptive Big Data Value (H2020 WP 2018–2019 calls)
- Phase III: Long-term Ecosystem Enablers (H2020 WP 2019–2020 calls)

The BDV SRIA has been regularly updated incorporating the multi-annual roadmap of the BDV PPP. BDV SRIA v4 (delivered at the end of 2017) provides direct input to the LEIT WP 2018–2020 as defined in its updated Phases II and III.

The BDV PPP projects cover Big Data technology, including Artificial Intelligence methods, and application research and innovation, new data-driven business models, data ecosystem support, data skills, regulatory and IPR requirements, and societal aspects. The value generated by Big Data technologies empowers Artificial Intelligence to foster linking, cross-cutting and vertical dimensions of value creation at the technical, business and societal level across many different sectors.

2.1 BDV PPP Vision and Objectives for European Big Data Value

The Big Data Value Association (BDVA) and the **BDV PPP** have pursued a common shared vision of positioning **Europe as the world leader in the creation of big data value**. The BDV PPP vision for Europe in 2020 has concerned the following aspects:

- **Data**: Zettabytes of useful public and private data will be widely and openly available. By 2020, smart applications such as smart grids, smart logistics, smart factories and smart cities will be widely deployed across the continent and beyond. Ubiquitous broadband access, mobile technology, social media, services and the IoT on billions of devices will have contributed to the explosion of generated data to a global total of 40 zettabytes. Much of this data will yield valuable information. Extracting this information and using it in intelligent ways will revolutionise decision-making in business, science and society, enhancing companies' competitiveness and leading to the creation of new industries, jobs and services.
- **Skills**: Millions of jobs will become established for data engineers and scientists, and the Big Data discipline will be integrated into technical and business degrees. The European workforce is increasingly data savvy, regarding data as an asset.
- **Legal**: Privacy and security can be guaranteed along the big data value chain. Data sharing and data privacy can be fully managed by citizens in a trusted data ecosystem.
- **Technology**: Real-time integration and interoperability among different multi-lingual, sensorial and non-structured datasets will be accomplished, and content will be automatically managed and visualised in real-time. By 2020, European

research and innovation efforts will have led to advanced technologies that make it significantly easier to use Big Data across sectors, borders and languages.

- **Application**: Applications using the BDV technologies can be built, which will allow anyone to create, use, exploit and benefit from Big Data. By 2020, thousands of specific applications and solutions will address data-in-motion and data-at-rest. There will be a highly secure and traceable environment supporting organisations and citizens, with the capacity to sustain various monetisation models.
- **Business**: One true EU single data market will be established, thus allowing EU companies to increase their competitiveness and become world leaders. By 2020 value creation from Big Data will have a disruptive influence on many sectors. From manufacturing to tourism, from healthcare to education, from energy to telecommunications services, from entertainment to mobility, big data value will be a key success factor in fuelling innovation, driving new business models, and supporting increased productivity and competitiveness.
- **Societal**: Societal challenges will be addressed through BDV systems, focusing on areas such as the high volume, mobility and variety of data.

The above-addressed aspects were planned to impact the European Union's priority areas as follows:

- **Economy**: The competitiveness of European enterprises will be significantly higher compared to their worldwide competitors, due to improved products and services and greater efficiency based on the value of Big Data. One true EU single data market will be established, allowing EU companies to increase their competitiveness and become world leaders.
- **Growth**: A flourishing sector of expanding new small and large businesses will result in a significant number of new jobs focusing on creating value out of data.
- **Society**: Citizens will benefit from better and more economical services in a stable economy where data can be shared with confidence. Privacy and security will be guaranteed throughout the life cycle of BDV exploitation.

These three factors were designed to support **the major EU pillars as stated in the Rome Declaration of March 2017** (European Council 2017): a safe and secure Europe; a prosperous and sustainable Europe; a social Europe; and a stronger Europe on the world stage.

2.2 Big Data Value Association (BDVA)

The BDVA is an industry-driven and fully self-financed international not-for-profit organisation under Belgian law. The BDVA has over 220 members all over Europe with a well-balanced composition of large, small and medium-sized industries (over 30% of SMEs), as well as research and user organisations. The Big Data Value Association is the private counterpart to the European Commission in implementing the BDV PPP.

BDVA members come together to collaborate on a joint mission: developing the European Big Data Value Ecosystem (BDVe) that will enable the data-driven digital transformation in Europe, delivering maximum economic and societal benefit, and achieving and sustaining Europe's leadership on big data value creation and Artificial Intelligence (Zillner et al. 2019). To achieve this mission, in 2017, the BDVA defined four strategic priorities:

- **Develop Data Innovation Recommendations**: Providing guidelines and recommendations on data innovation to the industry, researchers, markets and policymakers
- **Develop Ecosystem**: Developing and strengthening the European Big Data Value Ecosystem
- **Guiding Standards**: Driving Big Data standardisation and interoperability priorities, and influencing standardisation bodies and industrial alliances
- **Know-How and Skills**: Improving the adoption of Big Data through the exchange of knowledge, skills and best practices

Since 2017 the cross-technological nature of the data value chains, flowing across different technologies (IoT, Cloud, 5G, Cybersecurity, infrastructures, HPC, etc.), has triggered and accelerated the development of stronger collaborations between the BDV PPP/BDVA and other technological (cross-sectorial) sectorial communities and, in particular, other partnerships.

2.3 BDV PPP Objectives

As laid out in the Contractual Arrangement (CA) of the BDV PPP (*BDVPPP Contractual Arrangement* n.d.), the overarching general objectives are as follows:

- To foster European Big Data technology leadership in terms of job creation and prosperity by creating a Europe-wide technology and application base, and building up the competence and number of European data companies, including start-ups
- To reinforce Europe's industrial leadership and ability to compete successfully in the global data value solution market by advancing applications which can be converted into new opportunities, so that European businesses secure a 30% market share by 2020
- To enable research and innovation work, including activities related to interoperability and standardisation, and secure the future basis of Big Data Value creation in Europe
- To facilitate the acceleration of business ecosystems and appropriate business models, with a particular focus on SMEs, enforced by a Europe-wide benchmarking of usage, efficiency and benefits
- To provide and support successful solutions for major societal challenges in Europe, for example in the fields of health, energy, transport and the environment, and agriculture

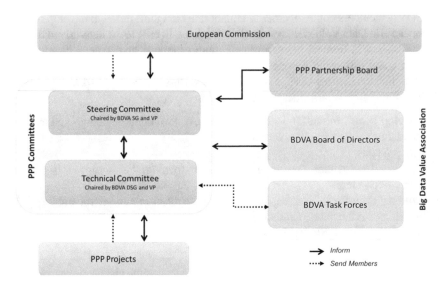

Fig. 1 BDV PPP governance structure

- To demonstrate the value of Big Data for businesses and the public sector and to increase citizens' acceptance levels by involving them as "prosumers" and accelerating take-up
- To support the application of EU data protection legislation and provide effective, secure mechanisms to ensure its enforcement in the Cloud and for Big Data, thus facilitating its adoption

2.4 BDV PPP Governance

The main governance structure of the BDV PPP (Fig. 1) was prepared and delivered at the beginning of the PPP to provide the framework for collaboration and alignment among all members of the PPP (EC, funded projects, the Association and its members).

The Cooperation Charter[4] was created by the Association as one of the key governance mechanisms to facilitate cooperation among the BDV PPP actions and the BDVA and has been updated every year accordingly.

---

[4]The Cooperation Charter was produced by the BDVA during 2016 and it has been integrated in the CAs or GAs of the Call 1 and Call 2 actions, thereby formalising the actions' commitment to supporting the cooperation within the BDV ecosystem. Latest version: http://www.bdva.eu/sites/default/files/BDV%20PPP%20COOPERATION%20CHARTER%20January%202019_approved.pdf

The BDVe project (CSA of the BDV PPP) has supported the implementation of the PPP projects governance structure by establishing the BDV PPP Steering Committee (SC) and the Technical Committee (TC). The Steering Committee (SC) provides executive-level steering and advice to ensure effective and efficient coordination and communication between the BDV PPP actions. The Technical Committee (TC) facilitates knowledge exchange and cooperation on the technical aspects, methodology and implementation of the BDV PPP programme. A non-formal Communication Committee was also established to support cooperation in Marketing and Communications.

The Board of Directors[5] (BoD) of the BDVA is selected by the General Assembly of the Association (2-year mandate) and is in charge of achieving the objectives of the association. It follows the resolutions, instructions and recommendations adopted by the General Assembly.

The Partnership Board (PB) is the monitoring body of the PPP formed by selected directors of the Board of the BDVA, and representatives of the European Commission. The PB meets approximately 1–2 times per year and complements this with regular bi-weekly exchanges of information. The European Commission is represented by DG Connect Directorate G (Unit G1 in particular).

2.5 BDV PPP Monitoring Framework

The BDVA leads the production of the Monitoring Report of the Big Data Value PPP as part of its contractual obligations in the PPP. The work is developed by the BDVA TF2 (impact). Since 2019 and in accordance with the European Commission, the full monitoring report of the partnership will only be submitted every 2 years. The most recent version was delivered in 2019 covering the period from beginning 2018 to beginning 2019. The list of key performance indicators (KPIs) for this PPP, description and target values are defined by the following documents:

1. A note released by Directorate-General for Research and Innovation at the European Commission (DG RTD) in February 2018, defining, describing and providing a methodology for the four common KPIs to all the PPPs
2. BDV PPP contractual agreement[6]
3. BDV PPP SRIA

To produce the monitoring reports the association gathers input from all the running and selected Big Data PPP projects, all for-profit project partners from the projects, the members of the BDVA, the BDVA Task Forces and the BDVA Office, the EC DG CNECT G1 Unit and the European Data Market Monitoring Tool.[7]

---

[5]List of BoD members: http://www.bdva.eu/board-members

[6]http://www.bdva.eu/sites/default/files/BDVPPP_Contractual_Arrangement_.pdf

[7]SMART 2016/0063 – Study "Update of the European data marketMOnitoring Tool", IDC and Lisbon Councils.

3 Main Activities and Achievements During 2018

The main achievements of the Big Data Value PPP during 2018 can be summarised as follows:

- Mobilised private investments since the launch of the PPP (and until end of 2018) of €1.57 billion (€468 million in 2018). Considering the amount of EU funding allocated to the PPP by the end of 2018 (€201.30 million), the BDV PPP ended that period with a **leverage factor of 7.8**, much higher than the leverage factor of 4 committed contractually.
- Forty-two projects were running at the beginning of January 2019, with 32 projects active during 2018 (contributing to many of the KPIs) and 10 additional projects selected for funding in 2018 and starting in 2019. Participation of SMEs over 20% and targeted data incubation activities for start-ups (25.3% SME participation in the call for proposals 2018).
- The BDV PPP organised 181 training activities involving over 18,300 participants during 2018. Projects have contributed with 85 training activities during 2018 involving over 9700 participants. BDVA members contributed with 96 training activities involving over 8500 participants. Projects have developed 16 interdisciplinary programmes during 2018 reaching 250 participants.
- Forty-eight job profiles identified by projects in 2018.
- Data skills activities, including the launch of the BDV PPP Educational Hub for European MSc programmes in Data Science and Data Analytics and the launch of the pilot on the skills recognition programme
- Organisation of 323 events reaching over 630,000 participants during 2018.
- One hundred and six innovations of exploitable value, 39 of which are significant (37%), delivered by running projects during 2018. The BDVe project launches the Big Data Value Marketplace[8] and the Big Data Value Landscape.[9]
- Seventy-seven per cent of the BDV PPP projects contributing to job creation by 2023, with an estimation in accumulated numbers of thousands. Estimated numbers surpass 7500 new jobs created by 2023 linked to project activities and much more considering indirect effect.
- Two patents, over 61 publications and 24 products or software components in the field of advanced privacy- and security-respecting solutions for data access, processing and analysis in 2018.
- Sixty-three new economically viable services of high societal value developed during 2018.
- 100% coverage of research priorities defined in SRIA, with 204 new systems and technologies developed in different sectors during 2018. The major focus of technical contributions lies in "Data Analytics".

---

[8]http://marketplace.big-data-value.eu/

[9]https://landscape.big-data-value.eu/

- Two hundred and twenty-four use cases and/or experiments conducted during 2018 by projects and 165 additional experiments conducted by BDVA i-Spaces.
- Eighty-two large-scale experiments were developed by the projects during 2018, 64 involving closed (private) data. BDVA i-Spaces also contributed to this KPI, reporting in total 38 large-scale experiments performed during 2018, 28 involving private data.
- Four major sectors (bio-economy; transport, mobility and logistics; healthcare; smart manufacturing) covered with close-to-market large-scale implementations, and over 15 different sectors covered in total including (in addition to the ones already mentioned) telecom, Earth observation, media, retail, energy, finance and banking, public services, water and natural resources, business services, smart cities insurance, public safety, personal security, public tenders, e-commerce, marketing, fashion industry, citizen engagement, ICT/Cloud services, social networks, procurement and legal services.
- 0,10696 Exabytes (106.73 Petabytes) of data made available for experimentation (86,25 Petabytes by projects, 20,71 Petabytes by i-Spaces).
- Evidence of contribution to the environmental KPIs, with some pilots showing 25% and 51% in energy reduction and improvements concerning CO_2 emissions, reaching up to 29% and 23% of emission reductions in general.
- "During 2018, 396 FTEs master and PhD students (60 masters and 136 PhD) were involved in PPP projects".
- SME turnover evolution increase of 60% with respect to 2014 and 17.7% in the last year. A positive trend in employment evolution with an average increase in employment for the SMEs that are part of the PPP is 75% with respect to 2014 and a growth of 11.83% in the last year.
- The European Data Incubators DataPitch and EDI have given support and new opportunities to 47 start-ups and entrepreneurs, creating an impact on revenues, jobs created and competitiveness, and supporting them to raise additional private finding.
- Third wave of BDVA i-Spaces labelling,[10] with 10 labelled i-Spaces selected during 2018.
- The BDVA joined the EuroHPC Joint Undertaking[11] as one of its private members, bringing synergy between Big Data, AI and HPC for industrial use-cases. The BDVA appoints two official representatives in the EuroHPC RIAG and two additional observers.
- During 2018 the BDVA developed collaborations with impact on technology integration, roadmapping and the digitisation of industry challenges.
- The BDVA delivered 7 strategic papers during 2018 supporting this strategic roadmap, and an additional 10 new papers were released in 2019 and early 2020,

---

[10]All Information about the i-Spaces labelling can be found on the BDVA website. General information: http://bdva.eu/I-Spaces. Labelling process: Information about labelled i-Spaces 2018: http://www.bdva.eu/node/1172

[11]https://eurohpc-ju.europa.eu/

including on essential topics such as data protection in the era of Artificial Intelligence and use of data in Smart Manufacturing.

- Official liaison with the ISO/JTC1/SC42 with the main objective of channelling European input (PPP) into global standards for AI and Big Data. In 2019 the BDVA was in the process to sign an MoU with CEN (European Committee for Standardization)/CENELEC (European Committee for Electromechanical Standardization) and ETSI (European Standards Telecommunications Institute).
- During 2018 the BDVA also developed strong foundations for the future, building upon the current BDV PPP by joining the EuroHPC Joint Undertaking (as a private member) and by driving (together with euRobotics) the future partnership on AI, data and Robotics. As of today, the BDVA is the main promoter of the AI, Data and Robotics partnership,[12] one of the candidates for European Partnerships in digital, industry and space in Horizon Europe, developed in collaboration with euRobotics and the AI Research communities CLAIRE, ELLIS and EurAI.
- The BDVe project has supported the collaboration of the Network of Centres of Excellence in Big Data and the establishment of a new Centre of Excellence in Bulgaria, the first such centre in Eastern Europe.

3.1 Mobilisation of Stakeholders, Outreach, Success Stories

The year 2018 was one of remarkable progress and advancements for the Big Data Value PPP and the BDVA. In its second year of operations, the PPP showed a great quantity and variety of success stories from projects and the association. The main success stories from the projects related to:

- The impact created in specific sectors (e.g. results in the Lighthouse projects TT and DataBio already reporting evidence on reduction of operation and production costs, reduction of emissions, improvements on energy efficiency, etc.)
- "Close to market" technology and solutions delivered (e.g. FLAIR (framework for Natural Language Processing developed by FashionBrain) already integrated into the PyTorch ecosystem, or SLIPO workbench already used by other PPP projects and in commercial settings in the PPP Point of Interest (POI) data sets on a global scale)
- Performance (e.g. in one of its pilots BigMedilytics achieved a better prediction of re-admission for chronic heart failure over 50%)
- Resources generated (new knowledge, new ontologies, datasets)
- Incubation of new data-driven businesses (47 start-ups in 2018 with individual success stories)
- Research excellence (publications and paper awards)
- Impact in Standardisation
- Strong foundations put in place for future activities

---

[12]https://ai-data-robotics-partnership.eu/home/

The European Data Incubators/accelerators DataPitch and EDI gave support and new opportunities to 47 start-ups and entrepreneurs, helping them to grow their business in the new Data Economy offering skills development, access to resources, data, infrastructure, ecosystem and additional private funding. This has generated a significant impact on revenues, jobs created and competitiveness.

It is important to highlight the **positive effect that participation in a more extensive programme has brought to individual projects**. Eighty per cent of the projects reported value created for their Research and Innovation projects by being part of the BDV PPP, e.g. facilitating collaboration and exchanges between projects, such as complementary functionalities (e.g. SLIPO and QROWD), reuse of projects outcomes (functionality, solutions or ontologies, data sharing[13] and specific know-how sharing). Additionally, the PPP is seeking to be effective in coordinating communication activities, providing new opportunities for start-ups, and providing a common framework and vocabulary to develop effective end-to-end ecosystems.

It is also quite remarkable to note the overall impact in communication and engagement of the PPP, with the estimated number of people outreached in dissemination activities around 7.8 million in 2018 with the objective of raising awareness about their different activities, to engage new stakeholders, and communicating the result. Additionally, the BDV PPP organised 181 training activities involving over 18,300 participants during 2018. The range and diversity of actors and stakeholders outreached is very broad, in alignment with the overall objectives of the PPP.

4 Monitored Achievements and Impact of the PPP

Enabled by the monitoring framework, as described above, the progress of the BDV PPP is continuously monitored. Below we report the key achievements and impacts in alignment with the development phases described in the SRIA that are backed by the monitoring data.

4.1 Achievement of the Goals of the PPP

According to the Big Data Value PPP SRIA v4,[14] the programme would develop the European data ecosystem in three distinct phases of development, each with a primary theme:

---

[13]Discussions going on between projects working in same sector.

[14]And Multi-Annual roadmap version 2017.

- Phase I: Establish the ecosystem (governance, i-Space, education, enablers) and demonstrate the value of existing technology in high-impact sectors (Lighthouses, technical projects) (Work Programme WP 16–17)
- Phase II: Pioneer disruptive new forms of Big Data Value solutions (Lighthouses, technical projects) in high-impact domains of importance for EU industry, addressing emerging challenges of the data economy (WP 18–19)
- Phase III: Develop long-term ecosystem enablers to maximise sustainability for economic and societal benefit (WP 19–20)

The PPP goals achieved are analysed based on the defined roadmap. The year 2018 lies between Phase I and Phase II, and thus the progress of the PPP is assessed considering the objectives of both phases.

Phase I: Establish an Innovation Ecosystem (WP 2016–17) focused on laying the foundations needed to establish a sustainable European data innovation ecosystem (Table 1).

Phase II: Pioneer disruptive new forms of Big Data Value solutions (Lighthouses, technical projects) in high-impact domains of importance for EU industry, addressing emerging challenges of the data economy (WP 18–19). According to the SRIA, this second phase is meant to build on the foundations established in Phase I and will have a primary focus on Research and Innovation (R&I) activities to deliver the next generation of Big Data Value solutions. Although the projects implementing Phase II started in 2019 (or 2020), there are some activities in 2018 supporting the implementation of this stage, in particular those listed in Table 2.

Phase III[15]: Develop long-term ecosystem enablers to maximise sustainability for economic and societal benefit (WP 19–20). This phase started in late 2019 and will continue until the end of the PPP. As this phase has only just started, the analysis can only be incomplete. Some ideas about possible achievements are provided in Table 3.

4.2   Progress Achieved on KPIs

4.2.1   Private Investments

Through this KPI, we attempt to understand and capture/show the level of industrial engagement within the BDV PPP. This KPI includes both direct and indirect leverage, as described in Fig. 2.

Two hundred and ninety-six companies representing all for-profit organisations participating in Big Data Value PPP projects active during 2018 (including not only project partners but also third parties engaged through cascade funding) and all

---

[15]Reported as part of the BDVA annual report 2019: https://bdva.eu/sites/default/files/BDVA%20-%20BDVA%20PPP%20Annual%20Report%202019_v1.1%20for%20publication.pdf

Table 1 Summary achievements of the goals of the BDVA PPP: Phase I of the roadmap

Expected PPP activities and outcomes for WP2016-17 according to BDV PPP SRIA/ Multi-Annual roadmap	Achievements
Establish a European network of i-Spaces for cross-sectorial and cross-lingual data integration, experimentation and incubation (ICT14 – 2016–17)	Fifteen projects were running in 2018 including two European Data Incubators Ten labelled BDVA i-Spaces providing data experimentation and data incubation capabilities for SMEs Over 15 sectors covered Eighty-two large-scale experiments were developed by the projects during 2018, 64 involving closed (private) data. BDVA i-Spaces also contributed to this KPI, reporting in total 38 large-scale experiments performed during 2018, 28 of them involving private data 0,10696 Exabytes (106,96 Petabytes) of data made available for experimentation (86,25 Petabytes by the projects, 20,71 Petabytes by i-Spaces)
Demonstrate Big Data Value solutions via large-scale pilot projects in domains of strategic importance for EU industry, using existing technologies or very near-to-market technologies (ICT15 – 2016–17)	Four Lighthouse projects running in 2018. Additional four HPC-Big Data-enabled Lighthouse projects and associated projects) selected in 2018 to start in 2019 Four major domains of strategic importance covered: bio-economy, transport, logistics and mobility, and healthcare and manufacturing
Tackle the main technology challenges of the data economy by improving the technology, methods, standards and processes for Big Data Value (ICT16 – 2017)	One hundred per cent of SRIA technical priorities covered in 2018 Seven technical projects running in 2018 and six additional projects funded (started in 2019) One hundred and thirty-two innovations of exploitable value (106 delivered in 2018), thirty-five per cent of which are significant innovations, including technologies, platforms, services, products, methods, systems, components and/or modules, frameworks/architectures, processes, tools/toolkits, spin-offs, datasets, ontologies, patents and knowledge Two hundred and four new systems and technologies developed in different sectors during 2018. The major focus of technical contributions lies in "Data Analytics" BDV PPP reference model (2017)
Advance state of the art in privacy-preserving Big Data technologies and explore the societal and ethical implications (ICT18 – 2016)	Four projects running in 2018 (1 focused on societal and ethical implications). Three additional projects selected in 2018 and starting in 2019 to scale solutions Two patents, over 61 publications and 24 products or software components in the field of advanced privacy- and security-respecting solutions for data access, processing

(continued)

Table 1 (continued)

Expected PPP activities and outcomes for WP2016-17 according to BDV PPP SRIA/ Multi-Annual roadmap	Achievements
	and analysis in 2018 BDVA TF5 (societal and ethical aspects of data, among other things)
Establish key ecosystem enablers, including programme support and coordination structures for industry skills and benchmarking (ICT17 – 2016-17)	BDV marketplace, BDV landscape, the education hub and the skills recognition programme The BDV PPP organised 181 training activities involving over 18,300 participants during 2018. Projects contributed with 85 training activities during 2018 involving over 9700 participants. BDVA members contributed with 96 training activities involving over 8500 participants. Projects developed 16 interdisciplinary programmes during 2018 outreaching 250 participants Centres of Excellence in Big Data EBDVF and BDV PPP meet-up/Summit DataBench project ongoing

for-profit organisation members of the BDVA were outreached to provide input to this KPI with an overall response rate of 40.9%.

Table 4 shows the evolution of the reported numbers in private investments from 2015 to 2018, as well as the EU contributions.

Aggregated to the numbers reported in 2015 (€280.9 million), 2016 (€338.5 million) and 2017 (€482.25 million), the amount of mobilised private investments since the launch of the PPP until the end of 2018 was 1569.1M€ (€1.57 billion). Considering the amount of EU funding allocated to the PPP by that time (€201.3 million), the BDV PPP ended 2018 with a leverage factor of 7.8, much higher than the leverage factor of 4 committed contractually.

4.2.2 Job Creation, New Skills and Job Profiles

Seventy-seven per cent of the BDV PPP projects indicated that their project would contribute to job creation by 2023, with an estimation in accumulated numbers of thousands. The estimated numbers surpass 7500 new jobs created by 2023 linked to project activities and many more considering indirect effect.

BDV PPP projects contribute to job creation in Europe by (1) increasing the market share of Big Data Technology providers in Europe; (2) developing new job profiles that generate new jobs… the creation; (3) developing new opportunities for entrepreneurs and start-ups in the new Data Economy; (4) generating job opportunities by increasing data sharing; (5) creating new jobs already during the lifetime of the project; and (6) forecasting jobs created as a follow-up of project results.

Table 2 Summary of achievements of the goals of the BDVA PPP: Phase II of the roadmap

Expected PPP activities and outcomes for WP2018-19 according to the BDV PPP SRIA/ Multi-Annual roadmap	Achievements
Supporting the emergence of the data economy with a particular focus on accelerating the progress of SMEs, start-ups and entrepreneurs, as well as best practices and standardisation (ICT-13-c)	Ten labelled BDVA i-Spaces providing data experimentation and data incubation capabilities for SMEs Two European Data Incubators (EDI and DataPitch) with 47 start-ups incubated during 2018 Data Market Services project started in 2019 (support to SMEs and Standards).
Pioneering disruptive new forms of Big Data Value solutions with the Cloud and HPC or the IoT via large-scale pilot projects in emerging domains of importance for EU industry using advanced platforms, tools and testbeds (ICT-11, DT-ICT-11-2019)	Four ICT-11-a-2018 projects (HPC and Big Data enabled large-scale testbeds and applications) funded in 2018 and started in 2019 (projects associated with the PPP). Cooperation established with the new projects Two additional projects selected in 2019 for IoT-Big Data (ICT-11-b-2018)
Tackling the next generation of Big Data research and innovation challenges for extreme-scale analytics (ICT-12-a)	Six technical projects selected in 2018 (started in 2018) with focus on extreme-scale analytics From the running projects in 2018, there is a clear trend to focus on technical contributions in the areas of "Data Analytics" and "Data Processing Architectures", thus supporting the explanation that a solid base of "Data Management" solutions will enable analytics and processing innovations
Addressing ecosystem roadblocks and inhibitors to the take-up of Big Data Value platforms for data ecosystem viability, including platforms for personal and industrial data (ICT-13)	Three projects selected for funding in 2018 (started in 2019) to advance state of the art in the scalability and computational efficiency of methods for securing desired levels of privacy of personal data and/or confidentiality of commercial data Call for proposals for ICT-13-a for setting up operating platforms for secure and controlled sharing of "closed data" (proprietary and/or personal data) closed in April 2019
Providing programme support (continuing), facilitating networking and cooperation among ecosystem actors and projects, and promoting community building between BDV, Cloud, HPC and IoT activities (ICT-12-b)	BDVA has built strong collaborations with the European Technology Platform for High Performance Computing (ETP4HPC) (for HPC), Alliance for Internet of Things Innovation (AIOTI) (for IoT), European Factories of the Future Research Association (EFFRA) and euRobotics BDVA has become a private member of the EuroHPC Joint Undertaking BDVe project supports collaborations

Table 3 Summary of achievements of the goals of the BDVA PPP: Phase III of the roadmap

Expected PPP activities and outcomes for WP2018-19 according to the BDV PPP SRIA/ Multi-Annual roadmap	Achievements
Sowing the seeds for long-term ecosystem enablers to ensure self-sustainability beyond 2020 (ICT-13)	Launch of nine ICT-13 projects under the mechanisms of Data Platform, as defined by the SRIA. These projects are very relevant from the perspective of the new EU Data Strategy and to establish the sectorial EU data spaces, which are planned. The projects will also help establish the link with the future AI, Data and Robotics Partnership. Organisation of an online workshop on the Role of Data Innovation Spaces and Data-Driven Innovation Hubs in the European digital transformation. The event addresses the question of sustainability and of the links between the Horizon 2020 initiatives (i.e. i-Spaces) and the Horizon Europe and Digital Europe Programme new mechanisms (i.e. testing and experimentation facilities, European Digital Innovation Hubs)
Creating innovation projects within a federation of i-Spaces (European Digital Innovation Hubs for Big Data) to validate and incubate innovative Big Data Value solutions and business models (DT-ICT-05-2020)	Launch of three new energy-related projects under DT-ICT-11 A full analysis of the projects that bring together i-Spaces and data incubators will be part of the full Monitoring Report on 2020 activities due in 2021
Ensuring continued support for technology outputs of PPP (Lighthouses, R&I, CSA), including non-technical aspects (training) beyond 2020 (i.e. Open Source Community, Technology Foundation)	The BDVe project will soon deliver an exploitation plan which hands over many important activities to the BDVA. The association will continue to support the ecosystem on technical and non-technical aspects beyond 2020 BDVA is strongly engaged in the discussions concerning the future AI, Data and Robotics Partnership and the strategy for the EuroHPC. The association has also already established many new collaborations for strengthening the impact of the PPP's outputs beyond 2020 (i.e. by partnering with standardisation associations)
Establishing a Foundation for European Innovation Spaces with a charter to continue collaborative innovation activity beyond 2020, in line with the concept of the European Digital Innovation Hub for Big Data	Launch of the EUHubs4Data project which will set up a European federation of Big Data Digital Innovation Hubs (DIHs), with the ambition of becoming a reference instrument for data-driven cross-border experimentation and innovation and will support the growth of European SMEs and start-ups in a global Data Economy Organisation of an online workshop in June 2019 on the Role of Data Innovation Spaces

(continued)

Table 3 (continued)

Expected PPP activities and outcomes for WP2018-19 according to the BDV PPP SRIA/ Multi-Annual roadmap	Achievements
	and Data-Driven Innovation Hubs in the European digital transformation (see above) Continued efforts of the Data Incubators projects to deliver results
Liaising with private funding (including Venture Capital) to accelerate entry into the market and socio-economic impacts, including the provision of ancillary services to develop investment-ready proposals and support scaling for BDV PPP start-ups and SMEs to reach the market	Private investors are a specific and very important target for the new AI, Data and Robotics Partnership As part of the BDVe project investors from different countries such as France, Germany, Luxembourg, Spain, and the UK have been identified that have demonstrated interest in investing in start-ups focusing on the data-driven economy. Those investors have been identified through our participation in main events. BDVe is currently working on linking those investors with BDV PPP start-ups such as start-ups from European Data to reach matchmaking
Tackling the necessary strategy and planning for the BDV Ecosystem until 2030, including the identification of new stakeholders, emerging usage domains, technology, business and policy roadmapping activity (ICT-13)	Establishment of a vision paper for the new AI, Data and Robotics Partnership in March 2019, together with euRobotics Finalisation of a stakeholder mapping exercise to engage relevant communities in the new Partnership, especially from a sectorial perspective

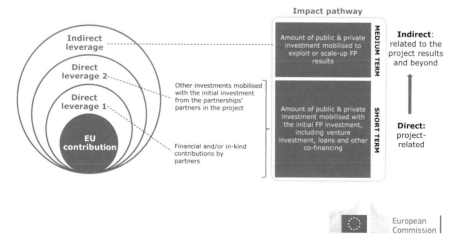

Fig. 2 Methodology and KPI structure proposed by EC for MR2018 (PPPs) (by European Commission licensed under CC BY 4.0)

Table 4 Evolution of private investments in BDV PPP over time

KPI	Description	Amounts in million €				
		2015	2016	2017	2018	Total
Indirect leverage 2	Estimated R&D expenses that are related to the BDV PPP but are not related to EC-funded projects (this excludes any expenses that are funded by the EC by definition)	245.80	289.10	388.67	370.33	1293.90
Indirect leverage 1	Estimated R&D expenses resulting from follow-up invest- ments of projects funded by the EC that are topic-wise related to the BDV PPP but initiated out- side the Big Data PPP (in FP7 or in H2020) (this excludes any expenses that are funded by the EC by definition)	35.10	49.40	83.55	70.40	238.45
Direct leverage2	Estimated R&D expenses resulting from follow-up invest- ments of BDV PPP projects (this excludes any expenses that are funded by the EC by definition)	N/A	N/A	1.09	7.57	8.67
Direct leverage 1	Additional investments in the execution of BDV PPP projects (this excludes any expenses that are funded by the EC by definition)	N/A	N/A	3.46	9.52	12.97
EU contribution	Annual private contribution (estimated for reporting period 2018)	N/A	N/A	5.48	9.64	15.12

Input to PPP project investment was 0 before 2017 as no projects had started. The number €12.4 million is calculated based on real input extrapolated from the percentage of responses and expected annual private investment

On the other hand, 40% of the BDVA members stated that their participation in the BDVA/BDV PPP had already contributed directly or indirectly to job creation, mainly because of the hiring of new experts to develop H2020 projects, start-ups created...), and new profiles hired to develop operations.

Projects reported that 48 job profiles were created or identified in 2018, and 106 new job profiles were reported as expected to be created from 2019 onwards and by the end of the project linked to the project activities.

Sixty-seven per cent of the projects running in 2018 reported contribution to the generation of new skills by the end of the project. In addition to the skills linked to the new job profiles, new skills are expected to be developed in cross-sectorial domains (e.g. in the form of "privacy-aware data processing" and "privacy-aware big data innovation" as reported by the SPECIAL project) and in specific sectors (e.g. analysis techniques using weather data, reported by the EW-SHOPP project). The BDV PPP incubators help start-ups to develop both the technical and non-technical skills needed to develop business in the Data Economy.

Among BDVA members, 51% of organisations reported contribution to the creation of new job profiles, and almost 60% contribute to the creation of new skills linked to the Big Data Value PPP. Finally, 60% of the projects and 51% of BDVA members have reported contributions to the Skills Agenda for Europe.

The BDV PPP organised 181 training activities involving over 18,300 participants during 2018. Projects contributed to this with 85 training activities during 2018 involving over 9700 participants. BDVA members reported 96 training activities involving over 8500 participants. Projects developed 16 interdisciplinary programmes during 2018 outreaching 250 participants.

During 2018, 396 equivalent FTEs masters and PhD students "(260 masters and 136 PhD) were involved in PPP projects, thereby collaborating with industrial players in developing industry-driven solutions and deploying experimentation testing scenarios. Contributing to raising awareness in professionals, users and the general public, the BDV PPP organised 323 events outreaching around 630,000 participants during 2018 contributing to raising awareness in professionals, users and the general public.

4.2.3 Impact of the BDV PPP on SMEs

Results of the Monitoring Report 2018 showed that a wide range of SMEs in Europe benefit from the Big Data Value PPP, considering the size (12% medium-size companies, 41% small companies and 48% micro-companies[16]), age (20% of the SMEs are 0 to 4 years old, 36% are 5 to 10 years old and 42% are 10 years old or older) and wide geographical distribution. SMEs play a variety of roles in the data value chain. SMEs participating in PPP projects clearly show a trend of an increase in turnover and in the number of employees. It is also important to mention that not all the SMEs involved in BDV PPP projects are technology companies but are also data users or providers, and the overall results and trend indicate an ongoing growth of turnover along the whole value chain.

Total turnover reported for SMEs in 2017 was €260.4 million.[17] In terms of turnover evolution, there is an increase in turnover in the SME companies that are part of the PPP with reported numbers of 60% increase in turnover with respect to 2014 and 17.7% in the last year. This number is in full alignment with the macro-economic numbers of data companies in Europe, and higher for some specific categories. In particular, young SMEs (5 and 10 years old) show on average the highest growth in turnover in relation to 2014 (up to 284%). The youngest companies (<5 years) show on average the highest growth in the last year (54.8%).

---

[16]Criteria for classification following EC rules: http://ec.europa.eu/growth/smes/business-friendly-environment/sme-definition_en

[17]Aggregated total of the companies.

In terms of employment evolution, the trend is also very positive in all companies that are part of the PPP, with an average increase in employment for the SMEs that are part of the PPP of 75% with respect to 2014 and a growth of 11.83% in the last year (2018 compared with 2017).

Special emphasis should be given to PPP instruments focused on supporting SMEs, in particular the Data Incubators and i-Spaces. The average age of the companies receiving cascade funding from the Data Incubators (DataPitch and EDI) is 4.9 years; 41% of those SMEs are younger than 5 years, 50% are between 5 and 10 years, and only 9% are older than 10 years old. Companies reported an increase in turnover of 315% for 2014 and 48.8% for 2017, and an 118.5% increase in employment for 2014 with a 22.4% increase in the last year.

4.2.4 Innovations Emerging from Projects

Innovations arising from the BDV PPP include:

- Specific project developments that have a marketable value, including Big Data products, processes, instruments, methods, systems and technologies, offering value to a wide variety of economic and industrial sectors
- Services of high societal value developed by projects
- Spin-offs arising from projects and start-ups incubated by the programme activities
- Patents and solutions enabling advanced privacy- and security-respecting solutions for data access, processing and analysis
- Contribution to Standards (individually as projects and coordinated activities at a programme level)
- Innovations resulting from cooperation between projects or programme-coordinated activities (e.g. advances in data sharing, innovative skill programmes, reuse of technical solutions across different sectors, etc.)
- Transformation of sectors of high economic value (led by the PPP Lighthouse projects, but also triggered by project cooperation): new business models and scaling innovations (advances in TRLs (technology readiness levels), cross-border solutions and bringing technology closer to the market, accelerating adoption)

In its second year of operation, the BDV PPP's 32 running projects reported 106 innovations of exploitable value as delivered in 2018: 63% have a medium impact and 37% are considered innovations of significant impact. Fifty per cent of the innovations delivered in 2018 are incremental innovations, 6% are architectural, 36% are disruptive and 1% are radical innovations.[18]

Ninety-three per cent of the innovations delivered in 2018 have an economic impact, and 48% have a societal impact.[19] Forty-one per cent are technologies

---

[18]Eight per cent are not included in any of these categories.

[19]Note that many innovations have both economic and societal impact.

(including platforms), 32% are services, 7% are products, 8% are methods, 8% are systems, 1% are software, 4% are components and/or modules and 11% are others, including frameworks/architectures, processes, tools and toolkits, spin-offs, datasets, ontologies, patents and knowledge.

Sixteen per cent of the innovations delivered in 2018 are fully cross-sectorial.

Sevety-five per cent provide solutions to the transport, mobility and logistics sector (the one with the best coverage in the PPP by the end of 2018); 20% of the innovations related to public services and smart cities; 19% to industry and manufacturing; 14% to bio-economy; 13% to the Telco sector; 12% marketing activities; 8% relate to health and healthcare; 8% to the ICT market; 7% to geospatial market; 5% to commerce; and 3% to others (including fashion, retail, business services, energy, media, compliance, etc.).

In relation to the maturity levels and TRLs, 7% of the innovations delivered are TRL 3 (experimental proof of concept), 10% are TRL 4 (technology validated in lab), 36% are TRL 5 or TRL 6 (technology validated in relevant environment, industrially relevant environment in the case of key enabling technologies), 32% are TRL 7 (system prototype demonstration in operational environment), 8% are TRL 8 (system complete and qualified) and 1% are TRL 9 (actual system proven in

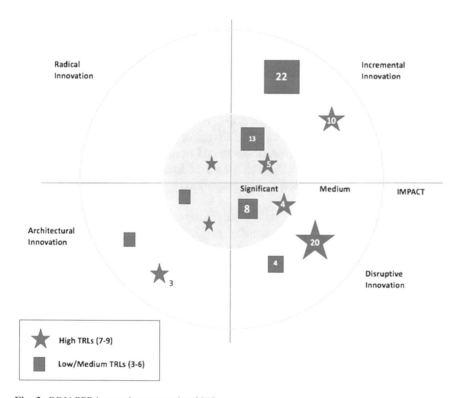

Fig. 3 BDV PPP innovations to market 2018

an operational environment—competitive manufacturing in the case of key enabling technologies—or in space).

Figure 3 provides a full overview of the innovations delivered by the BDV PPP during 2018, combining level of significance, type of innovation (incremental, disruptive, architectural or radical) and the TRLs. Although a large number of innovations are classified as incremental innovations of medium impact, it is remarkable to note the high percentage of significant innovations (and expected growth in the upcoming years), the high number of disruptive innovations and the high TRLs in some cases close to deployment. Although at a lower level, the BDV PPP is also delivering some architectural and radical innovations.

Sixty-three new economically viable services of high societal value were developed during 2018 as a result of the projects. Forty-seven per cent (over 30 projects) contributed to this KPI.

Projects reported 204 new systems and technologies developed during 2018. Many of them are already reported as part of the KPI "Significant Innovations to Market". Systems and technologies developed are not limited to one sector, and, in fact, the majority of the new systems and technologies can be utilised in different sectors/markets, thus stimulating the use of Big Data technologies in many areas.

Finally, many solutions and innovations arising from the Big Data PPP have been promoted in the BDV PPP Marketplace[20] developed by the BDVe CSA project to spread knowledge about the outcomes of the PPP.

4.2.5 Supporting Major Sectors and Major Domains with Big Data Technologies and Applications

The BDV PPP Lighthouse projects[21] active in 2018 focused on the bio-economy (agriculture, fisheries and forestry) (DataBio project), transport, mobility and logistics (transforming transport project), health and healthcare (BigMedilytics project) and manufacturing (BOOST4.0), with a total of four major sectors supported by Lighthouse projects and therefore widely supported by multiple use cases, scenarios and solutions.

Twenty per cent of the projects are fully cross-sectorial (their outcomes can be used in any sector or application domain) and 80% of the projects are working in more than one sector or application domain (this explains why the total is higher than 100% in Table 5). In particular, the BDV PPP projects address a wide variety of sectors[22], as shown in Table 5.

---

[20]http://marketplace.big-data-value.eu/

[21]Large-scale data-driven innovation and demonstration projects that aim at creating superior visibility, awareness and impact in specific relevant economic sectors.

[22]Grouped with a good level of alignment with the NACE registry. These categories are part of the information in the BDV PPP Marketplace that will be used for promoting all exploitable solutions coming out of the PPP projects (if needed, new categories can be added).

Table 5 Support to major sectors and domains

Sector/application domain	Projects addressing this sector/ application domain (% over active projects in 2018)	Innovations to market (delivered in 2018, % over the total)
Public services and smart cities	50%	20%
Transport, mobility and logistics	43%	75%
Retail	37%	20%
Business services	37%	20%
Health and healthcare	33%	8%
Manufacturing	23%	19%
Media	23%	0%
Finances and banking	23%	14%
Telecom	20%	13%
Energy	20%	0%
Bio-economy: agriculture, forestry and fisheries	17%	14%
Water and natural resources	17%	0%
Earth observation	13%	7%
Others	43%	27%

Others (43% of the projects) includes sectors such as insurance, public safety, personal security, public tenders, e-commerce, marketing, fashion industry, citizen engagement, ICT/Cloud services, social networks, procurement and legal domain.

Considering the whole project portfolio, the number of sectors supported is higher than 15, with a solid distribution of use cases, experiments, solutions and outreach activities among different sectors.

4.2.6 Experimentation

Projects reported 224 use cases and/or experiments conducted during 2018 with contributions from 18 different projects. This is an increase of 48.3% with respect to 2017 (151 experiments). The BDVA i-Spaces reported an additional 165 experiments with 6 i-Spaces contributing to this KPI.

Projects reported 82 large-scale experiments developed during 2018, 64 involving closed (private) data (78% of the total). Large-scale experiments either involve a large number of users with high TRLs or are developed in large geographical areas, in many cases involving a large number of users and actors or a combination of data volume, complexity and velocity; a large number of data sources; or integrated complex datasets flowing across borders. The BDVA i-Spaces also contributed to this KPI, reporting in total 38 large-scale experiments performed during 2018, 28 of them involving private data.

In relation to the amount of data made available for experimentation, reported information from projects and i-Spaces (members of BDVA) shows that the amount of data made available by the BDV PPP for experimentation in 2018 is 0.10696 Exabytes (106.96 Petabytes). A total of 0.08625 Exabytes (86.25 Petabytes) was reported by the projects.[23] It is important to note that some of the projects are not only providing internal access to diverse data sets from different sources but are also improving and creating new valuable datasets (e.g. of DataBio project). BDVA i-Spaces contributed to this KPI, reporting an additional 20.71 Petabytes of data for experimentation.

4.2.7   SRIA Implementation and Update

Concerning SRIA coverage, measured as "% of research priorities covered compared to the overall scope of research priorities defined in SRIA", projects have delivered contributions during 2018 already covering 100% of all the SRIA technical priorities. The major focus of technical contributions was "Data Analytics", followed at some distance by "Data Processing Architectures" and "Data Management". This is a significant change from the 2017 coverage, where "Data Management" was the top priority. A clear trend to focus on technical contributions in the areas of "Data Analytics" and "Data Processing Architectures" was anticipated in the BDV PPP Annual Monitoring Report 2017,[24] thus supporting our explanation that a solid base of "Data Management" solutions will enable analytics and processing innovations.

In relation to the BDV SRIA update, at the end of 2017 the BDVA released the BDV PPP SRIA v4.0 (detailed process and results reported in the 2017 Monitoring Report). This version was the basis to support the H2020 LEIT ICT WP2018–20. During 2018 a minor update, towards a version 4.1, was released in the community, crystallising in a series of individual deliverables in the format of vision, position or discussion papers that supported the transition towards the next framework programme and the creation of a new strategic agenda and roadmap.

In total, there were at least 12 events organised during 2018 that contributed to input in the BDVA strategic papers – multiple online meetings with a total of 2085 participants/contributions.

In total, since the launch of the BDV PPP, we can count 6422 potential contributions to the strategic roadmapping activities.

---

[23]Thirteen projects provided data for this KPI (Aegis, BigDataOcean, DataBio, euBusinessGraph, EW-Shopp, TT, QROWD, BigDataStack, BigMedilytics, Boost 4.0, CLASS, EDI, TheBuyForYou).

[24]http://www.bdva.eu/sites/default/files/MR2017_BDV_PPP_Main%20Report_September%202018_1.pdf

4.2.8 Technical Projects

The BDV PPP contributes to enabling advanced privacy- and security-respecting solutions for data access, processing and analysis. For 2018, 97 contributions were reported (2 patents,[25] 61 publications and 24 OSS/SW/Products).

Fifty per cent of the projects confirmed that they are assessing quality, diversity and value of data assets. These results show the intense usage of metrics to measure quality, diversity and value of data assets in projects, and some projects have developed specific metrics and methods to ensure quality, diversity and value in the data (e.g. I-BiDaaS has developed a Data Quality Assurance Process (DQAP) aiming at ensuring the high quality of the data generated/collected during the lifetime of the project). However, we cannot talk yet (2018) about the "PPP"-developed metric expected for 2019+.

Concerning the speed of data throughput, 40% of the projects reported that they expect the project to improve data throughput. Some projects, such as BigDataOcean and FashionBrain, measured improvements over 1000%. Others such as I-BiDaaS have specific objectives to develop data processing tools and techniques applicable in real-world settings and to demonstrate a significant increase in speed of data throughput and access.

4.2.9 Macro-economic KPIs

The monitoring of macro-economic KPIs is based on input from the European Data Market Monitoring Tool[26] as they are presented in the most recent report by IDC (https://www.idc.com/).[27]

Development of the market share of the European Union in the global Big Data Market. As an indicator, we compare the total revenues of EU Data Companies with other economies, i.e. the US, Japan and Brazil, as they are used as a benchmark in the IDC report.[28] The EU share of the total revenues in these economies the 2013 baseline was 27.7%. This share increased slightly to 27.9% in 2018, which is remarkable because the international indicators grew very fast in this period, but the EU kept pace with them. In absolute terms, the total revenue of US data companies in 2018 was approximately twice that of EU28 data companies in the same year (€162 billion vs. €77 billion). Seventy per cent of PPP projects active in 2018[29] reported contribution to increasing the revenue share of EU companies. Projects contributed by:

---

[25]Filled patents.

[26]SMART 2016/0063 – Study "Update of the European Data Market Monitoring Tool", IDC and Lisbon Councils.

[27]Gabriella Cattaneo, Giorgio Micheletti et al. "Update of the European Data Market Tool - Second report on Facts and Figures" April 2019 www.datalandscape.eu

[28]Gabriella Cattaneo et al., ibid. Chap. 10, pp. 129–142.

[29]Based on number of respondents.

- Accelerating adoption of new technologies
- Supporting EU data-driven companies to build innovative solutions that can be scaled internationally
- Developing innovative technologies to make European companies more competitive (e.g. news data protection approaches)
- Enabling industries to exploit their big data efficiently and therefore increase their market share and services provided to their customers

According to the most recent report,[30] the number of data companies increased to 283,100 by 2018, compared to 271,700 in 2017, with a growth rate of 4.2%. It should be noted that almost half of them are based in the UK, due to the high concentration of the ICT industry there. BDVA i-Spaces and Data Incubators (ICT 14-b projects, i.e. DataPitchand EDI projects) are in particular designed to contribute to this KPI as they support start-ups and entrepreneurs from early ideas to technical and business development until the go-to-market stage.[31] Seventy-seven per cent of the BDV PPP projects active in 2018[32] reported contribution towards increasing the number of European companies offering data technology and applications. The projects contributed in different ways, such as:

- Creating tools that will stimulate the creation of new companies
- Creating new companies as a result of a project (e.g. BigDataOcean)
- Supporting EU data-driven companies
- Building innovative solutions to solve data-related challenges
- Supporting companies in complying with the GDPR
- Lowering the threshold to create new business in a particular sector

In addition, 25% of BDVA members reported that their organisation ran or supported a programme that is specifically targeted at supporting start-ups or entrepreneurs in the field of Big Data.

The revenue of **data companies in the European Union**, according to the IDC report,[33] reached €77 billion in 2018 compared to €69 billion the previous year, with a growth rate of 12%. The revenue share of SMEs in 2018 amounts to €55.5 billion (72% of the total revenue), an absolute growth of €5.7 billion on the year before. The growth rate of revenue increases in proportion to company size, with the revenue of large companies with over 500 employees growing at 16% in 2018 over 2017. Seventy-seven per cent of the PPP projects active in 2018[34] reported contribution (or plan to contribute) to the revenue generated by European data companies. Project contribution to this KPI is mainly by:

---

[30]Gabriella Cattaneo et al., ibid.

[31]Further information can be found in Sect. 2.1 of this report.

[32]Based on number of respondents.

[33]Gabriella Cattaneo et al., ibid. pp. 89–97.

[34]Based on number of respondents.

- Opening up sectors to data-intensive companies
- Offering direct support and getting funding for data start-ups
- Making data processing easier and cheaper for companies
- Creating new opportunities through privacy-preserving analytics solutions
- Commercialising new services with a marketable value
- Creating opportunities for common exploitation based on joint Big Data technology pipelines
- Developing simplicity in some business ecosystems

The baseline for **data professionals in the European Union** in 2013 amounts to 5.77 million. The number of data professionals increased to a total of 7.2 million by 2018, resulting in an absolute growth rate of 1453 million professionals since 2013. The rate of growth of data professionals is increasing, with approximately 559,000 positions added in 2018 and an increase of 8.4% on the year before.[35] Eighty-seven per cent of the PPP projects active in 2018[36] reported contribution from their project to increase the number of data workers in Europe. Projects contribute to this KPI in different ways:

- New organisations created as a result of the projects hiring new data professionals
- Supporting growth of emerging start-ups
- Developing more data-driven services that will require new data workers
- Unlocking the value of data services by introducing privacy-preserving technologies
- Creating new job profiles
- Supporting the adoption of data solutions in different sectors
- Supporting education and training

4.2.10 Contributions to Environmental Challenges

Over 20% of the projects running in 2018 reported that they contribute to the reduction of energy, and 30% contribute to **reduction in CO_2 emission**. Quantitative results are provided by some projects, such as the Transforming Transport (TT) project that shows that in some specific monitored items improvements in efficiency range between 25% and 51% in energy reduction, and improvements concerning CO_2 emissions reach up to 29% and emission reductions in general (including PM and NOx) up to 23%.

The three Lighthouse projects running in 2018 (DataBio, Transforming Transport and Boost4.0) have reported contribution to **reduction in waste**. For example, in DataBio and in particular in forestry, although still with early data and experiments, the experience from customer cases shows a reduction in waste of up to 10%. Some pilot TT projects show approximately 25% improvement in the management of

---

[35]Ibid.

[36]Based on number of respondents.

assets, which can adequately demonstrate a relative high-level achievement in waste reduction at this final stage of the project.

Seventeen per cent of the projects running in 2018 have reported contribution to **reduction in the use of material resources**; e.g. BigMedilytics provides quantitative data in a particular scenario, reporting that the Asset Management pilot aims to reduce the number of unused mobile assets in hospitals by up to 20%.

Finally, in relation to **energy reduction in big data analytics**, there is no quantitative input in results provided by any project but, e.g., the E2Data project develops a framework that optimises calculations, leading to decreased use of energy.

4.2.11 Standardisation Activities with European Standardisation Bodies

During 2017, the BDVA and the BDV PPP set up some foundations defining priorities for the PPP in Big Data standardisation implemented during 2018 as follows:

- Establish an official liaison between the BDVA Standards Group and the AIOTI WG3; this activity was developed through different workshops during 2017 and implemented in 2018 with the signing of an MoU with AIOTI and common activities organised during the year.
- Further develop the BDVA Reference Model pursuing alignment with others, such as oneM2M, BDE Platform, AIOTI and RAMI 4.0, implemented through different workshops organised during 2017 and 2018.
- Open an official dialogue with CEN, CENELEC and ETSI on standards harmonisation, implemented through different workshops during 2017 and 2018. The BDVA intends to sign an MoU with CEN/CENELEC in 2019, and it is under discussion with ETSI.
- Create the BDVA Roadmap for Big Data standards harmonisation and industry engagement in Global Big Data standards development.

Thirty per cent of the projects running in 2018 reported that they perform activities leading to data/Big data standardisation. Three projects reported contribution to European standardisation bodies (ESBs) activities and reported 11 working items in ESBs. Twenty per cent of BDVA members reported that their organisations perform activities leading to data/Big data standardisation. In particular, BDVA members have reported contributions to IEC, DIN DKE and other consensus-based standardisation bodies; OPC foundation and other consortia-based standardisation bodies; OASIS; W3C committees and community group discussions; open data harmonisation national activities; ISO/IEC JTC1; and defining standards in georeferenced data for geoscience (Open Geospatial Consortium (OGC) and Commission for the Management and Application of Geoscience Information (IUGS/CGI)) and ETSI.

5 Summary and Outlook

The year 2018 was a transition year and an important inflexion point between the so-called Phase I (establishment of the ecosystem) and Phase II of the BDV PPP (pioneer disruptive new forms of big data value solutions). New calls for proposals were in place during 2018 and 2019 as part of the H2020 WP 2018–2020 (calls closing in April 2018, November 2018, April 2019 and November 2019) that brought new projects that enriched the BDV PPP portfolio, also increasing challenges of coordination, communication and cooperation. The year 2018 was also a transition year in defining the strategy and direction of the next partnership framework programme (2021–2028).

The increase in the quality and quantity of the data available for experimentation and the launch of the cross-border Industrial Data Platforms and Personal Data Platforms at the beginning of 2020, supported by other ecosystem enablers, have directed the final transition period towards Phase III as defined in SRIA v4. The BDV PPP projects starting in 2020 (e.g. EUHubs4Data project) are establishing a strong foundation for the next framework programme (deployment of data platforms, the federation of Big Data Innovation Hubs/data experimentation facilities, and advances in data and data-driven AI capabilities).

On 25 April 2018, the European Commission outlined a European strategy for AI to boost investment and set ethical guidelines. In its communication, the European Commission put forward a European approach to Artificial Intelligence based on three pillars: (i) "boosting financial support and encouraging uptake by public and private sectors", (ii) "preparing for socio-economic changes brought about by AI", and (iii) "ensuring an appropriate ethical and legal framework". The strategy acknowledged that member states had existing research and innovation objectives that focused on AI and encouraged alignment of individual roadmaps towards a European partnership. Also on 25 April the European Commission proposed a package of measures as a key step towards a common data space in the EU – a seamless digital area with a scale that will enable the development of new products and services based on data.

On 6 June 2018, the European Commission announced its proposal to create the first ever Digital Europe programme and invest €9.2 billion to align the next long-term EU budget 2021–2027 with increasing digital challenges. The Commission's proposal focused on five areas: supercomputers, Artificial Intelligence (AI) (including Data/European Data Space), cybersecurity and trust, digital skills, and ensuring a wide use of digital technologies across the economy and society.

On 7 June 2018, the European Commission announced Horizon Europe (research and innovation programme for the next long-term EU budget 2021–2027) with plans to bring a new generation of European Partnerships and increase collaboration with other EU programmes.

Towards the end of 2018, the BDVA committed its official participation as a private member of the EuroHPC Joint Undertaking aiming at bringing synergy between HPC, Big Data and Artificial Intelligence, and providing industry perspective.

Additionally, the BDVA and euRobotics officially joined forces at the end of 2018 and announced their intentions of working together in a future AI, Data and Robotics Partnership. On 7 December 2018, the European Commission presented a coordinated plan prepared with the members states to foster the development and use of AI in Europe. The plan proposes the development of a European AI public-private partnership building on the BDV PPP and SPARC PPPs.

During 2019 the BDVA and euRobotics developed a common vision paper and the first version of a common AI-PPP Strategic, Research Innovation and Deployment Agenda with strong involvement of ongoing PPP projects, members and many external communities. At the end of 2019, CLAIRE, ELLIS and EurAI joined forces with the BDVA and euRobotics, and the five organisations submitted a joint Partnership Proposal (Zillner et al. 2020). This document lays down the context, vision and objective, and suggests the impact of a possible Partnership of Data, AI and Robotics, building upon the strong assets developed by the BDV PPP and the SPARC PPP. During the first months of 2020, the member states and the European Commission carefully considered the Partnership Proposal and provided feedback for its improvement, which resulted in several updates of the document. On 22 September 2020, the joint release of the Strategic Research and Deployment Agenda (SRIDA v3.0) was published, paving the way towards the new Partnership for Horizon Europe and the Digital Europe Programme, bringing investments and new instruments to scale up the assets and impact of the current Big Data Value PPP.

References

BDVPPP Contractual Arrangement. (n.d.).

European Council. (2017). *The Rome Declaration.*

Zillner, S, Bisset, D., García Robles, A., Hahn, T., Lafrenz, R., Liepert, B., & Curry, E. (2019). *Strategic research, innovation and deployment agenda for an AI PPP: A focal point for collaboration on artificial intelligence, data and robotics.* Retrieved from https://www.bdva. eu/sites/default/files/AI%20PPP%20SRIDA-Consultation%20Version-June%202019%20-% 20Online%20version.pdf

Zillner, S., Bisset, D., Milano, M., Curry, E., Hahn, T., Lafrenz, R., et al. 2020. *Strategic research, innovation and deployment agenda - AI, data and robotics partnership. Third Release* (3rd). Brussels: BDVA, euRobotics, ELLIS, EurAI and CLAIRE.

Zillner, Sonja, Curry, E., Metzger, A., Auer, S., & Seidl, R. (Eds.). (2017). *European big data value strategic research & innovation agenda.* Retrieved from Big Data Value Association website www.bdva.eu

Part II
Research and Innovation Elements of Big Data Value

Technical Research Priorities for Big Data

Edward Curry, Sonja Zillner, Andreas Metzger, Arne J. Berre, Sören Auer,
Ray Walshe, Marija Despenic, Milan Petkovic, Dumitru Roman,
Walter Waterfeld, Robert Seidl, Souleiman Hasan, Umair ul Hassan, and
Adegboyega Ojo

Abstract   To drive innovation and competitiveness, organisations need to foster the
development and broad adoption of data technologies, value-adding use cases and
sustainable business models. Enabling an effective data ecosystem requires overcoming several technical challenges associated with the cost and complexity of
management, processing, analysis and utilisation of data. This chapter details a
community-driven initiative to identify and characterise the key technical research
priorities for research and development in data technologies. The chapter examines
the systemic and structured methodology used to gather inputs from over 200 stakeholder organisations. The result of the process identified five key technical research
priorities in the areas of *data management*, *data processing*, *data analytics*, *data*

E. Curry (✉) · S. Hasan · U. ul Hassan · A. Ojo
Insight SFI Research Centre for Data Analytics, NUI Galway, Galway, Ireland
e-mail: edward.curry@nuigalway.ie

S. Zillner
Siemens AG, Munich, Germany

A. Metzger
paluno, University of Duisburg-Essen, Duisburg, Germany

A. J. Berre · D. Roman
SINTEF Digital, Oslo, Norway

S. Auer
Leibniz Universität Hannover, Hannover, Germany

R. Walshe
ADAPT SFI Centre for Digital Content, Dublin City University, Dublin, Ireland

M. Despenic
ABN AMRO Bank, Amsterdam, the Netherlands

M. Petkovic
Philips and Eindhoven University of Technology, Eindhoven, the Netherlands

W. Waterfeld
Saarbrücken, Germany

R. Seidl
Nokia Bell Labs, Munich, Germany

visualisation and user interactions, and *data protection,* together with 28 sub-level challenges. The process also highlighted the important role of data standardisation, data engineering and DevOps for Big Data.

Keywords Research challenges · Data management · Data processing · Data analytics · Data visualisation · User interactions · Data protection · Data standardisation · Data ecosystem

1 Introduction

The expectations in refining data as the new oil of the twenty-first century are currently so high that virtually no business can afford not to have a big data project that 'unlocks' the value in their data (Chen et al. 2012). There is a noticeable increase in the adoption of data-driven business scenarios in sectors other than the web-based 'traditional' big data companies such as Google, Yahoo, Facebook and Twitter (Lavalle et al. 2011). However, many sectors still struggle with the adoption of data technologies, often due to a lack of expertise, regulatory barriers and unclear business value. This is especially true in non-IT-focused sectors, such as the energy sector that struggles with the adoption of data technologies (Rusitschka and Curry 2016). The benefits of sharing and linking data across domains and industry are apparent. Initiatives such as Smart Cities are showing how different sectors (i.e. energy and transport) can collaborate to maximise the potential for optimisation and value return (*Communication: A European strategy for data* 2020). The cross-fertilisation of stakeholders and datasets from different sectors is a key element for advancing the data economy.

To support the emergence of a data ecosystem, it was important that the different actors within the ecosystem 'define a shared vision and jointly identify gaps in the current data landscape' (DG Connect 2013). Data ecosystems face several problems such as data discovery, curation, linking, synchronisation, distribution, business modelling, sales and marketing (José María Cavanillas et al. 2016). To address these issues, the Big Data Value contractual Public-Private Partnership (BDV PPP) between the European Commission and the Big Data Value Association aimed to strengthen the data value chain (Curry 2016), foster cooperation in data research and innovation, enhance community building around data and set the groundwork for a thriving data-driven economy in Europe. The BDV PPP was driven by the conviction that research and innovation focusing on a combination of business and usage needs is the best long-term strategy to deliver value from big data and create jobs and prosperity. An essential requirement was to identify and characterise the key technical research challenges that need to be tackled to enable a data ecosystem.

This chapter identifies the key technical research priorities for research and development in data technologies. It presents the results of an investigation and consultation process that was conducted to capture the priorities for big data in public and private organisations across Europe. The chapter starts with an introduction to the methodology for the identification and prioritisation of the technical challenges for the adoption of data technologies. The chapter details the key challenges and outcomes needed in terms of data management, data processing, data analytics, data visualisation and user interaction, and data protection. It highlights the role of standardisation to further the development of data technology and the key role of data standards. Challenges with data engineering and DevOps for big data systems ensure productivity and quality are detailed. Finally, the chapter presents a scenario from the healthcare sector to emphasise the importance of adopting better big data strategies.

2 Methodology

In order to correctly identify the technical research priorities a systemic and structured methodology was needed to gather inputs from over 200 stakeholder organisations. The methodology built on and extended an established roadmapping methodology to gather consensus from a range of stakeholders (Curry et al. 2016). The key phases in the methodology, as illustrated in Fig. 1, are (a) technology state of the art and sector analyses, (b) subject matter expert interviews, (c) stakeholder workshops, (d) requirements consolidation and (e) community survey.

2.1 Technology State of the Art and Sector Analysis

The goal of the first phase was to identify the sectorial needs and requirements gathered from different stakeholders and the state of the art of data technologies, as well as identifying research challenges. As part of the investigation, application sectors expressed their need for the technology as well as possible limitations and expectations regarding its current and future deployment. The first step was to perform a systematic literature review based on the following activities:

- Identification of relevant type and sources of information
- Analysis of key information in each source
- Identification of key topics for each technical working group
- Identification of the key subject matter experts for each topic as potential interview candidates

Fig. 1 The workflow of research methodology

- Synthesisation of the key message of each data source into state-of-the-art descriptions for each identified topic

The following types of data sources were used: scientific papers published in workshops, symposia, conferences, journals and magazines, company white papers, technology vendor websites, open-source projects, online magazines, analysts' data, web blogs other online sources and interviews. The groups focused on sources that mention concrete technologies and analysed them concerning their values and benefits. The synthesis step compared the key messages and extracted agreed views. Topics were prioritised based on the degree to which they can address business needs.

2.2   Subject Matter Expert Interviews

The literature survey was complemented by a series of interviews with subject matter experts for relevant topic areas. Subject matter expert interviews are a technique well suited to data collection and particularly for exploratory research because it allows extensive discussions that illuminate factors of importance (Oppenheim 1992; Yin 2013). The information gathered is likely to be more accurate than information collected by other methods since the interviewer can avoid inaccurate or incomplete answers by explaining the questions to the interviewee (Oppenheim 1992). The interviews followed a semi-structured protocol. The topics of the interview covered different aspects of big data:

- Goals of big data technology
- Beneficiaries of big data technology
- Drivers and barriers for data technologies
- Technology and standards for data technologies

Interviewees were selected to be representative of the different stakeholders within the data ecosystem. The selection of interviewees covered (1) established providers of big data technology (typically MNCs), (2) innovative sectorial players who are successful at leveraging big data, (3) new and emerging SMEs in the big data space and (4) world-leading academic authorities in technical areas related to the big data value chain.

The data collection and the analysis strategy were inspired by the triangulation approach (Flick 2004). Reviewing and quantitatively assessing the high-level application scenarios derived a reliable analysis of user needs. Examinations of the likely constraints of big data applications helped to identify the relevant requirements that needed to be addressed.

2.3   Stakeholder Workshops

The third step involved a cross-check and validation of the initial results of the first two steps by involving stakeholders from multiple domains in dedicated workshops and webinars to discuss and review the outcomes. Multiple workshops and consultations took place to ensure the most comprehensive representation of views and positions, including the full range of public and private sector entities not only from technology provision but also technology adoption. Sectoral workshops were conducted in various fields: geospatial/environment, energy, media, mobility, manufacturing, retail, health and the public sector. The purpose was to identify the main priorities with approximately 200 organisations and other relevant stakeholders physically participating and contributing. A wide range of stakeholders contributed to the process with inputs and analysis from SMEs and large enterprises, public organisations, and research and academic institutions. They included suppliers and service providers, data owners and early adopters of big data in many sectors. Extensive analysis reports were then produced, which helped both formulate and reformulate the identified requirements. From the analysis of the results, it was clear that addressing the technical needs of these vertical application markets required a set of cross-sector technologies.

2.4   Requirement Consolidation

Comparison among the different sectors enabled the identification of commonalities and differences at multiple levels. The analysis was used to define integrated cross-sectorial priorities that provide a coherent, holistic view of the big data domain and establish a common understanding of requirements, as well as technology descriptions and terms used across domains. A consolidated description was established to align the sector-specific labelling of requirements. In doing so, each sector provided its requirements with the associated user needs. Thus, the initial list of 13 high-level requirements and 28 sub-level requirements could be reduced to 5 high-level requirements and 20 sub-level requirements.

2.5   Community Survey

The objective of the community survey was to engage with the broader community to ensure a comprehensive perspective concerning the technical and business impact of the identified technical priorities, as well as to identify emerging priorities with high impact for the European big data economy. An inclusive approach was taken to ensure stakeholder engagement, with inputs actively solicited from the wider community composed of experts in technical domains as well as in business sectors. The

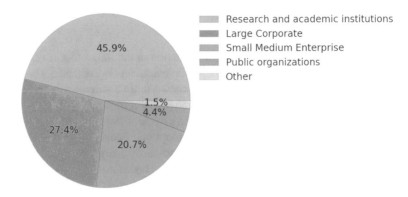

Fig. 2 Distribution of participants in terms of the type of organisation

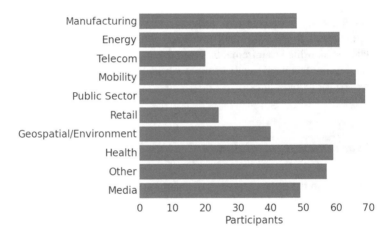

Fig. 3 Number of organisations associated with different sectors

survey received participation from a wide range of organisations. In total, 135 organisations responded to the survey through their representatives.

Figure 2 shows the distribution of participants in terms of the type of organisation. The majority of participant organisations (almost 95%) were either private companies or research and academic institutions. The response indicates a broader interest and contribution from stakeholders in shaping the future of the European big data community.

Figure 3 shows the number of organisations working in various sectors. In general, the organisations identified themselves as being active in multiple sectors, which underlines the cross-sectoral perspectives on the technical and non-technical priorities of big data as identified by the survey. Figure 4 shows that more than 70% of the participants chose two or more sectors. On average, more than three different sectors were chosen by participants to indicate the diversity of their portfolio. This

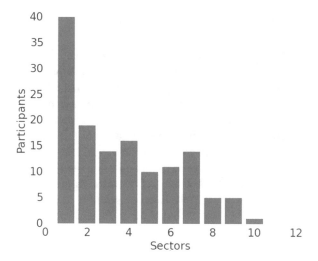

Fig. 4 Histogram of the number of sectors per organisation

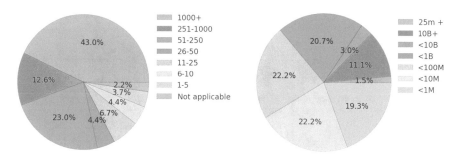

Fig. 5 Composition of participating organisations in terms of number of employees (left) and annual revenue (right)

also highlights the need to consider the multidisciplinary nature of the big data economy.

To quantify the size of the organisation, the survey participants were asked to indicate the number of employees (full-time equivalent) and annual revenue. Figure 5 summarises the composition of participating organisations in terms of employees and revenue. Primarily due to participation from the public sector and large corporates, the majority of organisations have more than 200 employees and revenue higher than 10 million. It should be noted that big data challenges for companies with more than 1000 employees are not only limited to their specific sectors but also in their day-to-day operations, such as human resource management and finance. The following section discusses the technical priorities for data technologies, in addition to their ranking based on the community survey.

3 Research Priorities for Big Data Value

The first three steps of the methodology produced a set of consolidated cross-sectorial technical research requirements. The result of this process was the identification of five key technical research priorities as illustrated in Fig. 6 (data management, data processing architectures, deep analytics, data protection and pseudonymisation, advanced visualisation and user experience), together with 28 sub-level challenges to delivering big data value. In this section, we report on the results of the survey to identify a prioritisation of the cross-sectorial requirements. As far as possible, the roadmaps were quantified using the results of the survey to allow for well-founded prioritisation and action plans, as illustrated in Fig. 7. The remainder of this chapter summaries the technical priorities as defined in the Strategic Research and Innovation Agenda (SRIA) of the BDVA (Zillner et al. 2017).

3.1 Priority 'Data Management'

More and more data are becoming available. This data explosion, often called a 'data tsunami', has been triggered by the growing volumes of sensor data and social data,

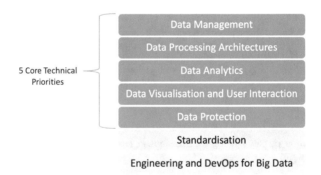

Fig. 6 High-level technical priorities for data technologies

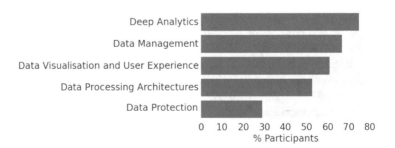

Fig. 7 Distribution of high-level technical priorities across participants

born out of Cyber-Physical Systems (CPS) and Internet of Things (IoT) applications. Traditional means for data storage and data management are no longer able to cope with the size and speed of data delivered in heterogeneous formats and at distributed locations.

Large amounts of data are being made available in a variety of formats ranging from unstructured to semi-structured to structured formats, such as reports, Web 2.0 data, images, sensor data, mobile data, geospatial data and multimedia data. For instance, important data types include numeric types, arrays and matrices, geospatial data, multimedia data and text. A great deal of this data is created or converted and further processed as text. Algorithms or machines are not able to process the data sources due to the lack of explicit semantics. In Europe, text-based data resources occur in many different languages, since customers and citizens create content in their local language. This multilingualism of data sources means that it is often impossible to align them using existing tools because they are generally available only in the English language. Thus, the seamless aligning of data sources for data analysis or business intelligence applications is hindered by the lack of language support and gaps in the availability of appropriate resources.

Isolated and fragmented data pools are found in almost all industrial sectors. Due to the prevalence of data silos, it is challenging to accomplish seamless integration with and smart access to the various heterogeneous data sources. And still today, data producers and consumers, even in the same sector, rely on different storage, communication and thus different access mechanisms for their data. Due to a lack of commonly agreed standards and frameworks, the migration and federation of data between pools impose high levels of additional costs. Without a semantic interoperability layer being imposed upon all these different systems, the seamless alignment of data sources cannot be realised.

To ensure a valuable big data analytics outcome, the incoming data has to be high quality; or, at least, the quality of the data should be known to enable appropriate judgements to be made. This requires differentiating between noise and valuable data, and thereby being able to decide which data sources to include and which to exclude to achieve the desired results.

Over many years, several different application sectors have tried to develop vertical processes for data management, including specific data format standards and domain models. However, consistent data lifecycle management – that is, the ability to clearly define, interoperate, openly share, access, transform, link, syndicate and manage data – is still missing. In addition, data, information and content need to be syndicated from data providers to data consumers while maintaining provenance, control and source information, including IPR considerations (data provenance). Moreover, to ensure transparent and flexible data usage, the aggregation and management of respective datasets enhanced by a controlled access mechanism through APIs should be enabled (Data-as-a-Service, or DaaS).

3.1.1 Challenges

As of today, collected data is rapidly increasing; however, the methods and tools for data management are not evolving at the same pace. From this perspective, it becomes crucial to have – at a minimum – good metadata, Natural Language Processing (NLP), and semantic techniques to structure the datasets and content, annotate them, document the associated processes, and deliver or syndicate information to recipients. The following research challenges have been identified:

- **Semantic annotation of unstructured and semi-structured data:** Data needs to be semantically annotated in digital formats, without imposing extra effort on data producers. In particular, unstructured data, such as videos, images or text in a natural language (including multilingual text), or specific domain data, such as Earth observation data, have to be pre-processed and enhanced with semantic annotation.
- **Semantic interoperability:** Data silos have to be unlocked by creating interoperability standards and efficient technologies for the storage and exchange of semantic data and tools to allow efficient user-driven or automated annotations and transformations.
- **Data quality:** Methods for improving and assessing data quality have to be created, together with curation frameworks and workflows. Data curation methods might include general-purpose data curation pipelines, online and offline data filtering techniques, improved human–data interaction, and standardised data curation models and vocabularies, as well as ensuring improved integration between data curation tools.
- **Data lifecycle management and data governance:** With the tremendous increase in data, integrated data lifecycle management is facing new challenges in handling the sheer size of data, as well as enforcing consistent quality as the data grows in volume, velocity and variability, including providing support for real-time data management and efficiency in data centres. Furthermore, as part of the data lifecycle, data protection and management must be aligned. Control, auditability and lifecycle management are key for governance, cross-sector applications and the General Data Protection Regulation (GDPR).
- **Integration of data and business processes:** This relates to a conceptual and technically sound integration of results from the two 'worlds' of analytics. Integrating data processes, such as data mining or business intelligence, on the one side, with business processes, such as process analysis in the area of Business Process Management (BPM), on the other side, is needed.
- **Data-as-a-Service:** The issue here is how to bundle both the data and the software and data analytics needed to interpret and process them into a single package that can be provided as an (intermediate) offering to the customer.
- **Distributed trust infrastructures for data management:** Mechanisms are required to enforce consistency in transactions and data management, for example, based on distributed ledger/blockchain technologies. Flexible data

management structures are based on microservices with the possibility of integrating data transformations, data analysis and data anonymisation, in a decentralised manner.

3.1.2 Outcomes

The main expected advances in data management are as follows:

- Languages, techniques and tools for measuring and ensuring data quality (such as novel data management processing algorithms and data quality governance approaches that support the specifics of big data) and for assessing data provenance, control and IPRs.
- Principles for a clear Data-as-a-Service (DaaS) model and paradigm fostering the harmonisation of tools and techniques with the ability to easily reuse, interconnect, syndicate, auto/crowd annotate and bring to life data management use cases and services across sectors, borders and citizens by decreasing the costs of developing new solutions. Furthermore, trusted and flexible infrastructures need to be developed for the DaaS paradigm, potentially based on technologies such as distributed ledgers, blockchains and microservices.
- Methods and tools for a complete data management lifecycle, ranging from data curation and cleaning (including pre-processing veracity, velocity integrity and quality of data) and using scalable big data transformations approaches (including aspects of automatic, interactive, sharable and repeatable transformations), to long-term data storage and access. New models and tools to check integrity and veracity of data, through both machine-based and human-based (crowdsourcing) techniques. Furthermore, mechanisms need to be developed for the alignment of data protection and management, addressing aspects such as control, auditability and lifecycle management of data.
- Methods and tools for the sound integration of analytics results from data and business processes. This relies on languages and techniques for semantic interoperability such as standardised data models and interoperable architectures for different sectors enriched through semantic terminologies. Particularly important are standards and multilingual knowledge repositories/sources that allow industries and citizens to seamlessly link their data with others. Mechanisms to deal with semantic data lakes and industrial data spaces and the development of enterprise knowledge graphs are of high relevance in this context.
- Techniques and tools for handling unstructured and semi-structured data. This includes natural language processing for different languages and algorithms for the automatic detection of normal and abnormal structures (including automatic measuring, tools for pre-processing and analysing sensor, social, geospatial, genomics, proteomics and other domain-orientated data), as well as standardised annotation frameworks for different sectors supporting the technical integration of different annotation technologies and data formats.

3.2   Priority 'Data Processing Architectures'

The Internet of Things (IoT) is one of the key drivers of the big data phenomenon. Initially, this phenomenon started by applying the existing architectures and technologies of big data that we categorise as data-at-rest, which is data kept in persistent storage. In the meantime, the need for processing immense amounts of sensor data streams has increased. This type of data-in-motion (i.e. non-persistent data processed on the fly) has extreme requirements for low-latency and real-time processing. What has hardly been addressed is the concept of complete processing for the combination of data-in-motion and data-at-rest.

For the IoT domain, these capabilities are essential. They are also required for other domains like social networks or manufacturing, where huge amounts of streaming data are produced in addition to the available big datasets of actual and historical data.

These capabilities affect all layers of future big data infrastructures, ranging from the specifications of low-level data, to flows with the continuous processing of micro-messages, to sophisticated analytics algorithms. The parallel need for real-time and large data volume capabilities is a key challenge for big data processing architectures. Architectures to handle streams of data, such as the lambda and kappa architectures, will be considered as a baseline for achieving a tighter integration of data-in-motion with data-at-rest.

Developing the integrated processing of data-at-rest and data-in-motion in an ad hoc fashion is, of course, possible, but only the design of generic, decentralised and scalable architectural solutions leverages their true potential. Optimised frameworks and toolboxes to enable the best use of both data-in-motion (e.g. data streams from sensors) and data-at-rest leverage the dissemination of reference solutions which are ready and easy to deploy in any economic sector. For example, proper integration of data-in-motion with the predictive models based on data-at-rest enable efficient, proactive processing (detection ahead of time). Architectures that can handle heterogeneous and unstructured data are also important. When such solutions become available to service providers, in a straightforward manner, they can focus on the development of business models.

The capability of existing systems to process such data-in-motion and answer queries in real time and for thousands of concurrent users is limited. Special-purpose approaches based on solutions like Complex Event Processing (CEP) are not sufficient for the challenges posed by the IoT in big data scenarios. The problem of achieving effective and efficient processing of data streams (data-in-motion) in a big data context is far from being solved, especially when considering the integration with data-at-rest and breakthroughs in NoSQL databases and parallel processing (e.g. Hadoop, Apache Spark, Apache Flink, Apache Kafka). Applications, for instance of Artificial Intelligence, are also required to fully exploit all the capabilities of modern and heterogeneous hardware, including parallelism and distribution to boost performance.

To achieve the agility demanded by real-time business and next-generation applications, a new set of interconnected data management capabilities is required.

3.2.1 Challenges

There have been several advances in big data analytics to support the dimension of big data volume. In a separate development, stream processing has been enhanced in terms of analytics on the fly to cover the velocity aspect of big data. This is especially important as business needs to know what is happening now. The main challenges to be addressed are:

- **Heterogeneity:** Big data processing architectures form places to gather and process various pieces of relevant data together. Such data can vary in several aspects, including different syntactic formats, heterogeneous semantic representations and various levels of granularity. In addition, data can be structured, semi-structured or unstructured, or multimedia, audio-visual or textual. Hardware can also be heterogeneous (CPUs, GPUs and FPGAs). Having the ability to handle big data's variety and uncertainty over several dimensions is a challenge for big data processing architectures.
- **Scalability:** Being able to apply storage and complex analytics techniques at scale is crucial to extract knowledge out of the data and develop decision-support applications. For instance, predictive systems such as recommendation engines must be able to provide real-time predictions while enriching historical databases to continuously train more complex and refined statistical models. The analytics must be scalable, with low latency adjusting to the increase of both the streams and volume of big datasets.
- **Processing of data-in-motion and data-at-rest:** Real-time analytics through event processing and stream processing, spanning inductive reasoning (machine learning), deductive reasoning (inference), high-performance computing (data centre optimisation, efficient resource allocation, quality of service provisioning) and statistical analysis, has to be adapted to allow continuous querying over streams (i.e. online processing). The scenarios for big data processing also require a greater ability to cope with systems which inherently contain dynamics in their daily operation, alongside their proper management, to increase operational effectiveness and competitiveness. Most of these processing techniques have only been applied to data-at-rest and in some cases to data-in-motion. A challenge here is to have suitable techniques for data-in-motion and also integrated processing for both types of data at the same time.
- **Decentralisation:** Big data producers and consumers can be distributed and loosely coupled as in the Internet of Things. Architectures have to consider the effect of distribution on the assumptions underlying them, such as loose data agreements and missing contextual data. The distribution of big data processing nodes poses the need for new big data-specific parallelisation techniques, and (at least partially) the automated distribution of tasks over clusters is a crucial

element for effective stream processing. Especially important is efficient distribution of the processing to the Edge (i.e. local data Edge processing and analytics), as a part of the ever-increasing trend of Fog computing.

- **Performance:** The performance of algorithms has to scale up by several orders of magnitude while reducing energy consumption compatible with the best efforts in the integration between hardware and software. It should be possible to utilise existing and emerging high-performance-computing and hardware-oriented developments, such as main memory technology, with different types of caches, such as Cloud and Fog computing, and software-defined storage with built-in functionality for computation near the data (e.g. Storlets). Also to be utilised are data availability guarantees to avoid unnecessary data downloading and archiving, and data reduction to support storing, sharing and efficient in-place processing of the data.

- **Novel architectures for enabling new types of big data workloads (hybrid big data and HPC architecture):** Some selected domains have shown a considerable increase in the complexity of big data applications, usually driven by computation-intensive simulations, which are based on complex models and generate enormous amounts of output data. On the other hand, users need to apply advanced and highly complex analytics and processing to this data to generate insights, which usually means that data analytics needs to take place in situ, using complex workflows and in synchrony with computing platforms. This requires novel big data architectures which exploit the advantages of HPC infrastructure and distributed processing, and includes the challenges of maintaining efficient distributed data access (enabling the scaling of deep learning applications) and efficient energy consumption models in such architectures.

- **The introduction of new hardware capabilities:** Computing capacity has become available to train larger and more complex models more quickly. Graphics processing units (GPUs) have been repurposed to execute the data and algorithm crunching required for machine learning at speeds many times faster than traditional processor chips. In addition, Field Programmable Gate Arrays (FPGAs) and dedicated deep learning processors are influencing big data architectures.

3.2.2   Outcomes

The main expected advances in data processing architectures are:

- **Techniques and tools for processing real-time heterogeneous data sources:** The heterogeneity of data sources for both data-at-rest and data-in-motion requires efficient and powerful techniques for transformation and migration. This includes data reduction and mechanisms to attach and link to arbitrary data. Standardisation also plays a key role in addressing heterogeneity.

- **Scalable and dynamic data approaches:** The capabilities for processing very large amounts of data in a very short time (in real-time applications and/or reacting to dynamic data) and analysing sizable amounts of data to update the analysis results as the information content changes. It is important to access only relevant and suitable data, thereby avoiding accessing and processing irrelevant data. Research should provide new techniques that can speed up training on large amounts of data, for example by exploiting parallelisation, distribution and flexible Cloud computing platforms, and by moving computation to Edge computing.

- **Real-time architectures for data-in-motion:** Architectures, frameworks and tools for real-time and on-the-fly processing of data-in-motion (e.g. IoT sensor data) and integrating it with data-at-rest. Furthermore, there is a need to dynamically reconfigure such architectures and dynamic data processing capabilities on the fly to cope with, for example, different contexts, changing requirements and optimisation in various dimensions (e.g. performance, energy consumption and security).
- **Decentralised architectures:** Architectures that can deal with the big data produced and consumed by highly decentralised and loosely coupled parties such as in the Internet of Things, with secure traceability such as blockchain. Additionally, architectures with parallelisation and distributed placement of processing for data-in-motion and its integration with data-at-rest.
- **Efficient mechanisms for storage and processing:** Real-time algorithms and techniques are needed for requirements demanding low latency when handling data-in-motion. Developing hardware and software together for Cloud and high-performance data platforms will, in turn, enable applications to run agnostically with outstanding reliability and energy efficiency.
- **Hybrid big data and high-performance computing architecture:** Efficient hybrid architectures that optimise the mixture of big data (i.e. Edge) and HPC (i.e. central) resources – combining local and global processing – to serve the needs of the most extreme and/or challenging data analytics at scale, called high-performance data analytics (HPDA).

3.3   Priority 'Data Analytics'

The progress of data analytics is key not only for turning big data into value but also for making it accessible to the wider public. Data analytics have a positive influence on all parts of the data value chain, and increase business opportunities through business intelligence and analytics while bringing benefits to both society and citizens.

Data analytics is an open, emerging field, in which Europe has substantial competitive advantages and a promising business development potential. It has been estimated that governments in Europe could save $149 billion (Manyika et al. 2011) by using big data analytics to improve operational efficiency. Big data analytics can provide additional value in every sector where it is applied, leading to more efficient and accurate processes. A recent study by the McKinsey Global

Institute placed a strong emphasis on analytics, ranking it as the main future driver for US economic growth, ahead of shale oil and gas production (Lund et al. 2013).

The next generation of analytics needs to deal with a vast amount of information from different types of sources, with differentiated characteristics, levels of trust and frequency of updating. Data analytics have to provide insights into the data in a cost-effective and economically sustainable way. On the one hand, there is a need to create complex and fine-grained predictive models for heterogeneous and massive datasets such as time series or graph data. On the other hand, such models must be applied in real time to large amounts of streaming data. This ranges from structured to unstructured data, from numerical data to micro-blogs and streams of data. The latter is exceptionally challenging because data streams, aside from their volume, are very heterogeneous and highly dynamic, which also calls for scalability and high throughput. For instance, data collection related to a disaster area can easily occupy terabytes in binary GIS formats, and real-time data streams can show bursts of gigabytes per minute.

In addition, an increasing number of big data applications are based on complex models of real-world objects and systems, which are used in computation-intensive simulations to generate new massive datasets. These can be used for iterative refinements of the models, but also for providing new data analytics services which can process massive datasets.

3.3.1 Challenges

Understanding data, whether it is numbers, text or multimedia content, has always been one of the most significant challenges for data analytics. Entering the era of big data, this challenge has expanded to a degree that makes the development of new methods necessary. The following list details the research areas identified for data analytics:

- **Semantic and knowledge-based analysis:** Improvements in the analysis of data to provide a near-real-time interpretation of the data (i.e. sentiment, semantics, etc.). Also, ontology engineering for big data sources, interactive visualisation and exploration, real-time interlinking and annotation of data sources, scalable and incremental reasoning, linked data mining and cognitive computing.
- **Content validation:** Implementation of veracity (source reliability/information credibility) models for validating content and exploiting content recommendations from unknown users.
- **Analytics frameworks and processing:** New frameworks and open APIs for the quality-aware distribution of batch and stream processing analytics, with minimal development effort from application developers and domain experts. Improvement in the scalability and processing speed of the algorithms mentioned above to tackle linearisation and computational optimisation issues.
- **Advanced business analytics and intelligence:** All of the above items enable the realisation of real and static business analytics, as well as business intelligence

empowering enterprises and other organisations to make accurate and instant decisions to shape their markets. The simplification and automation of these techniques are necessary, especially for SMEs.

- **Predictive and prescriptive analytics:** Machine learning, clustering, pattern mining, network analysis and hypothesis testing techniques applied on extremely large graphs containing sparse, uncertain and incomplete data. Areas that need to be addressed are building on the results of related research activities within the current EU work programme, sector-specific challenges and contextualisation combining heterogeneous data and data streams via graphs to improve the quality of mining processes, classifiers and event discovery. These capabilities open up novel opportunities for predictive analytics in terms of predicting future situations, and even prescriptive analytics providing actionable insights based on forecasts.
- **High-performance data analytics:** Applying high-performance computing techniques to the processing of extremely large amounts of data. Taking advantage of a high-performance infrastructure that powers different workloads and starting to support workflows that accelerate insights and lead to improved business results for enterprises. The goal is to develop new data analytics services with workloads typically characterised as follows: insights derived from analysis or simulations that are extremely valuable; the time-to-insight must be extremely fast; models and datasets are exceptionally complex.
- **Data analytics and Artificial Intelligence:** Machine-learning algorithms have progressed in recent years, primarily through the development of deep learning and reinforcement-learning techniques based on neural networks. The challenge is to make use of this progress in efficient and reliable data analytics processes for advanced business applications. This includes the intelligent distribution of the processing steps, from very close to data sources to Cloud (e.g. distributed deep learning). In addition, different techniques from AI can be used to enable better reasoning about data analytics' processes and outcomes.

3.3.2 Outcomes

The main expected advanced analytics innovations are as follows:

- **Improved models and simulations:** Improving the accuracy of statistical models by enabling fast non-linear approximations in very large datasets. Moving beyond the limited samples used so far in statistical analytics to samples covering the whole or the largest part of an event space/dataset.
- **Semantic analysis:** Deep learning, contextualisation based on AI, machine learning, natural language and semantic analysis in near real time. Providing canonical paths so that data can be aggregated and shared easily without dependency on technicians or domain experts. Enabling smart analysis of data across and within domains.

- **Event and pattern discovery:** Discovering and predicting rare real-time events that are hard to identify since they have a small probability of occurrence, but a great significance (such as physical disasters, a few costly claims in an insurance portfolio, rare diseases and treatments).
- **Multimedia (unstructured) data mining:** The processing of unstructured data (multimedia, text) Linking and cross-analysis algorithms to deliver cross-domain and cross-sector intelligence.
- **Deep learning techniques for business intelligence:** Coupled with the priorities on visualisation and engineering, providing user-friendly tools which connect to open and other datasets and streams (including a citizen's data), offering intelligent data interconnection for business- and citizen-orientated analytics, and allowing visualisation (e.g. diagnostic, descriptive and prescriptive analytics).
- **HPDA reference applications**: Well-defined processes for realising HPDA scenarios. Through enabling the combination of models (so-called Digital Twins) with the real-time operation of complex products/systems to more speedily project the inferences from (Big-Data-based) real-time massive data streams into (HPC-based) models and simulations (processing terabytes per minute/hour to petabytes of data per instance), the temporal delta between as-designed and as-operated can be reduced considerably.

3.4   Priority 'Data Visualisation and User Interaction'

Data visualisation plays a key role in effectively exploring and understanding big data. Visual analytics is the science of analytical reasoning assisted by interactive user interfaces. Data generated from data analytics processes need to be presented to end-users via (traditional or innovative) multi-device reports and dashboards which contain varying forms of media for the end-user, ranging from text and charts to dynamic 3D and possibly augmented-reality visualisations. For users to quickly and correctly interpret data in multi-device reports and dashboards, carefully designed presentations and digital visualisations are required. Interaction techniques fuse user input and output to provide a better way for a user to perform a task. Common tasks that allow users to gain a better understanding of big data include scalable zooms, dynamic filtering and annotation.

When representing complex information on multi-device screens, design issues multiply rapidly. Complex information interfaces need to be responsive to human needs and capacity (Raskin 2000). Knowledge workers need to be supplied with relevant information according to the just-in-time approach. Too much information, which cannot be efficiently searched and explored, can obscure the most relevant information. In fast-moving, time-constrained environments, knowledge workers need to be able to quickly understand the relevance and relatedness of information.

3.4.1 Challenges

In the data visualisation and user interaction domain, the tools that are currently used to communicate information need to be improved due to the significant changes brought about by the expanding volume and variety of big data. Advanced visualisation techniques must therefore consider the range of data available from diverse domains (e.g. graphs or geospatial, sensor and mobile data). Tools need to support user interaction for the exploration of unknown and unpredictable data within the visualisation layer. The following list briefly outlines the research areas identified for visualisation and user interaction:

- **Visual data discovery:** Access to information is at present based on a user-driven paradigm: the user knows what they need, and the only issue is to define the right criteria. With the advent of big data, this user-driven paradigm is no longer the most efficient. Data-driven paradigms are needed in which information is proactively extracted through data discovery techniques, and systems anticipate the user's information needs.
- **Interactive visual analytics of multiple-scale data:** There are significant challenges in visual analytics in the area of multiple-scale data. Appropriate scales of analysis are not always clear in advance, and single optimal solutions are unlikely to exist. Interactive visual interfaces have great potential for facilitating the empirical search for acceptable scales of analysis and the verification of results by modifying the scale and the means of any aggregation.
- **Collaborative, intuitive and interactive visual interfaces:** What is needed is an evolution of visual interfaces towards their becoming more intuitive and exploiting the advanced discovery aspects of big data analytics. This is required to foster effective exploitation of the information and knowledge that big data can deliver. In addition, there are significant challenges for effective communication and visualisation of big data insights to enable collaborative decision-making processes in organisations.
- **Interactive visual data exploration and querying in a multi-device context:** A key challenge is the provisioning of cross-platform mechanisms for data exploration, discovery and querying. Some difficult problems are how best to deal with uniform data visualisation on a range of devices and how to ensure access to functionalities for data exploration, discovery and querying in multi-device settings, requiring the exploration and development of new approaches and paradigms.

3.4.2 Outcomes

The main expected advances in visualisation and user experience are as follows:

- **Scalable data visualisation approaches and tools:** To handle extremely large volumes of data, the interaction must focus on aggregated data at different scales of abstraction rather than on individual objects. Techniques for summarising data in different contexts are highly relevant. There is a need to develop novel interaction techniques that can enable easy transitions from one scale or form of aggregation to another (e.g. from neighbourhood level to city level) while supporting aggregation and comparisons between different scales. It is necessary to address the uncertainty of the data and its propagation through aggregation and analysis operations.

- **Collaborative, 3D and cross-platform data visualisation frameworks:** Novel ways to visualise large amounts of possibly real-time data on different kinds of devices are required, including the augmented reality visualisation of data on mobile devices (e.g. smart glasses), as well as real-time and collaborative 3D visualisation techniques and tools.
- **New paradigms for visual data exploration, discovery and querying:** End-users need simplified mechanisms for the visual exploration of data, intuitive support for visual query formulation at different levels of abstraction, and tool-supported mechanisms for the visual discovery of data.
- **Personalised end-user-centric reusable data visualisation components:** Also useful are plug-and-play visualisation components that support the combination of any visualisation asset in real time and can be adapted and personalised to the needs of end-users. These also include advanced search capabilities rather than pre-defined visualisations and analytics. User feedback should be as simple as possible.
- **Domain-specific data visualisation approaches:** Techniques and approaches are required that support particular domains in exploring domain-specific data, for example innovative ways to visualise data in the geospatial domain, such as geo-locations, distances and space/time correlations (i.e. sensor data, event data). Another example is time-based data visualisation (it is necessary to take into account the specifics of time) – in contrast to common data dimensions which are usually 'flat'. Finally, the visualisation of interrelated/linked data that exploits graph visualisation techniques to allow easy exploration of network structures.

3.5 Priority 'Data Protection'

Data protection and anonymisation is a significant issue in the areas of big data and data analytics. With more than 90% of today's data having been produced in the last 2 years, a huge amount of person-specific and sensitive information from disparate data sources, such as social networking sites, mobile phone applications and electronic medical record systems, is increasingly being collected. Analysing this wealth and volume of data offers remarkable opportunities for data owners, but, at the same time, requires the use of state-of-the-art data privacy solutions, as well as the application of legal privacy regulations, to guarantee the confidentiality of individuals who are represented in the data. Data protection, while essential in the development of any modern information system, becomes crucial in the context of large-scale sensitive data processing.

Recent studies on mechanisms for protecting privacy have demonstrated that simple approaches, such as the removal or masking of the direct identifiers in a dataset (e.g. names, social security numbers), are insufficient to guarantee privacy. Indeed, such simple protection strategies can be easily circumvented by attackers who possess little background knowledge about specific data subjects. Due to the critical importance of addressing privacy issues in many business domains, the employment of privacy-protection techniques that offer formal privacy guarantees has become a necessity. This has paved the way for the development of privacy models and techniques such as differential privacy, private information retrieval, syntactic anonymity, homomorphic encryption, secure search encryption and secure multiparty computation, among others. The maturity of these technologies varies, with some, such as k-anonymity, more established than others. However, none of these technologies has so far been applied to large-scale commercial data processing tasks involving big data.

In addition to the privacy guarantees that can be offered by state-of-the-art privacy-enhancing technologies, another important consideration concerns the ability of the data protection approaches to maintain the utility of the datasets to which they are applied, to support different types of data analysis. Privacy solutions that offer guarantees while maintaining high data utility will make privacy technology a key enabler for the application of analytics to proprietary and potentially sensitive data.

There is a need for a truly modern and harmonised legal framework on data protection which has teeth and can be enforced appropriately to ensure that stakeholders pay attention to the importance of data protection. At the same time, it should enable the uptake of big data and incentivise privacy-enhancing technologies, which could be an asset for Europe as this is currently an underdeveloped market. In addition, users are beginning to pay more attention to how their data are processed. Hence, firms operating in the digital economy may realise that investing in privacy-enhancing technologies could give them a competitive advantage.

3.5.1   Challenges

In this perspective, the following main challenges have been identified:

- A more generic, easy-to-use and **enforceable data protection** approach suitable for large-scale commercial processing is needed. Data usage should conform to current legislation and policies. On the technical side, mechanisms are needed to provide data owners with the means to define the purpose of information gathering and sharing and to control the granularity at which their data will be shared with authorised third parties throughout the lifecycle of the data (data-in-motion and data-at-rest). Moreover, citizens should be able, for example, to have a say over the destruction of their personal data (the right to be forgotten). Data protection mechanisms also need to be 'easy', or at least capable of being used and understood with a reasonable level of effort by the various stakeholders, especially the end-users. Technical measures are also needed to enable and enforce the auditability of the principle that the data is only used for the defined purpose and nothing else – in particular, in relation to controlling the usage of personal information. In distributed settings such as supply chains, distributed trust technologies such as blockchains can be part of the solution.

- Maintaining robust **data privacy with utility guarantees** is a significant challenge and one which also implies sub-challenges, such as the need for state-of-the-art data analytics to cope with encrypted or anonymised data. The scalability of the solutions is also a critical feature. Anonymisation schemes may expose weaknesses exploitable by opportunistic or malicious opponents, and thus new and more robust techniques must be developed to tackle these adversarial models. Thus, ensuring the irreversibility of the anonymisation of big data assets is a key big data issue. On the other hand, encrypted data processing techniques, such as multiparty computation or homomorphic encryption, provide stronger privacy guarantees, but can currently only be applied to small parts of computation due to their large performance penalty. Also important are data privacy methods that can handle different data types as well as co-existing data types (e.g. datasets containing relational data together with sequential data about users), and methods that are designed to support analytic applications in different sectors (e.g. telecommunications, energy, and healthcare). Finally, preserving anonymity often implies removing the links between data assets. However, the approach to preserving anonymity also has to be reconciled with the needs for data quality, on which link removal has a very negative impact. This choice can be located on the side of the end-user, who has to balance the service benefits and possible loss of privacy, or on the side of the service provider, who has to offer a variety of added-value services according to the privacy acceptance of their customers. Measures to quantify privacy loss and data utility can be used to allow end-users to make informed decisions.

- Risk-based approaches calibrating information controllers' obligations regarding **privacy and personal data protection** must be considered, especially when dealing with the combined processing of multiple datasets. It has indeed been shown that when processing combinations of anonymised, pseudonymised and even public datasets, there is a risk that personally identifiable information can be retrieved. Thus, providing tools to assess or prevent the risks associated with such data processing is an issue of significant importance.

3.5.2   Outcomes

The main expected advances in data protection are as follows:

- **Complete data protection framework:** A good mechanism for data protection includes protecting the Cloud infrastructure, analytics applications and the data from leakage and threats, but also provides easy-to-use privacy mechanisms. Apart from the specification of the intended use of data, usage control mechanisms should also be covered.

- **Mining algorithms:** Developed privacy-preserving data mining algorithms.
- **Robust anonymisation algorithms:** Scalable algorithms that guarantee anonymity even when other external or publicly available data is integrated. In addition, algorithms that allow the generation of reliable insights by cross-referring data from a particular user in multiple databases, while protecting the identity of the user. Moreover, anonymisation methods that can guarantee a level of data utility to support intended types of analyses. Lastly, algorithms that can anonymise datasets of co-existing data types or generate synthetic data, which are commonly encountered in many business sectors, such as energy, healthcare and telecommunications.
- **Protection against reversibility:** Methods to analyse datasets to discover privacy vulnerabilities, evaluate the privacy risk of sharing the data and decide on the level of data protection that is necessary to guarantee privacy. Risk assessment tools to evaluate the reversibility of the anonymisation mechanisms.
- **Multiparty mining/pattern hiding:** Secure multiparty mining mechanisms over distributed datasets, so that data on which mining is to be performed can be partitioned, horizontally or vertically, and distributed among several parties. The partitioned data cannot be shared and must remain private, but the results of mining on the 'union' of the data are shared among the participants. The design of mechanisms for pattern hiding so that data is transformed in such a way that certain patterns cannot be derived (via mining) while others can.

4 Big Data Standardisation

Standardisation is a fundamental pillar in the construction of a Digital Single Market and Data Economy. It is only through the use of standards that the requirements of interconnectivity and interoperability can be ensured in an ICT-centric economy. Further development of technology and data standards for big data is needed by:

- Leveraging existing common standards as the basis for an open and thriving big data market
- Supporting standards development organisations (SDOs), such as ETSI, CEN-CENELEC, ISO, IEC, W3C, ITU-T and IEEE, by making experts available for all aspects of big data in the standardisation process
- Aligning the BDVA Big Data Reference Model with existing and evolving compatible architectures
- Liaising and collaborating with international consortia and SDOs through the TF6SG6 Standards Group and workshops

- Integrating national efforts on an international (European) level as early as possible
- Providing education and educational material to promote developing standards

Standards are the essential building blocks for product and service development as they define clear protocols that can be easily understood and adopted internationally. They are a prime source of compatibility and interoperability and simplify product and service development as well as speeding the time-to-market. Standards are globally adopted; they make it easier to understand and compare competing products, and thus drive international trade.

In the data ecosystem, standardisation applies to both the technology and the data.

Technology Standardisation Most technology standards for big data processing are de facto standards that are not prescribed (but are at best described after the fact) by a standards organisation. However, the lack of standards is a major obstacle. One example is the NoSQL databases. The history of NoSQL is based on solving specific technology challenges that lead to a range of different storage technologies. The broad range of choices, coupled with the lack of standards for querying the data, makes it harder to exchange data stores, as this may tie application-specific code to a specific storage solution. A pragmatic approach to standardisation is needed by influencing, in addition to NoSQL databases, the standardisation of technologies such as complex event processing for real-time data applications, languages to encode the extracted knowledge bases, Artificial Intelligence, computation infrastructure, data curation infrastructure, query interfaces and data storage technologies.

Data Standardisation The 'variety' of big data makes it very difficult to standardise. Nevertheless, there is a great deal of potential for data standardisation in the areas of data exchange and data interoperability. The exchange and use of data assets are essential for functioning ecosystems and the data economy. Enabling the seamless flow of data between participants (i.e. companies, institutions and individuals) is a necessary cornerstone of the ecosystem. Collaborative efforts are needed to support, where possible and pragmatic, the definition of semantic standardised data representation, ranging from domain (industry sector)-specific solutions, like domain ontologies, to general concepts such as Linked Open Data, to simplify and reduce the costs of data exchange.

5 Engineering and DevOps for Big Data

Big data technologies have gained significant momentum in research and innovation. However, mature, proven and empirically sound engineering methodologies for building next-generation big data value systems are not yet available. Also, we lack proven approaches for continuous development and operations (DevOps) of big data value systems. The availability of engineering methodologies and DevOps approaches – combined with adequate toolchains and big data platforms – will be essential for fostering productivity and quality. As a result, these methodologies and approaches will empower the new wave of data professionals to deliver high-quality next-generation big data value systems.

5.1 Challenges

Engineering and DevOps toolchains for big data value systems need to look at and systematically integrate a diverse set of aspects for: (1) system/software engineering, (2) development and operations and (3) quality assurance.

The main challenges to be addressed include:

- **Big data value engineering:** The engineering of big data value systems needs to be supported by targeted methodologies and tooling. Particularly important is significantly extending from online analytical processing (OLAP) systems to fully fledged frameworks which integrate data management, data analytics and data protection by bringing these data technologies into a unified systems perspective.
- **DevOps:** Integrated development and operations (DevOps) approaches need to be tailored to data systems. In particular, these approaches should align the work of data scientists (who develop data analytics solutions) and data engineers (who manage and curate data for and during operations).
- **Quality assurance:** Novel methods of quality assurance are required to deliver trustworthy and reliable big data value systems. Proven quality assurance techniques from software engineering, for example, can only be a starting point, as these techniques have to be significantly extended to cope with the values of big data. This may include generating (e.g. using simulation) sufficient and representative test data (e.g. incorporating extreme cases) to cover the volume and variety of big data. As testing may not scale to the ever-increasing size, velocity and variety of data, complementary (formal) verification techniques may be required to deliver confidence in the systems' quality. Also, to cope with velocity, existing monitoring techniques need to be extended to ensure the quality of big data value systems during their operation.
- **Considering multiple dimensions of big data value:** The design and advancement of methodologies, tooling and platforms should carefully consider the multifaceted issues of big data, such as real-time processing and analytics, as well as data veracity and variety.

5.2 Outcomes

The expected primary outcomes for engineering and DevOps are:

- Engineering principles, as well as fully integrated toolchains and frameworks, that significantly increase productivity in terms of developing and deploying big data value systems
- Testing, monitoring and verification tools and methodologies to significantly increase reliability, security, energy efficiency and quality of big data value systems
- Enhancing real-time capabilities of data systems and platforms to handle high-intensity and highly distributed data and event streams

6 Illustrative Scenario in Healthcare

This section illustrates how the technical priorities may help in delivering big data solutions for specific industry sectors. To this end, we present a scenario from the healthcare sector. A BDVA white paper collected and analysed the needs, opportunities and challenges for big data technologies in healthcare (TF7 Healthcare subgroup 2016).

There is a clear opportunity to transform healthcare by applying data technologies. To improve the productivity of the healthcare sector, it is necessary to reduce costs while maintaining or improving the quality of the care provided. The fastest, least costly and most effective way to achieve this is to use the knowledge that is hiding within the already existing large amounts of generated medical data. According to current estimates, medical data is already at the zettabyte scale and will soon reach the yottabyte (e.g. 1000 zettabytes, a billion petabytes) scale. While most of this data was previously stored in hard copy format, the current trend is towards digitisation of these large amounts of information, thus making them amenable to analysis, resulting in what is known as big data.

The challenges and needs for research and innovation in this illustrative scenario are quite evident for each of the technical priorities listed above. Let's consider them one by one, starting with data management.

- **Data management:** Access to high-quality, large healthcare datasets to optimise care processes, disease diagnosis, personalised care and the healthcare system in general. Furthermore, a real transformation of the healthcare sector can only be achieved if all stakeholders and verticals in the healthcare sector (the HealthTech industry, healthcare providers, pharma, and insurance) share data and allow free data flow. Topics such as data quality, semantic interoperability and data management lifecycles are of the utmost importance in breaking down data silos in healthcare.

- **Data processing:** Consequently, the data processing architecture needs to be able to deal with heterogeneous health data (medical records, medical images and lab results), ensuring scalability (e.g. to process millions of patient records to find a similar patient) and performance (e.g. for smart alarms in intensive care units).
- **Data analytics:** The main challenges arise in the field of data analytics. The core of healthcare transformation is expected to come from AI-based propositions to enable personalised medicine, clinical decision support, workflow optimisation, clinical research and, finally, better diagnosis and medical treatment for patients.
- **Data visualisation and user interaction:** An area closely related to analytics and data interpretation is data visualisation and user interaction. Visualising models obtained by machine learning, as well as effective and clear user interaction technologies, is of utmost importance for the acceptance of AI technologies in the healthcare sector.
- **Data protection:** The developing focus on data protection is especially important in the healthcare sector, which deals with sensitive health data. Robust data privacy and anonymisation techniques, privacy-preserving data mining, end-to-end security and consent management are significant challenges to be addressed.
- **Standards:** Finally, in the healthcare sector data is often fragmented or generated by different systems with incompatible formats. Therefore, interoperability and standardisation are key to deploying the full potential of data held.
- **Engineering and DevOps:** Linked to this are the engineering methodologies for building next-generation big data value systems in healthcare, which need to be correctly validated by clinical trials and regulatory approval. An interesting challenge is to create methodologies to regulate AI-based propositions more quickly and also address the liability and regulatory aspects of techniques such as continuous learning.

7 Summary

Enabling an effective data ecosystem requires overcoming several technical challenges associated with the cost and complexity of extracting value from data. This chapter identifies and characterises the key research areas. A systemic and structured methodology was used to gather inputs from over 200 stakeholder organisations. The results of this process, as illustrated in Fig. 8, identify the five technical research priorities together with 28 sub-challenges of big data. The requirement analysis was done in consultation with a community of stakeholders that included organisations for industry, research and government.

The results presented in this chapter provide a prioritised list of cross-sectorial business needs of data technologies and their impact in industry, research and government. These findings serve as a guide for directing the research and development efforts towards fostering a data ecosystem. The findings indicate that deep analytics and data management are viewed as the top two technical challenges for big data, with more than 60% of organisations prioritising them as having a high

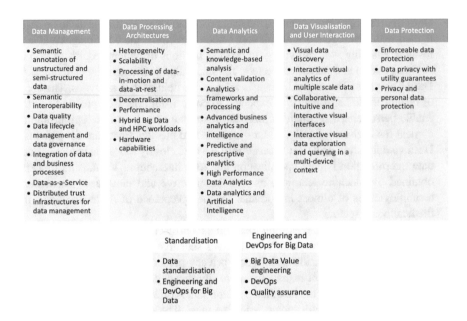

Fig. 8 High-level technical priorities and sub-challenges for big data value

impact on the data ecosystem. Although data privacy was considered a significant challenge, it was ranked lowest compared to other key challenges. This may be because not all data applications and domains have privacy implications and may focus on industrial/machine data.

Finally, these data research priorities have laid the foundations for a joint Strategic Research, Innovation and Deployment Agenda for an AI, Data and Robotics Partnership in Europe (Zillner et al. 2020) with the goal to unify the strategic focus of each of the three disciplines engaged in creating the Partnership.

Acknowledgements We greatly acknowledge the collective effort of the SRIA teams: Carlos A. Iglesias, Antonio Alfaro, Jesus Angel, Sören Auer, Paolo Bellavista, Arne Berre, Freek Bomhof, Stuart Campbell, Geraud Canet, Giuseppa Caruso, Edward Curry, Paul Czech, Davide Dalle Carbonare, Nuria de Lama, Stefano de Panfilis, Thomas Delavallade, Marija Despenic, Ana Garcia Robles, Wolfgang Gerteis, Aris Gkoulalas-Divanis, Nuria Gomez, Paolo Gonzales, Thomas Hahn, Souleiman Hasan, Jim Keneally, Bjarne Kjær Ersbøll, Bas Kotterink, Yannick Legré, Yves Mabiala, Julie Marguerite, Dirk Mayer, Ernestina Menasalves, Andreas Metzger, Elisa Molino, Thierry Nagellen, Dalit Naor, Maria Perez, Milan Petkovic, Roberta Piscitelli, Klaus-Dieter Platte, Pierre Pleven, Dumitru Roman, Titi Roman, Alexandra Rosén, Nikos Sarris, Stefano Scamuzzo, Simon Scerri, Corinna Schulze, Robert Seidl, Bjørn Skjellaug, Caj Södergård, Claire Tonna, Francois Troussier, Colin Upstill, Josef Urban, Meilof Veeningen, Tonny Velin, Ray Walshe, Walter Waterfeld, Stefan Wrobel, and Sonja Zillner.

References

Cavanillas, J. M., Curry, E., & Wahlster, W. (Eds.). (2016). *New horizons for a data-driven economy: A roadmap for usage and exploitation of big data in Europe* (pp. 1–303). New York: Springer. https://doi.org/10.1007/978-3-319-21569-3

Chen, H., Chiang, R. H. L., & Storey, V. C. (2012). Business intelligence and analytics: From big data to big impact. *MIS Quarterly, 36*(4), 1165. https://doi.org/10.2307/41703503

Communication: A European strategy for data. (2020). Retrieved from https://ec.europa.eu/info/sites/info/files/communication-european-strategy-data-19feb2020_en.pdf

Curry, E. (2016). The big data value chain: Definitions, concepts, and theoretical approaches. In J. M. Cavanillas, E. Curry, & W. Wahlster (Eds.), *New horizons for a data-driven economy: A roadmap for usage and exploitation of big data in Europe.* New York: Springer. https://doi.org/10.1007/978-3-319-21569-3_3

Curry, E., Becker, T., Munné, R., De Lama, N., & Zillner, S. (2016). The BIG project. In J. M. Cavanillas, E. Curry, & W. Wahlster (Eds.), *New horizons for a data-driven economy: A roadmap for usage and exploitation of big data in Europe.* New York: Springer. https://doi.org/10.1007/978-3-319-21569-3_2

DG Connect. (2013). *A European strategy on the data value chain.* Retrieved from http://ec.europa.eu/information_society/newsroom/cf/dae/document.cfm?doc_id=3488

Flick, U. (2004). Triangulation in qualitative research. In *A companion to qualitative research* (p. 432).

Lavalle, S., Lesser, E., Shockley, R., Hopkins, M. S., & Kruschwitz, N. (2011). Big data, analytics and the path from insights to value. *MIT Sloan Management Review, 52*(2), 21–32.

Lund, S., Manyika, J., Nyquist, S., Mendonca, L., & Ramaswamy, S. (2013). *Game changers: Five opportunities for US growth and renewal.*

Manyika, J., Chui, M., Brown, B., Bughin, J., Dobbs, R., Roxburgh, C., & Byers, A. H. (2011). *Big data: The next frontier for innovation, competition, and productivity.* Retrieved from McKinsey Global Institute website http://scholar.google.com/scholar.bib?q=info:kkCtazs1Q6wJ:scholar.google.com/&output=citation&hl=en&as_sdt=0,47&ct=citation&cd=0

Oppenheim, A. N. (1992). Questionnaire design, interviewing and attitude measurement. *Journal of Marketing Research, 30.* Retrieved from http://www.amazon.com/Questionnaire-Design-Interviewing-Attitude-Measurement/dp/1855670445

Raskin, J. (2000). *Humane interface, the: New directions for designing interactive systems.* Addison-Wesley Professional.

Rusitschka, S., & Curry, E. (2016). Big data in the energy and transport sectors. In J. M. Cavanillas, E. Curry, & W. Wahlster (Eds.), *New horizons for a data-driven economy: A roadmap for usage and exploitation of big data in Europe.* New York: Springer. https://doi.org/10.1007/978-3-319-21569-3_13

TF7 Healthcare subgroup. (2016). *Big Data Technologies in Healthcare.*

Yin, R. K. (2013). Case study research: Design and methods. *SAGE Publications, 26*(1), 93–96. https://doi.org/10.1017/CBO9781107415324.004

Zillner, S., Curry, E., Metzger, A., Auer, S., & Seidl, R. (Eds.). (2017). *European big data value strategic research & innovation agenda.* Retrieved from Big Data Value Association website www.bdva.eu

Zillner, S., Bisset, D., Milano, M., Curry, E., Hahn, T., Lafrenz, R., et al. (2020). *Strategic research, innovation and deployment agenda - AI, data and robotics partnership. Third Release* (3rd). Brussels: BDVA, euRobotics, ELLIS, EurAI and CLAIRE.

A Reference Model for Big Data Technologies

Edward Curry, Andreas Metzger, Arne J. Berre, Andrés Monzón, and
Alessandra Boggio-Marzet

Abstract The Big Data Value (BDV) Reference Model has been developed with
input from technical experts and stakeholders along the whole big data value chain.
The BDV Reference Model may serve as a common reference framework to locate
big data technologies on the overall IT stack. It addresses the main technical
concerns and aspects to be considered for big data value systems. The BDV
Reference Model enables the mapping of existing and future data technologies
within a common framework. Within this chapter, we detail the reference model in
more detail and show how it can be used to manage a portfolio of research and
innovation projects.

Keywords Reference model · Big data technologies · Data management · Data
processing · Data analysis · Data visualisation · Data protection

1 Introduction

The Big Data Value (BDV) Reference Model has been developed with input from
technical experts and stakeholders along the whole big data value chain. The BDV
Reference Model may serve as a common reference framework to locate big data
technologies on the overall IT stack. It addresses the main concerns and aspects to be
considered for big data value systems. Within this chapter, we detail the reference

E. Curry (✉)
Insight SFI Research Centre for Data Analytics, NUI Galway, Galway, Ireland
e-mail: edward.curry@nuigalway.ie

A. Metzger
paluno, University of Duisburg-Essen, Duisburg, Germany

A. J. Berre
SINTEF Digital, Oslo, Norway

A. Monzón · A. Boggio-Marzet
Universidad Politécnica de Madrid, Madrid, Spain

© The Author(s) 2021
E. Curry et al. (eds.), *The Elements of Big Data Value*,
https://doi.org/10.1007/978-3-030-68176-0_6

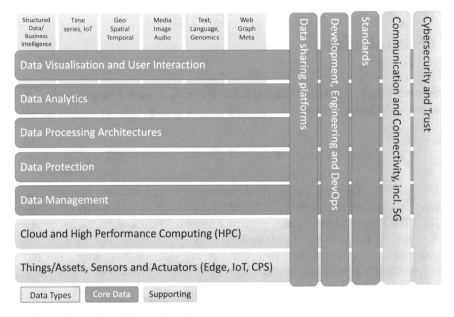

Fig. 1 Big Data Value Reference Model

model in more detail and show how it can be used to manage a portfolio of research and innovation projects. Section 2 details the Reference Model with its horizontal and concerns. Section 3 describes the use of the Reference Model within large-scale data projects to map projects' technical outcomes. Finally, Sect. 4 concludes the chapter.

2 Reference Model

An overview of the BDV Reference Model is shown in Fig. 1. It distinguishes between two different elements. On the one hand, it describes the elements that are at the core of the BDVA (also see Chap. "The European Big Data Value Ecosystem"); on the other, it outlines the features that are developed in strong collaboration with related European activities.

The BDV Reference Model has been developed by the Big Data Value Association (BDVA), taking into account input from technical experts and stakeholders along the whole big data value chain, as well as interactions with other related public-private partnerships (PPPs) (Zillner et al. 2017). The BDV Reference Model may serve as a common reference framework to locate big data technologies on the overall IT stack. It addresses the main concerns and aspects to be considered for big data value systems.

The BDV Reference Model is structured into horizontal and vertical concerns.

- Horizontal concerns cover specific aspects along the data processing chain, starting with data collection and ingestion, and extending to data visualisation. It should be noted that the horizontal concerns do not imply a layered architecture. As an example, data visualisation may be applied directly to collected data (the data management aspect) without the need for data processing and analytics.
- Vertical concerns address cross-cutting issues, which may affect all the horizontal concerns. In addition, vertical concerns may also involve non-technical aspects.

It should be noted that the BDV Reference Model has no ambition to serve as a technical reference architecture. However, it is compatible with such reference architectures, most notably the emerging ISO JTC1 WG9 Big Data Reference Architecture.

The following elements as expressed in the BDV Reference Model are elaborated in the remainder of this section.

2.1   Horizontal Concerns

Horizontal concerns cover specific aspects of a big data system. On the one hand, they cover the different elements of the data processing chain, starting from data collection and ingestion up to data visualisation and user interaction. On the other hand, they cover elements that facilitate deploying and operating big data systems, including Cloud and HPC, as well as Edge and IoT.

2.1.1   Data Visualisation and User Interaction

This concern covers advanced visualisation approaches for improved user experience. Data visualisation plays a key role in effectively exploring and understanding big data. Visual analytics is the science of analytical reasoning assisted by interactive user interfaces. Data generated from data analytics processes need to be presented to end-users via (traditional or innovative) multi-device reports and dashboards which contain varying forms of media for the end-user, ranging from text and charts to dynamic, 3D and possibly augmented-reality visualisations. In order for users to quickly and correctly interpret data in multi-device reports and dashboards, carefully designed presentations and digital visualisations are required. Interaction techniques fuse user input and output to provide a better way for a user to perform a task. Common tasks that allow users to gain a better understanding of big data include scalable zooms, dynamic filtering and annotation.

When representing complex information on multi-device screens, the design issues multiply rapidly. Complex information interfaces need to be responsive to human needs and capacity (Raskin 2000). Knowledge workers need to be supplied with relevant information according to the just-in-time approach. Too much information, which cannot be efficiently searched and explored, can obscure the information that is most relevant. In fast-moving time-constrained environments, knowledge workers need to be able to quickly understand the relevance and relatedness of information.

2.1.2 Data Analytics

This concern covers data analytics, which ranges from descriptive analytics ("What happened and why?") through predictive analytics ("What will happen and when?") to prescriptive analytics ("What is the best course of action to take?"). The progress of data analytics is key not only for turning big data into value but also for making it accessible to the wider public. Data analytics will have a positive influence on all parts of the data value chain (Cavanillas et al. 2016) and increase business opportunities through business intelligence and analytics while bringing benefits to both society and citizens.

Data analytics is an open, emerging field, in which Europe has strong competitive advantages and a promising business development potential. It has been estimated that governments in Europe could save \$149 billion (Manyika et al. 2011) by using big data analytics to improve operational efficiency. Big data analytics can provide additional value in every sector where it is applied, leading to more efficient and accurate processes. A study by the McKinsey Global Institute placed a strong emphasis on analytics, ranking it as the main future driver for US economic growth, ahead of shale oil and gas productions (Lund et al. 2013).

The next generation of analytics will be required to deal with a vast amount of information from different types of sources, with differentiated characteristics, levels of trust and frequency of updating. Data analytics will have to provide insights into the data in a cost-effective and economically sustainable way. On the one hand, there is a need to create complex and fine-grained predictive models for heterogeneous and massive datasets such as time series or graph data. On the other hand, such models must be applied in real time to large amounts of streaming data. This ranges from structured to unstructured data, from numerical data to micro-blogs and streams of data. The latter is exceptionally challenging because data streams, in addition to their volume, are very heterogeneous and highly dynamic, which also calls for scalability and high throughput. For instance, data collection related to a disaster area can easily occupy terabytes in binary GIS formats, and real-time data streams can show bursts of gigabytes per minute.

In addition, an increasing number of big data applications are based on complex models of real-world objects and systems, which are used in computation-intensive simulations to generate new huge datasets. These can be used for iterative refinements of the models, but also for providing new data analytics services which can process extremely large datasets.

2.1.3   Data Processing Architectures

This concern covers optimised and scalable architectures for analytics of both data-at-rest and data-in-motion, thereby delivering low-latency real-time analytics.

The Internet of Things (IoT) is one of the key drivers of the big data phenomenon. Initially, this phenomenon started by applying the existing architectures and technologies of big data that we categorise as data-at-rest, which is data kept in persistent storage. In the meantime, the need for processing immense amounts of sensor data streams has increased. This type of data-in-motion (i.e. non-persistent data processed on the fly) has extreme requirements for low-latency and real-time processing. What has hardly been addressed is the concept of complete processing for the combination of data-in-motion and data-at-rest.

For the IoT domain, these capabilities are essential. They are also required for other domains like social networks or manufacturing, where huge amounts of streaming data are produced in addition to the available big datasets of actual and historical data.

These capabilities will affect all layers of future big data infrastructures, ranging from the specifications of low-level data flows with the continuous processing of micro-messages, to sophisticated analytics algorithms. The parallel need for real-time and large data volume capabilities is a key challenge for big data processing architectures. Architectures to handle streams of data such as the lambda and kappa architectures will be considered as a baseline for achieving a tighter integration of data-in-motion with data-at-rest.

Developing the integrated processing of data-at-rest and data-in-motion in an ad hoc fashion is of course possible, but only the design of generic, decentralised and scalable architectural solutions will leverage their true potential. Optimised frameworks and toolboxes allowing the best use of both data-in-motion (e.g. data streams from sensors) and data-at-rest will leverage the dissemination of reference solutions which are ready and easy to deploy in any economic sector. For example, proper integration of data-in-motion with predictive models based on data-at-rest will enable efficient, proactive processing (detection ahead of time). Architectures that can handle heterogeneous and unstructured data are also important. When such solutions become available to service providers, in a straightforward manner, they will then be free to focus on the development of business models.

The capabilities of existing systems to process such data-in-motion and answer queries in real time and for thousands of concurrent users are limited. Special-purpose approaches based on solutions like Complex Event Processing (CEP) are not sufficient for the challenges posed by the IoT in big data scenarios. The problem of achieving effective and efficient processing of data streams (data-in-motion) in a big data context is far from being solved, especially when considering the integration with data-at-rest and breakthroughs in NoSQL databases and parallel processing (e.g. Hadoop, Apache Spark, Apache Flink, Apache Kafka). Applications, for instance of Artificial Intelligence, are also required to fully exploit all the capabilities of modern and heterogeneous hardware, including parallelism and distribution to boost performance.

To achieve the agility demanded by real-time business and next-generation applications, a new set of interconnected data management capabilities is required.

2.1.4 Data Protection

This concern covers privacy and anonymisation mechanisms to facilitate data protection. This is shown related to data management and processing as there is a strong link here, but it can also be associated with the area of cybersecurity.

Data protection and anonymisation is a major issue in the areas of big data and data analytics. With more than 90% of today's data having been produced in the last 2 years, a huge amount of person-specific and sensitive information from disparate data sources, such as social networking sites, mobile phone applications and electronic medical record systems, is increasingly being collected. Analysing this wealth and volume of data offers remarkable opportunities for data owners, but, at the same time, requires the use of state-of-the-art data privacy solutions, as well as the application of legal privacy regulations, to guarantee the confidentiality of individuals who are represented in the data. Data protection, while essential in the development of any modern information system, becomes crucial in the context of large-scale sensitive data processing.

Recent studies on mechanisms for protecting privacy have demonstrated that simple approaches, such as the removal or masking of the direct identifiers in a dataset (e.g. names, social security numbers), are insufficient to guarantee privacy. Indeed, such simple protection strategies can be easily circumvented by attackers who possess little background knowledge about specific data subjects. Due to the critical importance of addressing privacy issues in many business domains, the employment of privacy-protection techniques that offer formal privacy guarantees has become a necessity. This has paved the way for the development of privacy models and techniques such as differential privacy, private information retrieval, syntactic anonymity, homomorphic encryption, secure search encryption and secure multiparty computation, among others. The maturity of these technologies varies, with some, such as k-anonymity, more established than others. However, none of these technologies has so far been applied to large-scale commercial data processing tasks involving big data.

In addition to the privacy guarantees that can be offered by state-of-the-art privacy-enhancing technologies, another important consideration concerns the ability of the data protection approaches to maintain the utility of the datasets to which they are applied, with the goal of supporting different types of data analysis. Privacy solutions that offer guarantees while maintaining high data utility will make privacy technology a key enabler for the application of analytics to proprietary and potentially sensitive data.

A truly modern and harmonised legal framework on data protection which has teeth and can be enforced appropriately will ensure that stakeholders pay attention to the importance of data protection. At the same time, it should enable the uptake of big data and incentivise privacy-enhancing technologies, which could be an asset for Europe as this is currently an underdeveloped market. In addition, users are beginning to pay more attention to how their data are processed. Hence, firms operating in the digital economy may realise that investing in privacy-enhancing technologies could give them a competitive advantage.

2.1.5 Data Management

This concern covers principles and techniques for data management, including data ingestion, sharing, integration, cleansing and storage. More and more data are becoming available. This data explosion, often called a "data tsunami", has been triggered by the growing volumes of sensor data and social data, born out of Cyber-Physical Systems (CPS) and Internet of Things (IoT) applications. Traditional means for data storage and data management are no longer able to cope with the size and speed of data delivered in heterogeneous formats and at distributed locations.

Large amounts of data are being made available in a variety of formats – ranging from unstructured to semi-structured to structured – such as reports, Web 2.0 data, images, sensor data, mobile data, geospatial data and multimedia data. Important data types include numeric types, arrays and matrices, geospatial data, multimedia data and text. A great deal of this data is created or converted and further processed as text. Algorithms or machines are not able to process the data sources due to the lack of explicit semantics. In Europe, text-based data resources occur in many different languages, since customers and citizens create content in their local language. This multilingualism of data sources means that it is often impossible to align them using existing tools because they are generally available only in the English language. Thus, the seamless aligning of data sources for data analysis or business intelligence applications is hindered by the lack of language support and gaps in the availability of appropriate resources.

Isolated and fragmented data pools are found in almost all industrial sectors. Due to the prevalence of data silos, it is challenging to accomplish seamless integration with and smart access to the various heterogeneous data sources. And still today, data producers and consumers, even in the same sector, rely on different storage, communication and thus different access mechanisms for their data. Due to the lack of commonly agreed standards and frameworks, the migration and federation of data between pools impose high levels of additional costs. Without a semantic interoperability layer being imposed upon all these different systems, the seamless alignment of data sources cannot be realised.

In order to ensure a valuable big data analytics outcome, the incoming data has to be of high quality, or, at least, the quality of the data should be known to enable appropriate judgements to be made. This requires differentiating between noise and valuable data, and thereby being able to decide which data sources to include and which to exclude to achieve the desired results.

Over many years, several different application sectors have tried to develop vertical processes for data management, including specific data format standards and domain models. However, consistent data lifecycle management – that is, the ability to clearly define, interoperate, openly share, access, transform, link, syndicate and manage data – is still missing. In addition, data, information and content need to be syndicated from data providers to data consumers while maintaining provenance, control and source information, including IPR considerations (data provenance). Moreover, to ensure transparent and flexible data usage, the aggregation and management of respective datasets enhanced by a controlled access mechanism through APIs should be enabled (Data-as-a-Service).

2.1.6   Cloud and High-Performance Computing (HPC)

Efficient big data processing, data analytics and data management require the effective use of Cloud and High-Performance Computing infrastructures to address the computational resource and storage needs of big data systems.

Cloud   Data ecosystems, promoted by the BDVA, should include strong links to scientific research that is becoming predominantly data driven. The BDVA is in a strong position to nurture such links as it has established strong relationships with European big data academia. However, a lack of access, trust and reusability prevents European researchers in academia and industry from gaining the full benefits of data-driven science. Most datasets from publicly funded research are still inaccessible to the majority of scientists in the same discipline, not to mention other potential users of the data, such as company R&D departments. Approximately 80% of research data is not in a trusted repository. However, even if the data openly appears in repositories, this is not always enough. As a current example, only 18% of the data in open repositories is reusable.[1] This leads to inefficiencies and delays; in recent surveys, the time reportedly spent by data scientists in collecting and cleaning data sources made up 80% of their work (G. Press 2016).

In response to these challenges, the Commission has launched a large effort to create "a European Open Science Cloud to make science more efficient and productive and let millions of researchers share and analyse research data in a trusted environment across technologies, disciplines and borders"[1]. The initial outline for the European Open Science Cloud (EOSC) was laid out in the report from the High-Level Expert Group.[2] The report advised the Commission on several measures needed to implement the governance and the financial scheme of the European Open Science Cloud, such as being based on a federated system of existing and emerging research (e-)infrastructures operating under light international governance with well-defined Rules of Engagement for participation. Machine understanding of

---

[1]"Are FAIR data principles FAIR?" LIBER Webinar by Alastair Dunning, 10.03.2017.

[2]Realising the European Open Science Cloud, 2016, https://ec.europa.eu/research/openscience/pdf/realising_the_european_open_science_cloud_2016.pdf

data – based on common or widely used data standards – is required to handle the exponential growth in publications. Attractive career paths for data experts should be created through proper training and by applying modern reward and recognition practices. This should help to satisfy the growing demand for data scientists working together with substance scientists. Turning science into innovation is emphasised, and alongside this there is a need for industry, especially SMEs and start-ups, to be able to access the appropriate data resources.

A first phase aims at establishing a governance and business model that sets the rules for the use of the EOSC, creating a cross-border and multi-disciplinary open innovation environment for research data, knowledge and services, and ultimately establishing global standards for the interoperability of scientific data.

The EU has already initiated and will go on to launch several more infrastructure projects, such as EOSC-hub, within H2020 for implementing and piloting the EOSC. In addition to these projects, Germany and the Netherlands, among other countries, are promoting the GO FAIR initiative (Germany and the Netherlands 2017). The FAIR principles aim to ensure that Data and Digital Research Objects are Findable, Accessible, Interoperable and Reusable (FAIR) (Wilkinson et al. 2016). As science becomes increasingly data driven, making data FAIR will create real added value since it allows for combining datasets across disciplines and across borders to address pressing societal challenges that are mostly interdisciplinary.

The GO FAIR initiative is a bottom-up, open-to-all, cross-border and cross-disciplinary approach aiming to contribute to a broad involvement of the European science community as a whole, including the "long tail" of science.

The EOSC initiative is aligned with the BDVA agenda, as both promote data accessibility, trustworthiness and reproducibility over domains and borders. In the BDVA, this mainly applies to the i-Spaces and Lighthouse instruments, where the interoperability of datasets is central. Data standardisation is a self-evident topic for cooperation, but there are also common concerns in non-technical priorities – most notably skills development (relating to data-intensive engineers and data scientists). Both industry and academia benefit from findable, accessible, interoperable and reproducible data.

High-Performance Computing In some sectors, big data applications are expected to move towards more computation-intensive algorithms to reap deeper insights across descriptive (explaining what is happening), diagnostic (exploring why it happens), prognostic (predicting what can happen) and prescriptive (proactive handling) analysis. The adoption of specific HPC-type capabilities by the big data analytics stack is likely to be of assistance where big data insights will be of the utmost value. Faster decision-making is crucial and extremely complex datasets are involved – i.e. extreme data analytics.

The Big Data and HPC communities (through BDVA and ETP4HPC collaboration[1]) have recognised their shared interests in strengthening Europe's position regarding extreme data analytics. Recent engagements between PPPs have focused on the relevant issues of looking at how HPC and Big Data platforms are implemented, understanding the platform requirements for HPC and Big Data workloads, and exploring how the cross-transfer of certain technical capabilities belonging to either HPC or big data could benefit each other. For example, the application of deep learning is one such workload that readily stands to benefit from certain HPC-type capabilities regarding optimising and parallelising difficult optimisation problems.

Major technical requirements include highly scalable performance, high memory bandwidth, low power consumption and excellent short arithmetic performance. Additionally, more flexible end-user education paths, utilisation and business models will be required to capitalise on the rapidly evolving technologies underpinning extreme data analytics, as well as continued support for collaboration across the communities of both big data and HPC to jointly define the way forward for Europe.

2.1.7 IoT, CPS, Edge and Fog Computing

The main source of big data is sensor data from an IoT context and actuator interaction in Cyber-Physical Systems. To meet real-time needs, it will often be necessary to handle big data aspects at the edge of the system. This area is separately elaborated further in collaboration with the IoT (Alliance for Internet of Things Innovation (AIOTI)) and CPS communities.

Internet of Things (IoT) technology, which enables the connection of any type of smart device or object, will have a profound impact on many sectors in the European economy. Fostering this future market growth requires the seamless integration of IoT technology (such as sensor integration, field data collection, Cloud, Edge and Fog computing) and big data technology (such as data management, analytics, deep analytics, edge analytics and processing architectures).

The mission of the Alliance of Internet of Things Innovation (AIOTI) is to foster the European IoT market uptake and position by developing ecosystems across vertical silos, contributing to the direction of H2020 large-scale pilots, gathering evidence on market obstacles for IoT deployment in the Digital Single Market context, championing the EU in spearheading IoT initiatives, and mapping and bridging global, EU and Members States' IoT innovation and standardisation activities. AIOTI working groups cover various vertical markets from smart farming to smart manufacturing and smart cities, and specific horizontal topics on standardisation, policy, research and innovation ecosystems. The AIOTI was launched by the European Commission in 2015 as an informal group and established as a legal entity in 2016. It is a major cross-domain European IoT innovation activity.

Close cooperation between the AIOTI and the BDVA is seen as being very beneficial for the BDVA. The following areas of collaboration are of particular interest to the BDVA:

- Alignment of high-level reference architectures: A common understanding of how the AIOTI High-Level Architecture (HLA) and the BDVA Reference Model are related to each other enables well-grounded decisions and prioritisations related to the future impact of technologies.

- Deepening the understanding about sectorial needs: Through the mutual exchange of roadmaps, accompanied by insights about sectorial needs in the various domains, the BDVA will receive additional input about drivers for and constraints on the adoption of big data in the various sectors. In particular, insights about sector-specific user requirements as well as topics related to the BDV strategic research and innovation roadmap will be fed back into our ongoing updating process.
- Standardisation activities: To foster the seamless integration of IoT and big data technologies, the standardisation activities of both communities should be aligned whenever technically required. In addition, the BDVA can benefit from the already established partnerships between the AIOTI and standardisation bodies to communicate big-data-related standardisation requirements.

Aligning Security Efforts The efforts to strengthen security in the IoT domain will have a huge impact on the integrity of data in the big data domain. When IoT security is compromised, so too is the generated data. By developing a mutual understanding on security issues in both domains, trust in both technologies and their applications will be increased.

2.2 Vertical Concerns

Vertical concerns address cross-cutting issues, which are relevant and may affect more than one of the horizontal concerns. They may not be purely technical and also involve some non-technical aspects.

2.2.1 Big Data Types and Semantics

One specific vertical concern defined by the BDV Reference Model is data types. Different data types may require the use of different techniques and mechanisms in the horizontal concerns, for instance for data analytics and data storage.

The following six big data types have been identified as the main relevant data types used in big data systems: (1) structured data, (2) time series data, (3) geospatial data, (4) media data (image, video, audio, etc.), (5) text data (including natural language data and genomics representations) and (6) graph or network data. In addition, it is important to support both the syntactical and semantic aspects of data for all big data types, in particular, considering metadata.

2.2.2 Standards

This concern covers the standardisation of big data technology areas to facilitate data integration, sharing and interoperability.

Standardisation is a fundamental pillar in the construction of a Digital Single Market and Data Economy. It is only through the use of standards that the requirements of interconnectivity and interoperability can be ensured in an ICT-centric economy. The PPP will continue to lead the way in the development of technology and data standards for big data by:

- Leveraging existing common standards as the basis for an open and successful big data market
- Supporting standards development organisations (SDOs), such as ETSI, CEN-CENELEC, ISO, IEC, W3C, ITU-T and IEEE, by making experts available for all aspects of big data in the standardisation process
- Aligning the BDV Reference Model with existing and evolving compatible architectures
- Liaising and collaborating with international consortia and SDOs through the TF6SG6 Standards Group and Workshops
- Integrating national efforts on an international (European) level as early as possible
- Providing education and educational material to promote developing standards

Standards are the essential building blocks for product and service development as they define clear protocols that can be easily understood and adopted internationally. They are a prime source of compatibility and interoperability and simplify product and service development as well as speeding the time-to-market. Standards are globally adopted; they make it easier to understand and compare competing products, and thus drive international trade.

In the data ecosystem, standardisation applies to both the technology and the data.

Technology Standardisation Most technology standards for big data processing are de facto standards that are not prescribed (but are at best described after the fact) by a standards organisation. However, the lack of standards is a significant obstacle. One example is the NoSQL databases. The history of NoSQL is based on solving specific technology challenges that lead to a range of different storage technologies. The broad range of choices, coupled with the lack of standards for querying the data, makes it harder to exchange data stores, as this may tie application-specific code to a specific storage solution. The PPP is likely to take a pragmatic approach to standardisation and look to influence, in addition to NoSQL databases, the standardisation of technologies such as complex event processing for real-time big data applications, languages to encode the extracted knowledge bases, Artificial Intelligence, computation infrastructure, data curation infrastructure, query interfaces and data storage technologies.

Data Standardisation The "variety" of big data makes it very difficult to standardise. Nevertheless, there is a great deal of potential for data standardisation in the areas of data exchange and data interoperability. The exchange and use of data assets are essential for functioning ecosystems and the data economy. Enabling the seamless flow of data between participants (i.e. companies, institutions and individuals) is a necessary cornerstone of the ecosystem.

To this end, the PPP is likely to undertake collaborative efforts to support, where possible and pragmatic, the definition of semantic standardised data representation, ranging from the domain (industry sector)-specific solutions, like domain ontologies, to general concepts, such as Linked Open Data, to simplify and reduce the costs of data exchange.

In line with JTC1 Directives Clause 3.3.4.2, the Big Data Value Association (BDVA) requested the establishment of a Category C liaison with the ISO/IEC JTC1/WG9 Big Data Reference Architecture. This request was processed at the August Plenary meeting of ISO IEC JTC1 WG9, and the recommendation was unanimously approved by the working group. This liaison moves the BDVA work forward from a technology standardisation viewpoint, and now the BDVA Big Data Reference Model is closely aligned with the ISO Big Data Reference Architecture, as described in ISO IEC JTC1 WG9 20547-3. The BDVA TF6SG6 Standardisation Group is now also in the process of using the WG9 Use Case Template to extract data from the PPP Projects to extend the European use case influence on the ISO big data standards.

As the data ecosystem overlaps with many other ecosystems, such as Cloud computing, IoT, smart cities and Artificial Intelligence, the PPP will continue to be a forum for bringing together industry stakeholders from across these other domains to collaborate. These fora will continue to drive interoperability within the big data domain but will also extend this activity across the other technological ecosystems.

2.2.3 Communication and Connectivity

This concern covers effective communication and connectivity mechanisms, which are necessary for providing support for big data. This area is separately further elaborated, along with various communication communities, such as the 5G community.

The 5G PPP will deliver solutions, architectures, technologies and standards for the ubiquitous next generation of communication infrastructures in the coming decade. It will provide 1000 times higher wireless area capacity by facilitating very dense deployments of wireless communication links to connect over 7 trillion wireless devices serving over 7 billion people. This guarantees access to a wider panel of services and applications for everyone, everywhere.

5G provides the opportunity to collect and process big data from the network in real time. The exploitation of Data Analytics and big data techniques supports Network Management and Automation. This will pave the way to monitoring users' Quality of Experience (QoE) and Quality of Service (QoS) through new metrics combining network and behavioural data while guaranteeing privacy. 5G is also based on flexible network function orchestration, where machine learning techniques and approaches from big data handling will become necessary to optimise the network.

Turning to the IoT arena, the per-bit value of IoT is relatively low, while the value generated by holistic orchestration and big data analytics is enormous. Combinations of 5G infrastructure capabilities, big data assets and IoT development may help to create more value, increased sector knowledge and ultimately more ground for new sector applications and services.

On the agenda of 5G PPP is the realisation of prototypes, technology demos, and pilots of network management and operation, Cloud-based distributed computing, edge computing and big data for network operation – as is the extension of pilots and trials to non-ICT stakeholders to evaluate the technical solutions and their impact on the real economy.

The aims of 5G PPP are closely related to the agenda of the BDVA. Collaborative interactions involving both ecosystems (e.g. joint events, workshops and conferences) could provide opportunities for the BDVA and 5G PPP to advance understanding and definition in their respective areas. The 5G PPP and BDVA ecosystems need to increase their collaboration with each other, and in so doing could develop joint recommendations related to big data.

2.2.4 Cybersecurity

This concern covers security and trust elements that go beyond privacy and anonymisation. The aspect of trust frequently has links to trust mechanisms such as blockchain technologies, smart contracts and various forms of encryption.

Cybersecurity and big data naturally complement each other and are closely related, for instance in using cybersecurity algorithms to secure a data repository, or reciprocally, using big data technologies to build dynamic and smart responses and protection from attacks (web crawling to gather information and learning techniques to extract relevant information).

By its nature, any data manipulation presents a cybersecurity challenge. The issue of Data Sovereignty perfectly illustrates the way in which both technologies can be intertwined. Data Sovereignty consists in merging personal data from several sources, always allowing the data owner to retain control over their data, be it by partial anonymisation, secure protocols, smart contracts or other methods. The problem as a whole cannot be solved by considering each of these technologies separately, especially those relevant to cybersecurity and big data. The problem has to be solved globally, taking a functionally complete and secure-by-design approach.

In the case of personal data space, both security and privacy should be considered. For industrial dataspaces, the challenges relate more to the protection of IPRs, the protection of data at large and the secure processing of sensitive data in the Cloud.

In terms of research and innovation, several topics have to be considered, for example homomorphic encryption, threat intelligence and how to test a learning process, assurance in gaining trust, differential privacy techniques for privacy-aware big data analytics and the protection of data algorithms.

Artificial Intelligence could be used and could even be more efficient in attacking a system rather than protecting it. The impact of falsified data, and trust in data, should also be considered. It is essential to define the concepts of measurable trust and evidence-based trust. Data should be secured at rest and in motion.

The European Cyber Security Organisation (ECSO) represents the contractual counterpart to the European Commission for the implementation of the Cybersecurity contractual Public-Private Partnership (PPP)[1]. A collaboration with ECSO, supporting the Cybersecurity PPP, has been initiated and further steps planned.

2.2.5 Engineering and DevOps for Building Big Data Value Systems

This concern covers methodologies for developing and operating big data systems.

While big data technologies gain significant momentum in research and innovation, mature, proven and empirically sound engineering methodologies for building next-generation big data value systems are not yet available. Moreover, we lack proven approaches for continuous development and operations (DevOps) of big data value systems. The availability of engineering methodologies and DevOps approaches – combined with adequate toolchains and big data platforms – will be essential for fostering productivity and quality. As a result, these methodologies and approaches will empower the new wave of data professionals to deliver high-quality next-generation big data value systems.

2.2.6 Marketplaces, Industrial Data Platforms and Personal Data Platforms (IDPs/PDPs), Ecosystems for Data Sharing and Innovation Support

This concern covers data platforms for data sharing, which include, in particular, IDPs and PDPs, but also other data sharing platforms such as Research Data Platforms (RDPs), Data Platforms for Smart Environments (Curry 2020) and Urban/City Data Platforms (UDPs). These platforms facilitate the efficient usage of a number of the horizontal and vertical big data areas, most notably data management, data processing, data protection and cybersecurity.

Data sharing and trading are seen as essential ecosystem enablers in the data economy, although closed and personal data present particular challenges for the free flow of data (Curry and Ojo 2020). The following two conceptual solutions – Industrial Data Platforms (IDPs) and Personal Data Platforms (PDPs) – introduce new approaches to addressing this particular need to regulate closed proprietary and personal data.

3   Transforming Transport Case Study

This section illustrates the use of the BDV Reference Model within the large-scale European big data project TransformingTransport (http://www. transformingtransport.eu). The model was used to structure systematically, map, coordinate and align the project's technical outcomes, thereby also serving to distil lessons learned for the different technical concerns.

The TransformingTransport project demonstrated in a realistic, measurable and replicable way the transformations that big data can bring to the mobility and logistics market (Castiñeira and Metzger 2018; Metzger et al. 2019a). Structured into 13 different pilots, which cover areas of major importance for the mobility and logistics sector in Europe, TransformingTransport validated the technical and economic viability of big data for reshaping transport processes and services. To this end, TransformingTransport exploited access to industrial data sets from over 160 data sources, totalling 410,000 GB.

TransformingTransport ran from January 2017 to July 2019 and brought together knowledge, solutions and impact potential of major European ICT and big data technology providers with the competence and experience of key European industry players and public bodies in the mobility and logistics domain. TransformingTransport was one of the first two Lighthouse projects of the European Big Data Value Public-Private Partnership (http://www.big-data-value. eu/) funded by the European Commission within the framework of the Horizon 2020 programme.

TransformingTransport addresses 13 pilots in seven highly relevant pilot domains within mobility and transport that will benefit from big data solutions and the increased availability of data. The seven pilot domains and 13 pilots are shown in Fig. 2. For each pilot, TransformingTransport explored innovative use cases and engaged key players in the sector to demonstrate the transformative nature that big data technologies can bring about.

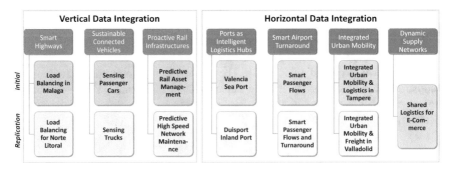

Fig. 2 Thirteen pilots in seven pilot domains

	Smart Highways		Sustainable Connected Vehicles		Proactive Rail Infrastructures		Ports as Intelligent Logistics Hubs		Smart Airport Turnaround		Integrated Urban Mobility		Dynamic Supply Networks
Data Management													
Semantic Annotation of unstructured and semi-structured data	2	2	3	3	3	3	3	3	3	3	4	4	3
Semantic interoperability	3	3	3	4	3	3	3	3	3	3	4	4	3
Data quality	3	3	4	4	2	2	4	4	4	4	4	4	4
Data lifecycle management and data governance	4	4	4	4	4	4	4	4	3	3	3	3	3
Integration of data and business processes	3	2	3	4	4	4	4	4	4	4	4	4	4
Data-as-a service	4	4	4	4	4	4	4	4	4	4	4	4	3
Distributed trust infrastructures for data management	4	4	4	4	4	4	4	4	4	4	4	4	4
Data Processing Architectures													
Heterogeneity	4	4	4	4	4	4	4	4	4	4	3	3	4
Scalability	3	3	3	3	3	3	3	3	3	3	3	3	3
Processing of data-in-motion and data-at-rest	4	4	4	4	4	4	4	4	4	4	4	4	4
Decentralizatrion	4	4	4	4	4	4	4	4	4	4	4	4	4
Performance	4	4	4	4	4	4	4	4	4	4	4	4	4
Novel architectures for enabling new types of big data workloads	3	3	4	4	4	4	4	4	4	4	4	4	4
Introduction of new hardware capabilities	4	4	4	3	4	4	4	4	4	4	4	4	3
Data Analytics													
Semantic and knowledge-based analysis	3	2	3	3	2	2	3	3	2	2	2	2	2
Content validation	4	4	4	4	3	3	4	4	3	3	4	4	4
Analytics frameworks & processing	2	3	3	3	3	3	3	3	3	3	3	3	3
Advanced business analytics and intelligence	3	2	2	1	1	1	2	2	3	3	2	2	2
Predictive and prescriptive analytics	1	1	1	2	1	1	1	1	1	1	1	1	1
High Performance Data Analytics (HPDA)	2	2	2	2	1	1	2	2	2	2	3	3	2
Data analytics and Artificial Intelligence	4	4	4	3	4	4	4	4	4	4	4	4	3
Data Protection													
Generic and easy to use data protection approaches	4	4	4	4	4	4	4	4	4	4	4	4	4
Robust Data privacy (incl. multi-party computation)	4	4	4	4	4	4	4	4	4	4	4	4	4
Risk based approaches	4	4	4	4	4	4	4	4	4	4	4	4	4
Data Visualisation and User Interaction													
Visual data discovery	3	3	3	2	2	2	3	3	3	3	3	3	3
Interactive visual analytics of multiple scale data	2	2	3	2	2	2	2	2	3	3	2	2	2
Collaborative, intuitive and interactive visual interfaces	2	2	2	2	2	2	2	2	3	3	2	2	2
Interactive visual data exploration and querying in a multi-device context	2	2	2	2	2	2	2	2	3	3	2	2	2

Fig. 3 Coverage of Big Data Value Reference Model (1 = Main focus; 2 = Topic addressed, but not main focus; 3 = Topic marginally addressed; 4 = Topic not addressed)

Figure 3 shows how the different pilots contributed to the different horizontal concerns of the Big Data Value Reference Model (as introduced in Sect. 2), breaking down their contributions to different technical priorities per concern. The numbers indicate the focus of the pilots on the respective technical priorities.

As can be seen, the most relevant horizontal concerns of TransformingTransport were (1) Data Analytics, (2) Data Visualisation and (3) Data Management, which we elaborate below together with lessons learned from the project. We then elaborate on how the impact of big data solutions on key business outcomes can be measured to assess the usefulness of these techniques, and then conclude the use case with some final observations.

3.1 Data Analytics

The key enabling analytics technology employed by TransformingTransport is predictive data analytics. Predictive analytics is a significant next step from descriptive analytics. While descriptive analytics answers the question "What happened and why?", predictive analytics attempts to answer the question "What will happen and when?" (see Sect. 2.1.2). For example, predictive analytics may help predict whether there may be a delay in a transport process, helping transport operators to be proactive and take action to decrease or prevent delays (Metzger et al. 2019a).

A case in point is the Smart Passenger Flows pilot at Athens Airport. With passenger demand increasing annually, the challenge for Athens Airport has been to identify intelligent ways to improve and streamline the flow of people through the airport, i.e., increase throughput, while at the same time ensuring the safety and the experience of passengers (Feltus et al. 2018). Increasing throughput requires sophisticated data analysis to build powerful big data models that can segment passengers and identify patterns and trends that will lead to actionable strategies on behalf of the airport.

Lessons learned in data analytics include:

- **Data quality**: Among the most universally accepted principles of analytics is "**Garbage in – Garbage out**", which refers to the quality of the data in the training models. It means that if poor-quality data enter the system, no matter how trendy the software for the analysis, the output value is expected to be of low quality too. To overcome this, checking and coping with missing data, data accuracy, data timelines, different time-zones (clocks), etc., is a must; so is assigning "data owners" that understand data and its field (domain) being able to be in the care of data quality.
- Using **Deep Learning** and Neural Networks helps to create more efficient development and engineering. They have been proven to work well even without extensive hyper-parametrisation, provided that enough good-quality data is available. This means that the time- and resource-consuming step of extensive experimentation with hyper-parameters may be skipped, leading to a more efficient development and deployment process of big data applications (Palm et al. 2020).
- **Data accuracy**: Operators benefit from information about data accuracy. This results in improved decision-making and helps to determine when to trust a prediction. Augmenting the quality of data (live or predicted) with confidence intervals, error ranges or reliability estimates allows operators to acquaint themselves with the most realistic situation.
- **Time series models** can be successfully approached by traditional **machine learning techniques**. It has been verified that machine learning techniques and Arima models are quite similar in short-term predictions, while the former tend to be more accurate as the time to be predicted increases. Not only are predictive models useful to improve a process, but it is also necessary to have teams with *enough experience* to select the most suitable alternative (descriptive or predictive). Another lesson learned is that external variables are easily included in the modelisation.

- **Historical data**: Regarding data analytics, pilots found it useful to **keep histor-ical non-reproducible data** and, when possible, in **raw format**. Several reasons support this method, such as possible errors or improvements in the code that do not allow rebuilding of processed data if the original data is deleted. If one substitutes raw data with processed data, and there are no possible mechanisms to reverse the process, important information can be missed in ulterior processing stages. A drawback in maintaining unprocessed raw data could be the need for increased storage capacity. Raw historical data can also be used for training in machine learning algorithms. The main idea is to keep the complete historical data since some bits of previously untreated information can be very important for future analyses.

3.2   *Data Visualisation*

As the project concluded, one of the most useful and profitable visualisation techniques that was considered as a "*key success factor*" was **cockpit** for data visualisation and real-time control. Cockpit is a flexible human-machine interface (HMI) designed to help operators in day-to-day monitoring, where pilots have shared their knowledge to gain the most valuable insights from these tools.

A case in point is developed as part of Dusiport inland port pilot. This cockpit exploits advanced data processing, predictive analytics capabilities and interactive visualisation to support terminal operators in proactive decision-making and process adaptation (Metzger et al. n.d.). In addition to raising alarms in the case of a predicted delay, the terminal productivity cockpit also shows a reliability estimate for the predicted delay. The reliability estimate gives the probability (in %) of whether the alarm is indeed a true alarm. Reliability estimates facilitate distinguishing between more and less reliable predictions on a case-by-case basis (Metzger et al. 2019b).

Lessons learned in data visualisation include the following:

- Despite being an excellent tool to see what is happening around the pilot, a cockpit should not be exhaustive in relation to the amount of information displayed, which can lead to cognitive overload due to information overflow. There are three main requisites. First, the **information** must be shown **hierar-chically** from top to bottom interface, enabling making summaries with the most relevant details. Second, widgets must be **intuitive, simple and "clean"** for the user and allow for quick handling to easily grasp the information shown. Third, cockpit should **only display critical and sufficiently well-validated events**, in order to avoid overloading the interface with superfluous warnings and focus the attention on the most important ones.
- Static user interfaces (UI) may be limiting. Providing **dynamic customisation** of UI from simple multi-option dropdowns to more complex interchangeable

requests could boost the efficiency of the analysis, adapting itself to specific user and operator needs.

- Visualisation helps to take decisions, with synthetic and clear results. Implications of the **human factors** team were found to be useful in understanding these aspects. Moreover, involving them in the early stages of the project also helped to gain a better perspective of the demonstrator.
- It is relevant to address the **right customer or user** who is going to work with the visualised data. In day-to-day business, there is often not enough time to only look at visualisations without an explicit added value. Yet, if the cockpit also serves as a decision-making tool, e.g. to plan routes, or has other technical implementations, it provides more added value. Another group to be approached could be **decision-makers** who can use these cockpits for strategic planning purposes.
- The goal of data visualisation is to make the data easily understandable and usable by the operators. To accomplish this, visualisations beyond just showing the quantitative data in big tables must be developed, thereby enabling the users to make a qualitative assessment of quantitative data intuitively. The terminal operators must be sure that the **data** is current. However, only knowing the current state is not sufficient for the operator. In addition, the **date and time** of the last critical event were perceived as important, to allow the operator to visualise/search for anomalies around the fault in historical data, and not only rely on the prediction algorithm. To enable the user to recognise critical trends more easily, it is recommended that spaces above and below certain thresholds be **colour-coded**.
- As it turns out, cockpits are an excellent means to gain a clear perception of the current status of activities. Nevertheless, **excessive overload in the presentation of the results can be risky** for a good understanding of the actual and relevant situation.

3.3   Data Management

Data collection, integration and quality requires significant effort and time in TransformingTransport. It has been estimated at around 80% by some pilots. Access to the data sources has turned out to be much more complicated than expected due to the following reasons: first, the number of **different sources** and data production and storage systems; secondly, the **access characteristics of data sources** – from a technical point of view, some of these sources and systems did not have the optimal flexibility. Using domain-specific data platforms (such as the BDV data platform project DataPorts: http://dataports-project.eu/) together with domain-specific machine learning components could significantly increase productivity in developing and deploying data analytics solutions.

Further lessons learned for data management are as follows:

- Concerning real-time analysis, tools have in many cases been implemented not as pure real-time but as **near-real-time** systems adapting the reaction time of the tools to the more lagged data-producing process. This is an important lesson because expecting pure real-time systems is nowadays far from easy due to ageing systems in several cases. This technology should be updated for further replications, mainly concerning big data projects, to take advantage of the new technologies.
- In order to provide services in real time, **extra storage** is required (which should be considered in the dimensioning phase of the system). This means that special care must be taken in defining optimised structures derived from the raw data that allows lower latency to process data. Additionally, in the case of databases, it is important to define appropriate indexes, reaching a compromise between the speed of writing in the database and reading from it. It has also been found that non-relational databases are more appropriate than traditional relational databases for evolving systems. Relational databases are more restrictive in their structure and do not allow rapid changes, offering advantages such as flexible schemas and better scaling (e.g. when new datasets are added and more fields in a table – or collection – are necessary, the addition is much easier in a non-relational database).
- One of the research goals was to identify **valuable data sources** that support the understanding of the different transport domains. Therefore, many different types of data and data sources were part of the pilots. These data sources differed in terms of format, timely availability and geographical spread (for pilots with large areas of action). One of the first things that many pilots learned was to abandon the idea of a holistic technical integration of all data sources. Data can also provide valuable insights when considered separately to some extent. Concerning visualisation, it was important to develop good use cases and to define the right data for them. Therefore, only useful data were used and further processed, which finally reduced complexity and increased understandability.
- The management of data required, in many cases, two approaches depending on whether processes required the use of raw datasets or processed datasets. Raw data were stored in file structures which were accessible to all workers. The parallelisation of computations was then organised such that each process task would use a different file from other processes, resulting in a mitigation of file access conflicts. Results were then stored in a variety of data structures that were capable of receiving data very quickly from multiple sources and enabled very fast search and retrieval times for records. Key was to have access to people who really know the data, because standardisation has not always been completed.
- **Data availability and fit for purpose:** Having data available on day 1 of the project does not mean it is fit for purpose (enough to answer the addressed business or operational needs) since technical access (interfaces) and organisational access (ownership) may require time to resolve. Because of this, first data analytics and visualisation goals must be defined and then it must be determined which data needs to be accessed and how, or vice versa.

3.4 Assessing the Impact of Big Data Technologies

As reported above, different lessons learned were collected for different technical concerns. However, such lessons learned were mostly qualitative. In order to complement these qualitative insights with quantitative measurements, TransformingTransport followed a stringent KPI measurement regime to demonstrate the transformative effects that big data could have on the transport sector through pilot projects in different countries, locations, transport modes and operating conditions. It applied big data for reshaping transport processes and services, increasing operational efficiency, improving customer experience and fostering new business models. As previously mentioned, data collection, integration and quality require significant effort and time, estimated at around 80% by some pilots mainly due to difficulties to be faced such as different data sources and storage characteristics. In this context, good and consistent data management is essential to improve operations.

A multi-criteria analysis (MCA) was designed specifically to assess the multiple impact levels of big data technologies implemented in the 13 different pilot cases of the project. The use of MCA appears to be an adequate option for simultaneously evaluating a certain number of both quantitative and qualitative criteria, some incommensurable, that ultimately need to be aggregated. MCA arose in the context of operations research (Charnes and Cooper 1977) and assessed alternatives on a set of criteria reflecting the decision-makers objectives, ranked based on an aggregation procedure. The scores achieved do not need to be translated into monetary terms but can simply be expressed in physical units or in qualitative terms (de Brucker et al. 2011). To make this method possible, a set of "Key Performance Indicators" (KPIs) were selected, defined as measurable figures able to shed light on how effective a certain application is. Applying the groundings of MCA, which enables the combination of both qualitative and quantitative aspects, TransformingTransport developed a methodology of assessing a high number of indicators pertaining to entirely different transport sectors (Velazquez et al. 2018) and Assessment Categories of major relevance, i.e. operational efficiency, asset management, environmental quality, energy consumption and safety. These categories have been used to perform a complete assessment of the different pilots and manage data collected through pilot-only evaluation and then – in a transversal way across pilots – a comparison between them.

The large differences among pilots and domains have led to the creation of a specific methodology out of which the analysis of results showed the impacts of the tested technological improvements. Throughout several consciously selected KPIs, it has been possible to assess the benefits of big data implementation on the transportation sector. Then, a four-level assessment was carried out. The first level consists of the evaluation of each pilot individually for each of the Assessment Categories, after an aggregation process. The second level goes through the analysis of the aggregated achievements within the same pilot domain, comparing the performance of the pilots within the domain. Therefore, the effects of big data in the same mode in different settings and conditions are analysed. The third level of the evaluation is the transversal assessment of the pilots for each category; the goal was to perform a comparative analysis through the different pilots on each of the aspects, e.g. how operational efficiency or energy savings vary among them. The fourth assessment is the strategic level, for which only the most relevant KPIs for each pilot are considered (Vázquez et al. 2020).

The evaluation procedure analyses the impact of the big data implementation over different transport sectors, by comparing KPI final measurements with the original ones. There is thus a four-level assessment comparison between two scenarios: the reference scenario before leveraging the big data technology (baseline or ex ante scenario) and the scenario once the technologies have been introduced (big data technology scenario) (Velazquez et al. 2018). The results of this assessment reveal improvements of around 40-60% regarding the operative cost, energy consumption, environmental quality and enhancement of the predictive maintenance of assets, among others. Big data technologies have demonstrated their usefulness when it comes to gaining deeper insights from the huge quantity of data to boost the different transport processes.

Effective and consistent data management is essential to improve transport operations. A further lesson learned from TransformingTransport is that due to the huge volume and variety of data and data sources, a coherent, in-depth and integrated approach for data management and analysis is necessary.

3.5   Use Case Conclusion

As can be concluded from the use case presented above, big data technologies promise to deliver profound economic and societal impact in mobility and logistics. TransformingTransport pursued big data use cases in all areas of major importance for the mobility and logistics sector in Europe, demonstrating the technical and economic viability of big data for reshaping transport processes and services. TransformingTransport employed predictive data analytics and predictive maintenance as the key enabling big data technologies to bring about this transformation.

The significant growth of transport data volumes and the rates at which such data is generated will be an important driver for the next level of technology innovation in transport: data-driven Artificial Intelligence (AI). Data-driven AI has a tremendous potential to benefit European citizens, economy and society (Sonja Zillner et al. 2018; Zillner et al. 2020). From an industrial point of view, AI means algorithm-based and data-driven computer systems that enable machines and people with digital capabilities such as perception, reasoning, learning and even autonomous decision-making. AI will facilitate software to draw conclusions, learn, adapt and adjust parameters accordingly. With recent advances in computing power, connectivity and algorithms, AI is making great strides. With today's promising results in using AI technology, we can expect the next level of efficiency and operational improvements in the mobility and transport sectors in Europe.

4   Summary

The Big Data Value Reference Model has been developed with input from technical experts and stakeholders along the whole big data value chain. The BDV Reference Model may serve as a common reference framework to locate big data technologies on the overall IT stack. This chapter elaborated the various elements (both horizontal and vertical) of the framework and illustrated how it might be used to map technical elements stemming from research and innovation projects. Complementing this application of the reference model, it has also been used to systematically monitor the technical progress of the Big Data Value PPP. To determine how well the technical priorities and challenges are covered by ongoing research and innovation activities, the BDVA performed a systematic collection of data, where the BDV Reference Model provided the structure for a common data collection template and frame for data analysis.

Acknowledgements Research leading to these results received funding from the European Union's Horizon 2020 research and innovation programme under grant agreement nos. 732630 (BDVe), 731932 (TransformingTransport) and 871493 (DataPorts). This publication has emanated from research supported in part by a research grant from Science Foundation Ireland (SFI) under grant no. SFI/12/RC/2289_P2, co-funded by the European Regional Development Fund.

References

Castiñeira, R., & Metzger, A. (2018, April). *The transforming transport project – Mobility meets big data.* https://doi.org/10.5281/zenodo.1484954
Cavanillas, J. M., Curry, E., & Wahlster, W. (Eds.). (2016). *New horizons for a data-driven economy: A roadmap for usage and exploitation of big data in Europe.* https://doi.org/10.1007/978-3-319-21569-3
Charnes, A., & Cooper, W. W. (1977). Goal programming and multiple objective optimisations: Part 1. *European Journal of Operational Research.* https://doi.org/10.1016/S0377-2217(77)81007-2
Curry, E. (2020). *Real-time linked dataspaces.* https://doi.org/10.1007/978-3-030-29665-0
Curry, E., & Ojo, A. (2020). Enabling knowledge flows in an intelligent systems data ecosystem. In *Real-time Linked Dataspaces* (pp. 15–43). https://doi.org/10.1007/978-3-030-29665-0_2
de Brucker, K., Macharis, C., & Verbeke, A. (2011). Multi-criteria analysis in transport project evaluation: An institutional approach. *European Transport - Trasporti Europei.*
Feltus, C., Proper, H., Metzger, A., Lopez, J., & Castineira, R. (2018). *Value CoCreation (VCC) Language design in the frame of a smart airport network case study* (pp. 858–865). https://doi.org/10.1109/AINA.2018.00127
G. Press. (2016). *Cleaning big data: Most time-consuming, least enjoyable data science task, survey says.*
Germany and the Netherlands. (2017). *Joint position paper on the European open science cloud.*
Lund, S., Manyika, J., Nyquist, S., Mendonca, L., & Ramaswamy, S. (2013). *Game changers: Five opportunities for US growth and renewal.*
Manyika, J., Chui, M., Brown, B., Bughin, J., Dobbs, R., Roxburgh, C., & Byers, A. H. (2011*). Big data: The next frontier for innovation, competition, and productivity.* Retrieved from McKinsey

Global Institute website http://scholar.google.com/scholar.bib?q=info:kkCtazs1Q6wJ:scholar.google.com/&output=citation&hl=en&as_sdt=0,47&ct=citation&cd=0

Metzger, A., Thornton, J., Valverde, F., Lopez, J. F. G., & Rublova, D. (2019a). *TransformingTransport, Predictive analytics and predictive maintenance innovation via big da-ta: The case of TransformingTransport.* In 13th Intelligent Transport Systems - European Congress (ITS Europe), Brainport-Eindhoven, The Netherlands, June 3–6.

Metzger, A., Neubauer, A., Bohn, P., & Pohl, K. (2019b). Proactive process adaptation using deep learning ensembles. In P. Giorgini & B. Weber (Eds.), *Advanced information systems engineering* (pp. 547–562). Cham: Springer.

Metzger, A., Franke, J., & Jansen, T. (n.d.). Data-driven deep learning for proactive terminal process management. In J. V. Brocke, J. Mendling, & M. Rosemann (Eds.), *Business process management cases* (Vol. 2). New York: Springer.

Palm, A., Metzger, A., & Pohl, K. (2020). Online reinforcement learning for self-adaptive information systems. In S. Dustdar, E. Yu, C. Salinesi, D. Rieu, & V. Pant (Eds.), *Advanced Information systems engineering* (pp. 169–184). Cham: Springer.

Raskin, J. (2000). *Humane interface, the: New directions for designing interactive systems.* Addison-Wesley Professional.

Vázquez, P., Monzon, A., Boggio-Marzet, A., & Corral, V. (2020). Assessing the impact of big data for improving transport efficiency: A cross-modal approach. In *8th Transport Research Arena TRA 2020.* Helsinki, Finland.

Velazquez, G., Monzon, A., & Roman, A. (2018). Big Data value for improving transport performance in all modes, an assessment methodology. In *7th Transport Research Arena TRA 2018.* Vienna, Austria.

Wilkinson, M. D., Dumontier, M., Aalbersberg, IJ., Appleton, G., Axton, M., Baak, A., et al. (2016). The FAIR guiding principles for scientific data management and stewardship. *Scientific Data, 3,* 160018. https://doi.org/10.1038/sdata.2016.18

Zillner, S, Curry, E., Metzger, A., & Auer, S. (Eds.). (2017). *European big data value strategic research & innovation agenda.* Retrieved from Big Data Value Association website www.bdva.eu

Zillner, S, Gómez, J. A., Robles, A. G., & Curry, E. (2018). *Data for artificial intelligence for European economic competitiveness and societal progress – BDVA position statement.*

Zillner, S., Bisset, D., Milano, M., Curry, E., Hahn, T., Lafrenz, R., et al. (2020). *Strategic research, innovation and deployment agenda - AI, data and robotics partnership. Third Release (3rd).* Brussels: BDVA, euRobotics, ELLIS, EurAI and CLAIRE.

Data Protection in the Era of Artificial Intelligence: Trends, Existing Solutions and Recommendations for Privacy-Preserving Technologies

Tjerk Timan and Zoltan Mann

Abstract This chapter addresses privacy challenges that stem particularly from working with big data. Several classification schemes of such challenges are discussed. The chapter continues by classifying the technological solutions as proposed by current state-of-the-art research projects. Three trends are distinguished: (1) putting the end user of data services back as the central focal point of Privacy-Preserving Technologies, (2) the digitisation and automation of privacy policies in and for big data services and (3) developing secure methods of multi-party computation and analytics, allowing both trusted and non-trusted partners to work together with big data while simultaneously preserving privacy. The chapter ends with three main recommendations: (1) the development of regulatory sandboxes; (2) continued support for research, innovation and deployment of Privacy-Preserving Technologies; and (3) support and contribution to the formation of technical standards for preserving privacy. The findings and recommendations of this chapter in particular demonstrate the role of Privacy-Preserving Technologies as an especially important case of data technologies towards data-driven AI. Privacy-Preserving Technologies constitute an essential element of the AI Innovation Ecosystem Enablers (Data for AI).

Keywords Data protection · Artificial Intelligence · Big data · Challenges · Future directions

T. Timan (✉)
Strategy, Analysis & Policy Department, TNO, The Hague, The Netherlands
e-mail: tjerk.timan@tno.nl

Z. Mann
paluno, University of Duisburg-Essen, Essen, Germany

© The Author(s) 2021
E. Curry et al. (eds.), *The Elements of Big Data Value*,
https://doi.org/10.1007/978-3-030-68176-0_7

153

1   Introduction

One of the challenges of big data analytics is to maximise utility whilst protecting human rights and preserving meaningful human control. One of the main questions in this regard for policymakers and lawmakers is to what extent they should allow for automation of (legal) protection in an increasingly digital society. This chapter contributes to this debate by looking into different technical solutions developed by the projects of the Big Data Value Public-Private Partnership (BDV PPP) that aim to protect both privacy and confidentiality whilst allowing for big data analytics. Such Privacy-Preserving Technologies are aimed at building privacy-by-design from the start into the back end and front end of digital services. They make sure that data-related risks are mitigated both at design time and run time, and they ensure that data architectures are safe and secure. In this chapter, we discuss recent trends in the development of tools and technologies that facilitate secure and trustworthy data analytics and provide recommendations based on the insights and outcomes of the projects of the BDV PPP and from the task forces of the Big Data Value Association (BDVA), combined with insights from recent debates and the literature.

1.1   *Aim of the Chapter*

The aim of this chapter is to provide an overview of trends in Privacy-Preserving Technologies and solutions as currently developed by research projects that are part of the Big Data Value Public-Private Partnership (BDV PPP). In the chapter, we focus on providing an overview of technical solutions for privacy and data protection challenges posed by Big Data and AI developments. The main particularity of big data is the number of data sources and the heterogeneousness of these sources. In many cases this leads to a mix of datasets that contain both personal and non-personal data. Combinations and aggregations of datasets in turn lead to new data. Mixing and reusing data on a large scale and at high velocity makes many forms of protection of data difficult, and enforcement of data protection laws challenging. In addition to legal, ethical, institutional and organisational checks and balances surrounding privacy rights, technological solutions to mitigate privacy issues caused by large-scale use of personal data are multiple, and rapidly developing. This chapter provides a selection of the many technologies aimed at protecting privacy while upholding the benefits of big data analytics. We hope the chapter serves policymakers, technology developers and other relevant audiences interested in Privacy-Preserving Technologies.

A note: Many solutions deal with mitigating risks of personal data breaches as a result of big data analytics. However, many of these solutions are equally applicable to the case of sharing non-personal data between parties.[1] As such, there is a difference between "privacy preservation" when talking about personal data, and "confidentiality preservation" when dealing with non-personal yet confidential data, although the techniques for the two can be the same. For the sake of simplicity, we will refer to solutions as "Privacy-Preserving Technologies", irrespective of whether they are applied to personal or non-personal data.

1.2   Context

Recent news about data leaks,[2] (the lack of) control over content and political influence of social networks has provided an increasing awareness of how social media platforms (mis)use personal data, which in turn has had an effect on the level of trust users have in such platforms and digital services (Newman et al. 2017). Many social media platforms get their (economic) value from capturing visitors' behaviour either directly (via services offered) or indirectly (by tracking users' online behaviour). With the migration from laptop- or PC-based browsing via web browsers to consuming media on mobile devices and via dedicated apps, it has become possible to collect far more types of data surrounding this behaviour in a far more targeted manner, even in near real time (Patent No. 9,720,569 2017). Combining places where people go digitally with where they are physically offers many possibilities, but also brings about many new privacy risks. Although location data is explicitly categorised as personal data in the GDPR (e.g. De Hert et al. 2018), it is not always clear what kinds of risks such data poses, specifically in combination with other types of personal or non-personal data. Debates on what personal data exactly entails (Purtova 2018) and how to apply personal data protection in the context of large-scale data analytics are even more pressing in the current landscape of data protection regulation.[3] Slowly but surely, companies and governments deploying big data analytics and processing personal data are applying (and complying with) the GDPR. Beyond the growing awareness of the need to comply (the first case of a

---

[1]Which can lead to personal data afterwards. For example, by processing data from a machine, an algorithm could identify the operator based on the consumption of electrical power of the machine. This then becomes related to personal data and could therefore be relevant to the EU General Data Protection Regulation (GDPR).

[2]While there are many data breaches on a corporate level that are often not mentioned or don't make headline news, a rather (in)famous one was the data breach of a company whose secrecy and data protection were part of its core value proposition: https://www.theguardian.com/technology/2016/feb/28/what-happened-after-ashley-madison-was-hacked

[3]For an overview of the current data regulatory landscape, see a recent deliverable by the LeMo project: https://lemo-h2020.eu/newsroom/2018/11/1/deliverable-d22-report-on-legal-issues

GDPR fine was issued in 2018[4]), there is a wider societal need for trust in digital environments.[5]

The question of how to foster trust in digital systems is a complex and multifaceted one. Many recent research projects are engaged directly or indirectly in (re)-building trust in digital environments, via different approaches, ranging from technical to social, ethical and organisational. Going beyond mere compliance with the GDPR and other data privacy laws (Gellert n.d.) (sometimes dubbed "phase 1" of privacy protection in data analytics), the main aim of many current research projects that deal with Privacy-Preserving Technologies is to explore how privacy can be utilised as an asset, as a competitive advantage or as a unique selling point (sometimes dubbed "phase 2"). One of the challenges of arriving at a fully functional digital single market is to take human rights as a starting point while also offering a unique environment for innovation, to offer framework conditions that allow companies to reach this phase 2. In this chapter, we highlight projects that are developing solutions to bridge the gap between utility and privacy and that offer a positive-sum outcome, instead of a zero-sum outcome (Cavoukian 2008), when it comes to privacy and security of data. We provide recommendations for policy concerning the development of Privacy-Preserving Technologies and the uptake of such technologies by different markets or sectors. Scalability of solutions is marked as one of the main barriers in this regard, especially when cryptographic techniques are used at any point along the analysis pipeline.

2   Challenges to Security and Privacy in Big Data

What is it about big data that makes for specific data protection challenges that need addressing, and how can we address them? The challenges of protection of personal data in the context of big data analytics (BDA) mainly connect to concepts such as profiling and prediction based on large datasets of personal data. A secondary result of big data analytics is that combinations of non-personal data (according to the definition provided in the GDPR (Zarsky n.d.)) can still lead to the identification of persons and/or other sensitive information (Kerr 2012), rendering many current pseudonymisation and anonymisation approaches insufficient. A dilemma put forward by data science is that data protection and data-driven innovation have diverging, even opposite, premises: the former requires a clear and defined purpose for any type of processing, whereas the latter is often based on exploration of data in order to find a purpose. While this dichotomy is not new, the increasing scale, speed and

---

[4]https://iapp.org/news/a/portugal-fines-hospital-400k-euros-for-gdpr-violation/
[5]See, for instance, https://medium.com/ipg-media-lab/how-tech-companies-are-failing-the-trust-test-1f1057de9317

complexity of current data analytics reinforce it.[6] We need to look for new ways to guarantee the protection of personal data while retaining the potential benefits of big data analytics. The BDVA subgroup on Data Protection and Pseudonymisation Mechanisms summarised current challenges in the most recent BDVA Strategic Research and Innovation Agenda (SRIA) (Zillner et al. 2017), including:

- A general, easy-to-use and enforceable data protection approach suitable for large-scale commercial processing[7]
- Maintaining robust data privacy with utility guarantees, also implying the need for state-of-the-art data analytics to cope with encrypted or anonymised data[8,9]
- Risk-based approaches calibrating data controllers' obligations regarding privacy and personal data protection[10]
- Combining different techniques for end-to-end data protection (Mann et al. 2018; Stojmenovic et al. 2016)

The last point has also been observed by the E-SIDES project, who have investigated a wide range of technologies for privacy preservation in big data: "In practice, the technologies need to be combined to be effective and there is no single most important class of technologies".[11]

Another challenge when designing privacy solutions for big data is the number of data sources, which can result in different settings where stakeholders can have varying degrees of access to the processed data. In the case of a single data owner, the data owner may encrypt their data with their own keying material and may apply data analytics on the encrypted data either locally or by offloading to a third-party platform. On the other hand, nowadays data is being collected by a vast range of

---

[6]See E-SIDES Deliverable D4.1, section 3.2. See also the ENISA report on privacy in the era of big data (https://www.enisa.europa.eu/publications/big-data-protection), in which the novelty is described as follows: "Therefore, the new thing in big data is not the analytics itself or the processing of personal data. It is rather the new, overwhelming and increasing possibilities of the technology in applying advanced types of analyses to huge amounts of continuously produced data of diverse nature and from diverse sources. The data protection principles are the same. But the privacy challenges follow the scale of big data and grow together with the technological capabilities of the analytics" (p. 22).

[7]For an elaborate overview of different types of measures, both technical and non-technical, see E-SIDES project Deliverable D4.1, section 4 and D3.2, section 4.4: https://e-sides.eu/assets/media/e-sides-d4.1-ver.-1.0-1540563562.pdf

[8]This is one of the goals of the MOSAICrOWN project, a recently started H2020 project which aims to enable data sharing and collaborative analytics in multi-owner scenarios in a privacy-preserving way, ensuring proper protection of private/sensitive/confidential information. https://mosaicrown.eu

[9]See e-sides Deliverable 3.2, in which a Privacy-Preserving Technologies uptake gap analysis is provided. https://e-sides.eu/resources/deliverable-d32-assessment-of-existing-technologies

[10]A risk-based tool featuring a didactic interface to carry out Data Protection Impact Assessment according to GDPR is available from the French data protection authority CNIL at: https://www.cnil.fr/en/open-source-pia-software-helps-carry-out-data-protection-impact-assesment

[11]See E-SIDES Deliverable D3.2, conclusions. https://e-sides.eu/resources/deliverable-d32-assessment-of-existing-technologies

applications and services, by different kinds of organisations. This data is often subject to deep analysis in order to infer valuable information for these organisations. Nevertheless, restrictions on data access and sharing (such as using traditional encryption techniques) can render data analytics less effective, in the sense that without access to high volumes of data, applications that rely on analytics cannot maintain a good level of accuracy of their analytical models.

The ability to train an accurate model depends on the diversity of training data. With more diverse data collected from different sources, analytical models can be increasingly accurate. However, recent privacy-related regulations or business interests inhibit data producers from sharing (sensitive) data with third parties. As a consequence, organisations are not benefiting from employing collaborative large-scale analytics and from deriving more accurate global analytical models. Privacy-preserving data analytics should consider the case of data coming from multiple sources while enabling collaborative analytics without compromising the privacy of the different data subjects involved.[12]

In this regard, two main approaches can be identified. The first one aims at providing means to protect the data, establishing trust among partners (e.g. possibly by encrypting the data or adding a perturbation under Differential Privacy principles), such that data can be outsourced and processed elsewhere, even by third parties. This approach requires a very strong level of protection, since the variety of manipulations/attacks is potentially very large. Such strong protection also imposes strong restrictions: limited types of operations on the data (possibly enforced by a usage control policy), presence of distortions that may bias the results, very high computational requirements and loss of control on the ultimate data usage. A second approach relies on the deployment of a controlled processing environment where the participants are expected, or forced, to operate under specific predetermined rules and protocols. In this scenario, the data does not leave the owner facilities, and the process of training relies on secure operations on the data following pre-specified protocols. Instances of this approach are the environments known as Industrial Data Platforms (IDP) and Personal Data Platforms (PDP). This approach has been adopted, for instance, in the Musketeer project,[13] as described in the next section. Several techniques of pseudonymisation and anonymisation have also been utilised in the Transforming Transport project in the context of an e-commerce pilot, the urban pilot in the city of Tampere (Finland) and several airport pilots.[14] Finally, one may also allow an authorised third party to make analytical queries over the collected data.

---

[12]This is the main goal of the Musketeer project, an H2020 project that has recently started, which aims at developing an Industrial Data Platform (IDP) facilitating the combination of information from multiple sources without actually exchanging raw data (thereby protecting privacy/confidentiality) such that, eventually, better machine learning models are obtained.

[13]Machine Learning to Augment Shared Knowledge in Federated Privacy-Preserving Scenarios. EU H2020 Research and Innovation Action – grant No. 824988. http://musketeer.eu

[14]See Transforming Transport newsletters here: https://transformingtransport.eu/downloads/newsletters

In short, the role of Privacy-Preserving Technologies is to establish trust in a digital world, in a digital way. Although some of the above-mentioned challenges also require non-technical solutions (organisational measures, ethical guidelines on data analytics and AI,[15] increased education, etc.), in the following we focus mostly on the technical solutions in the making.

3 Current Trends and Solutions in Privacy-Preserving Technologies

Different activities in Europe on data protection, such as works on privacy standards, privacy engineering and awareness-raising events, have been developed over recent decades.[16] However, while the field of privacy engineering is ever-evolving in research labs and universities, for the translation into applications and services their maturity level (sometimes also referred to as Technology-Readiness Level – TRL) is important. We need to better understand the current maturity levels and types of solutions available for a specific challenge or issue (sometimes referred to as Best Available Techniques), but also an overview in general about the available technological solutions. Companies, governments or other institutions might require different levels of maturity for a particular Privacy-Preserving Technology, depending on what kind of big data processes they are involved in. ENISA, the EU Agency for Cybersecurity, developed a portal[17] that provides an assessment methodology for determining the readiness of these solutions for certain problems or challenges.[18] For the classification of Privacy-Preserving Technologies, a first point of departure can be found in Jaap-Henk Hoepman's Blue Book on privacy-by-design strategies (Hoepman 2020). Here, an overview is provided in terms of how and where different privacy-by-design strategies can be applied. He distinguishes the following strategies, divided into data-related and process-related tasks around privacy protection (Gürses et al. 2006) (Table 1):

There are some parts of this structure that might overlap when it comes to Privacy-Preserving Technologies, especially if the notion of Privacy-Preserving Technologies is taken broadly, to include any technology that can aid in the protection of privacy or support Privacy-Preserving Data Processing activities.

---

[15]See, for instance, https://algorithmwatch.org/en/project/ai-ethics-guidelines-global-inventory/

[16]See https://edps.europa.eu/data-protection/ipen-internet-privacy-engineering-network_en and https://ipen.trialog.com/wiki/Wiki_for_Privacy_Standards

[17]https://www.enisa.europa.eu/events/personal-data-security/pets-maturity

[18]Sometimes also referred to as "best available technique", or BAT. The EDPS (European Data Protection Supervisor) describes BATs for data protection as follows: "the most effective and advanced stage in the development of activities and their methods of operation, which indicates the practical suitability of particular techniques for providing the basis for complying with the EU data protection framework. They are designed to prevent or mitigate risks to privacy, personal data and security" (see EDPS opinion, p. 10).

Table 1 Privacy strategies according to Hoepman

Data-related tasks	
Minimise	Limit as much as possible the processing of personal data.
Separate	Separate the processing of personal data as much as possible from the data itself.
Abstract	Limit as much as possible the detail in which personal data is processed.
Hide	Protect personal data, or make it unlinkable or unobservable. Make sure it does not become public or known.
Process-related tasks	
Inform	Inform data subjects about the processing of their personal data in a timely and adequate manner.
Control	Provide data subjects adequate control over the processing of their personal data.
Enforce	Commit to processing personal data in a privacy-friendly way, and enforce this adequately.
Demonstrate	Demonstrate that you are processing personal data in a privacy-friendly way.

Privacy-Enhancing Technologies, which precede the use of Privacy-Preserving Technologies as a term, are somewhat different: Privacy-Enhancing Technologies are aimed at improving privacy in existing systems, whereas Privacy-Preserving Technologies are mainly aimed at the design of novel systems and technologies in which privacy is guaranteed. Therefore, Privacy-Preserving Technologies adhere more strongly to the principle of "privacy-by-design".[19] When looking at some of the organisational aspects, we see that developments in big data and AI have also opened new avenues for pushing forward new modes of automated compliance, for instance via sticky policies and other types of scalable and policy-aware privacy protection.[20,21,22]

Other attempts have recently been made to create meaningful overviews or typologies of Privacy-Preserving Technologies, mainly with the goal to create clarity for the industry itself (e.g. via ISO standards) and/or to aid policymakers and SMEs.[23] Approaches are data-centred ("What is the data and where is it?"), actor-centred ("Whose data is it, and/or who or what is doing something with the data?") or risk-based[24] ("What are the likelihood and impact of a data breach?"). The ISO 20889 standard, which strictly limits[25] itself to tabular datasets and the

---

[19]We thank Freek Bomhof (TNO) for this point.

[20]This is one of the main aims of the SPECIAL project.

[21]The BOOST project is developing a European Industrial Data Space based on the IDSA framework, which promotes trust and sovereignty based on certification and usage control policies attached to datasets: https://boost40.eu/

[22]The RestAssured project uses sticky policies to capture user requirements on data protection, which are then enforced using run-time data protection mechanisms. More details can be found at https://restassuredh2020.eu/

[23]See, for instance, the E-SIDES project and the recently started SMOOTH platform project.

[24]See E-SIDES D3.2, page 10.

[25]See ISO standard 20889, introduction (p. VI).

de-identification of personally identifiable information (PII), distinguishes, on the one hand, privacy-preserving techniques such as statistical and cryptographic tools and anonymisation, pseudonymisation, generalisation, suppression and randomisation techniques, and, on the other hand, privacy-preserving models, such as differential privacy, k-anonymity and linear sensitivity. The standard also mentions synthetic data as a technique for de-identification.[26] In many such classifications, there are obvious overlaps, yet we can see some recurring patterns, for example in terms of when in the data value chain certain harms or risks can occur.[27] Such classifications aim to somehow prioritise and map technological and non-technological solutions. Recently, the E-SIDES project has proposed the following classification of solutions to data protection risks that stem from big data analytics: anonymisation, sanitisation, encryption, multi-party computation, access control, policy enforcement, accountability, data provenance, transparency, access/portability and user control.[28] When looking at technical solutions, they are aimed at preserving privacy at the source, during the processing of data or at the outcome of data analysis, or they are necessary at each step in the data value chain (Heurix et al. 2015).

Acknowledging both the needs and the challenges in making such solutions more accessible and implementable (Hoepman et al. 2016), we want to show how some current EU projects are contributing to both the state of the art and to the accessibility of their solutions. A number of research projects in the Horizon 2020 funding programme are working on technical measures that address a variety of data protection challenges. Among others, they work on the use of blockchain for patient data, homomorphic encryption, multi-party computation, privacy-preserving data mining (PPDM[29]), and non-technical measures and approaches such as ethical guidelines, and the development of Data Privacy Vocabularies and Controls Community Group (see W3C working group DPVCG).[30] Moreover, they explore ways of making use of data that are not known to the data provider before sharing them, based on usage policies and clearing house concepts.[31] Table 2 gives an overview of the types of challenges recognised by the BDV PPP projects and the BDVA Strategic Research and Innovation Agenda (SRIA), and the (technological) solutions connected to these challenges.

The following overview provides an insight into current trends and developments in Privacy-Preserving Technologies that have been or are being explored by recent

---

[26] See also https://project-hobbit.eu/mimicking-algorithms/#transport

[27] Although the assumption that data processing activities take place in a sequential way is contestable.

[28] E-SIDES D3.2, page 21.

[29] See, for example, https://web.stanford.edu/group/mmds/slides/mcsherry-mmds.pdf

[30] https://www.w3.org/community/dpvcg/

[31] See IDSA Reference Architecture Model: https://www.internationaldataspaces.org/wp-content/uploads/2019/03/IDS-Reference-Architecture-Model-3.0.pdf

Table 2 Challenges identified by BDVA members

Type of challenge	Solutions
Contradiction between big data innovation and data protection	Linked data, sticky policies
Secure and trusted personal data sharing	(Secure) multi-party computation, self-sovereign identity management, data governance
Processing sensitive (health) data	Blockchain, multi-party computation
(Limits of) anonymisation and pseudonymisation	Homomorphic encryption, differential privacy, data wrapping
Dealing with multiple data sources and untrusted parties	Multi-party computation, data sanitisation and wrapping techniques
Maintaining robust data privacy with utility guarantees	Multi-party computation, federated learning approaches, distributed ledger technologies
Risk-based approaches calibrating data controllers' obligations	Automated compliance, risk assessment tools
Combining different techniques for end-to-end data protection	Integration of approaches, toolboxes, overviews and repositories of Privacy-Preserving Technologies (such as ENISA's self-assessment kit)

research projects and that we see as being key for the future research and development of Privacy-Preserving Technologies.

3.1   Trend 1: User-Centred Data Protection

For many years, the main ideas of what data is or who it belongs to and who controls access to it have been predominantly aimed at service providers, data stores and sector-specific data users (scientific and/or commercial). The end user and/or data subject was (and predominantly still is) taken on board merely by ticking a consent box on a screen, or is denied a service if not complying or if personal data is not provided, via, for instance, forcing users to make an account or to accept platform lock-in conditions. An increasing data-scandal-fed dissatisfaction can be witnessed in society, which in turn also demands different models or paradigms on how we think about and deal with personal data. Technologically, this means that data architectures and logics need to be overhauled. Some of the trends we see revolve around (end) user control. The notion of control in itself is a highly contested concept when it comes to data protection and ownership, as it remains unclear what "exercising control" over one's personal data should actually entail (Schaub et al. 2017). Rather, novel approaches "flip" the logic of data sharing and access, for instance by actualising dynamic consent and by introducing self-sovereign identity schemes based on distributed ledger technologies.[32] Moreover, there are

---

[32]See, for instance, International Data Spaces Association. https://www.internationaldataspaces.org/publications/infografic/

developments to make digital environments more secure by making compliance with digital regulation more transparent and clear. Within the Transforming Transport[33] project, the pilot studies suggested that extra training or assistive tools (i.e. an electronic platform or digital service) should be utilised. These tools and training material will be characterised by a user-friendly natural language on the provided definitions on questions raised. Moreover, the explanations to be offered to everyday users should be easily digestible in comparison to the current legalistic and lengthy documents offered by national authorities, which still do not cover cases extensively. For example, the SPECIAL project aims to help data controllers and data subjects alike with new technical means to remain on top of data protection obligations and rights. The intent is to preserve informational self-determination by data subjects (i.e. the capacity of an individual to decide how their data is used), while at the same time unleashing the full potential of big data in terms of both commercial and societal innovation. In the SPECIAL project, the solution lies in the development of technologies that allow the data controller and the data subject to interact in new ways, and technologies[34] that mediate consent between them in a non-intrusive manner. MOSAICrOWN is another H2020 project that aims at a user-centred approach for data protection. This project aims to achieve its goal of empowering data owners with control over their data in multi-owner scenarios, such as data markets, by providing both a data governance framework, able to capture and combine the protection requirements that can possibly be specified by multiple parties who have a say over the data, and effective and efficient protection techniques that can be integrated in current technologies and that enforce protection while enabling efficient and scalable data sharing and processing. Another running H2020 project, MyHealthMyData (MHMD), aims at fundamentally changing the way sensitive data is shared. MHMD is poised to be the first open biomedical information network, centred on the connection between organisations and individuals, encouraging hospitals to make anonymised data available for open research, while prompting citizens to become the ultimate owners and controllers of their health data. MHMD is intended to become a true information marketplace, based on new mechanisms of trust and direct, value-based relationships between citizens, hospitals, research centres and businesses. The main challenge is to open up data silos in healthcare that are sealed at the moment for various reasons, one of them being that the protection of privacy of individual patients cannot be guaranteed otherwise. As stated by the research team, the "MHMD project aims at fundamentally changing this paradigm by improving the way sensitive data are shared through a decentralised data and transaction management platform based on blockchain technologies".[35] Building on the underlying principle of smart contracts, solutions are being developed that can connect different stakeholders of medical data,

---

[33] See https://transformingtransport.eu

[34] https://www.specialprivacy.eu/images/documents/SPECIAL_D1.7_M17_V1.0.pdf, p 36.

[35] http://www.myhealthmydata.eu/wp-content/themes/Parallax-One/deliverables/D1.1_Initial-List-of-Main-Requirements.pdf, p 6.

allowing for control and trust via a private ledger.[36] The idea behind using blockchain is that it allows for a shared and distributed trust model while also allowing for more dynamic consent and control for end users about how and for which (research) purposes their data can be used.[37] By interacting intensively with the different stakeholders within the medical domain, the MHMD project has developed an extensive list of design requirements for the different stakeholders (patients, hospitals, research institutes and businesses) to which their solutions should (in part) adhere.[38] While patient data is particular, both in sensitivity and in the fact that it also falls under specific healthcare regulations, some of these developments also allow for more generic solutions to alleviate user control. The PAPAYA project is developing a specific component to alleviate user control, named Privacy Engine (PE).[39] The PE provides the data subject with mechanisms to manage their privacy preferences and to exercise their rights derivative from the GDPR (e.g. the right to erasure of their personal data). In particular, the Privacy Preferences Manager (PPM) allows the data subject to capture their privacy preferences on the collection and use of their personal data and/or special categories of personal data for processing in privacy-preserving big data analytics tasks. The Data Subject Rights Manager (DSRM) provides to the data subjects the mechanism for exercising their rights derivative from the current legislation (e.g. GDPR, Article 17, Right to erasure or "right to be forgotten"). In order to do so, the PE allows data controllers to choose how to react to data subject events (email, publisher/subscriber pattern, protection orchestrator). For data subjects, the PE provides a user-centric Graphical User Interface (GUI) to easily exercise their rights. A related technical challenge is how to furnish back-end Privacy-Preserving Technologies with usable and understandable user interfaces. One underlying challenge is to define and design meaningful human control and to find a balance between cognitive load and opportunity costs. This challenge is a two-way street: on the one hand, there is a boundary to be sought in terms of explaining data complexities to wider audiences, and on the other hand there is a "duty of care" in digital services, meaning that technology development should aid human interaction with digital systems, not (unnecessarily) complicate them. Hence, the avenue of automating data regulation (Bayamlıoğlu and Leenes 2018) is of relevance here.

---

[36]http://www.myhealthmydata.eu/wp-content/themes/Parallax-One/deliverables/D6.8_Blockchainanalytics(1).pdf

[37]http://www.myhealthmydata.eu/wp-content/uploads/2018/06/ERRINICTWGBLOCKCHAIN_130618_MHMD_AR_FINAL.pdf

[38]http://www.myhealthmydata.eu/wp-content/themes/Parallax-One/deliverables/D1.1_Initial-List-of-Main-Requirements.pdf from page 15 onwards.

[39]https://www.papaya-project.eu/sites/default/files/papaya/public/content-files/deliverables/PAPAYA_D4_1_Platform_Design_and_Development.pdf, p 113 and onwards.

3.2   Trend 2: Automated Compliance and Tools for Transparency

Some legal scholars argue that the need to automate forms of regulation in a digital world is inevitable (Hildebrandt 2015), whereas others have argued that hardcoding laws is a dangerous route, because laws are inherently argumentative, and change along with society's ideas of what is right, or just (Koops and Leenes 2013). While the debate about the limits and levels of techno-regulation is ongoing, several projects actively work on solutions to harmonise and improve certain forms of automated compliance. When working with personal data, or sharing personal data, different steps in the data value chain (Curry 2016) can be automated with respect to preserving privacy. Data sharing in itself should not be interpreted as unprotected raw data exchange, since there are many steps to be taken in preparing the exchange (such as privacy protection). Following this premise, there are three main possible scenarios for sharing of personal data. The first one proposes to share data to be processed elsewhere, possibly protected using a Privacy-Preserving Technology (e.g. outsourced encrypted data to be processed in a cloud computing facility under Fully Homomorphic Encryption (FHE) principles). The second scenario proposes an information exchange, without ever communicating any raw data, to be gathered in a central position to build improved models (e.g. interaction among different data owners under Secure Multi-party Computations to jointly derive an improved model/analysis that could benefit them all). The third scenario relies on data description exchange at first. Then, when two stakeholders agree on exchanging data upon the description of a dataset (available in a broker), the exchange occurs directly between the two parties in accordance with the usage control policy (e.g. applying restrictions and pre-processing) attached to the dataset as presented by the International Data Spaces Association (IDSA) framework, for instance.[40] Furthermore, it is important to be aware of the trade-offs among data utility, privacy risk, algorithmic complexity and interaction level. The Best Available Technique concept cannot be defined in absolute terms, but rather in relation to a particular task and user context.

One of the challenges in automating compliance is the harmonisation of privacy terminology, both in the back end and the front end of information systems. The SPECIAL project focuses on sticky policies, developing a standard semantic layer for privacy terminology in big data, and dynamic user consent as a solution domain for dealing with the intrinsic challenge of obtaining consent from end users when dealing with big data. Basing their project on former work on architectures for big, open and linked data, they propose a specific architecture. Their approach to user control is via managing lifted semantic metadata[41]: "SPECIAL tries to leverage existing policy information into the data flow, thus recording environmental

---

[40]https://www.internationaldataspaces.org/wp-content/uploads/2019/03/IDS-Reference-Architecture-Model-3.0.pdf

[41]See https://www.specialprivacy.eu/images/documents/SPECIAL_D21_M12_V10.pdf

information at collection time with the information. This is more constraint than the semantic lifting of arbitrary data in the data lake. SPECIAL will therefore not only develop the semantic lifting further, but also develop ways how to register, augment and secure semantically lifted data".[42] The project is investigating the use of blockchain as a ledger to check and verify data(sets) on their compliance to several regulations and data policies. As they state: "The SPECIAL transparency and compliance framework needs to be realised in the form of a scalable architecture, which is capable of providing transparency beyond company boundaries. In this context, it would be possible to leverage existing blockchain platforms [...] each have their own strengths and weaknesses".[43] Building on existing platforms and solutions, they contribute by looking into automation and formalisation of policy and the coupling of different formal policies semantically. The challenge is, on the one hand, to make end-user rights (rights of companies or individuals) manageable in the context of big data, and, on the other hand, to explore the limits of policy formalisation and machine-readable policies (technically, legally and semantically). Other solutions for automated compliance can be found in, for instance, the PAPAYA project mentioned earlier, in which a privacy engine transforms high-level descriptions to computer-oriented policies, allowing their enforcement in subsequent processes to only permit the processing of the data already granted by the data subject (e.g. filtering and excluding certain personal attributes). BOOST is another example of a project developing automated compliance (once stakeholders are certified) and transparency tools (dynamic management of participant attributes, clearing house) based on the IDSA framework. BOOST aims to construct a European Industrial Data Space (EIDS), enabling companies to use and exchange more industrial data to foster the introduction of big data in the factory.[44] The EIDS relies on secured and monitored connectors deployed on every participant's facilities where data is hosted and made available for exchange.

All such solutions aim to translate and automate legal text into computer language, and then back again to some form of human control or intervention to tweak parameters in the computer language translation of legal requirements of compliance. This is a highly complex task, and, as we have seen with the cookie-law example (Leenes and Kosta 2015), not always easily implemented or well received. Yet we need to keep pushing such efforts in order to better understand the interaction between big data utility, human experience and interpretation of what personal data and privacy mean, and current and future privacy regulation.[45]

---

[42]https://www.specialprivacy.eu/images/documents/SPECIAL_D3.1_M6_V10.pdf, p. 12.

[43]See https://www.specialprivacy.eu/images/documents/SPECIAL_D2.4_M14_V10.pdf, p. 8.

[44]https://boost40.eu/wp-content/uploads/2018/02/boost_leaflet.pdf

[45]See also the DECODE project: https://decodeproject.eu/

3.3   Trend 3: Learning with Big Data in a Privacy-Friendly and Confidential Way

Several projects are working on ways to cooperate without actually sharing data. Projects such as Bigmedilytics, SODA (Scalable Oblivious Data Analytics) and Musketeer are developing and/or applying approaches to data analytics that fall under the header of (secure) Multi-party Computation. Although multi-party computation is not one technology, but rather a toolbox of different technologies, the main idea of multi-party computation is to share analytics or outcomes of analytics rather than to share data. This can be achieved by developing trust mechanisms based on encryption or data transformation to create a shared computational space that acts as a trusted third party. Where formerly such a third party needed to be some form of a legal entity, now this third party can be a computational, transformed space. The advantage of such a space is that only aggregated data or locally computed analyses are shared; this makes it possible to work together with trusted and less trusted parties without sharing one's data. There are downsides as well at the moment: multi-party computation does not work well for all data manipulations and it negatively affects performance.

One of the projects working on multi-party computation is PAPAYA. The main aim of the PAPAYA project is to make use of advanced cryptographic tools such as homomorphic encryption, secure two-party computation, differential privacy and functional encryption, to design and develop three main classes of big data analytics operations. The first class is dubbed privacy-preserving neural networks, in which PAPAYA makes use of two-party computation and homomorphic encryption to enable a third-party server to perform neural network-based classification over encrypted data. The underlying neural network model is customised in order to support the actual cryptographic tools: the number of neurons is optimised and the underlying operations mainly consist of linear operations and some minor comparison. Although the developed model differs from the original one, it is ready to support cryptographic tools in order to ensure data privacy while still keeping a good accuracy level. Furthermore, the project also focuses on the training phase and investigates a collaborative neural network training solution based on differential privacy. A second proposed solution is privacy-preserving clustering: PAPAYA investigates algorithms that consist of regrouping data items in k clusters without disclosing the content of the data. The project specifically focuses on trajectory clustering algorithms. Partially homomorphic encryption and secure two-party computation are the main building blocks to develop privacy-preserving variants of such clustering algorithms. The third area is privacy-preserving basic statistics. The project is developing privacy-preserving counting modules which make use of functional encryption to enable a server to perform the counting without discovering the actual numbers. The result can only be decrypted by authorised parties.

The SODA (Scalable Oblivious Data Analytics) project[46] aims to enable practical privacy-preserving analytics of information from multiple data assets, also making

---

[46] https://www.soda-project.eu/

use of multi-party computation techniques. The main problems addressed include privacy protection of personal data and protection of confidentiality for sensitive business data in analytics applications. This means that data does not need to be shared, only made available for encrypted processing. So far, SODA has been working on pushing forward the field of multi-party computation. In particular, they work on enabling practical privacy-preserving data analytics by developing core multi-party computation protocols and multi-party computation-enabled machine learning algorithms. The project also considers the combination of multi-party computation and Differential Privacy to enable the protection of (intermediate) results of multi-party computation. The aforementioned innovations are incorporated in multi-party computation frameworks and proof of concepts. They address underlying challenges such as compliance with privacy legislation (GDPR) requirements, willingness of individuals and organisations to share data, and reputation and liability risk appetite of organisations. SODA analyses user and legal aspects of big data analytics, using multi-party computation as a technical security measure under the GDPR, whereby encrypted data is to be considered de-identified data.

The Musketeer project aims at developing an open-source Industrial Data Platform (IDP) instantiated in an interoperable, highly scalable, standardised and extendable architecture, efficient enough to be deployed in real use cases. It incorporates an initial set of analytical (machine learning) techniques for privacy-preserving distributed model learning such that the usage of every user's data fully complies with the current legislation (such as the GDPR) or other industrial or legal limitations of use. Musketeer does not rely on a single technology; rather, different Privacy Operation Modes will be implemented. Machine learning algorithms will be developed on the basis of different Privacy Operation Modes. These Privacy Operation Modes have been designed to remove some privacy barriers and each one describes a potential scenario with different privacy preservation demands and with different computational, communication, storage and accountability features. To develop the Privacy Operation Modes, a wide variety of standard Privacy-Preserving Technologies will be used, such as federated machine learning, homomorphic encryption, differential privacy or multi-party computation, also aiming at developing new ones or incorporating others from third parties in the future. Upon definition of a given analytic task, the platform will help to identify the Best Available Technique to be selected among the Privacy Operation Modes, thereby facilitating the usage of the platform especially for non-expert users and SMEs. Security and robustness against attacks will be ensured, not only with respect to threats external to the data platform, but also internal threats, through early detection and diminishment of the potential misbehaviours of IDP members. To further foster the development of a user data economy based on the data value (ultimately enabling data- and AI-driven digital transformation in Europe), the project will explore reward models capable of estimating the contribution of a user's data to the improvement of a given task, such that a fair monetisation scheme becomes possible.

Having provided an overview of cutting-edge trends and directions of the field of Privacy-Preserving Technologies, we will now mention some key challenges regarding the development, scaling and uptake of solutions developed by these projects.

3.4 Future Direction for Policy and Technology Development: Implementing the Old & Developing the New

Looking at the origins of Privacy-Preserving Technologies, they are technologies to re-establish trust that was broken by technology in the first place. There are inherent risks in technological "solutionism", such as getting into an arms race between novel harm-inducing technologies and trying to find remedies. Also, many technological solutions for data protection themselves need personal data or some form of data processing in order to protect that same data and/or data subject. This bootstrapping problem is well known, and hence other solution domains have gained traction (such as organisational, ethical and legal measures[47]). Yet here there is also an increased interaction with, and demand for, novel remedying technologies: the GDPR has placed unique demands on implementing privacy-by-design and privacy-by-default solutions, which are entirely or in part technological. In the wake of AI, we also see the field of explainable AI (XAI[48]) turning to technical measures to explain or make apparent automated decision-making. In short, we need technical solutions to fix what is broken in present-day information societies, and/or to prevent novel harm. In the wake of recent H2020 calls, the timing seems adequate to take stock of what is already available and what is being developed for the near future. Moreover, the work needed in the research, development, implementation and maintenance of Privacy-Preserving Technologies reflects a growing market and an increased number of stakeholders working in the field of privacy and data protection.

The GDPR requires national data protection authorities from every EU member state to consult and agree as a group on cases for using specific datasets required by big data technologies. Several pilots that are running in the Transforming Transport project came across fragmented policies regarding GDPR across Europe, and thus they experienced an imbalance between the different interpretations of (the protection of) privacy rights. It is currently difficult for the industry to define personal data and the appropriate levels of privacy protection needed in a sample dataset. Such pilots provide the opportunity to give feedback to policymakers and influence the next version of the GDPR and other data regulations. Uncertainty about the interpretation of the GDPR also affects service operators in acquiring data for accurate situational awareness, for example. For instance, vehicle fleet operators may be

---

[47] See also E-Sides deliverable 3.2: https://e-sides.eu/resources/deliverable-d32-assessment-of-existing-technologies

[48] See, for instance, https://www.darpa.mil/program/explainable-artificial-intelligence

reluctant to provide data on their fleet to service operators since they are not certain which of the data is personal data (e.g. truck movements include personal data when the driver takes a break).[49] Due to such uncertainties, many potentially valuable services are not developed and data resources remain untapped.

There is an inherent paradox in privacy preservation and innovation in big data services: start-ups and SMEs need network effects, and thus more (often personal) data, in order to grow, but also have in their start-up phase the fewest means and possibilities to implement data protection mechanisms, whereas larger players tend to have the means to properly implement Privacy-Preserving Technologies, but are often against such measures (at the cost of fines that, unfortunately, do not scare them much so far). In order to make the Digital Single Market a space for human values-centric digital innovation, Privacy-Preserving Technologies need to become more widespread and easier to find, adjust and implement. Thus, we need to spend more effort in "implementing the old". While many technological solutions developed by the projects mentioned above are state of the art, there are Privacy-Preserving Technologies that have existed for longer and that are at a much higher level of readiness.

Many projects aim to develop a proof of principle within a certain application domain or case study, taking into account the domain-specificity of the problem, also with the aim of collecting generalisable experience that will lead to solutions that can be taken up in other sectors and/or application domains as well. The challenges of uptake of existing Privacy-Preserving Technologies can be found in either a lack of expertise or a lack of matchmaking between an existing tool or technology for privacy preservation and a particular start-up or SME looking for solutions while developing a data-driven service. A recent in-depth analysis has been made by the E-SIDES project on the reasons behind such a lack of uptake, and what we can do about it.[50] They identify two strands of gaps: issues for which there is no technical solution yet, and issues for which solutions do exist but implementation and/or uptake is lagging behind.[51] In addition to technical expertise, budget limitations or concerns that may prevent the implementation of Privacy-Preserving Technologies play a major role, as well as cultural differences in terms of thinking about privacy, combined with the fact that privacy outcomes are often unpredictable and context-dependent. The study of E-SIDES emphasises that the introduction of privacy-preserving solutions needs to be periodically reassessed with respect to their use and implications. Moreover, the ENISA self-assessment kit still exists and should perhaps be overhauled and promoted more strongly.[52]

When it comes to protecting privacy and confidentiality in big data analytics without losing the ability to work with datasets that hold personal data, the group of

---

[49]See, for example, https://www.big-data-value.eu/transformingtransport-session-and-policy-work shop-at-the-ebdvf-2018/

[50]https://e-sides.eu/assets/media/e-sides-d4.1-ver.-1.0-1540563562.pdf

[51]See https://e-sides.eu/resources/white-paper-privacy-preserving-technologies-are-not-widely-inte grated-into-big-data-solutions-what-are-the-reasons-for-this-implementation-gap

[52]https://www.enisa.europa.eu/publications/pets-controls-matrix/pets-controls-matrix-a-system atic-approach-for-assessing-online-and-mobile-privacy-tools

technologies that falls under multi-party computation seems a fruitful contender. However, at the moment, the technology remains at the lower ends of TRL levels. As one SODA project member outlined, uptake of multi-party computation solutions in the market is slow. Many activities in the project are aimed at increasing uptake of multi-party computation solutions: "To bring results to the market we incorporate them in the open source FRESCO multi-party computation framework[53] and other software and we use them in our SME institute consulting business or spinoff thereof. Thirdly, we adopt them internally in our large medical technology enterprise partner, and we advocate multi-party computation potential and progress in the state of the art to target audiences in areas of data science, business, medical and academia". The main barriers the project sees for adoption of multi-party computation solutions on a large commercial scale relate to, among others, "the relative newness of the technology (e.g. unfamiliarity, software framework availability and maturity) as well as the state of the technology that needs to develop further (e.g. performance, supported programming constructs and data types, technology usability)". As a main message to policymakers, they state that: "Policy makers should be aware that different Privacy-Preserving Technologies are in different phases of their lifecycle.[54] Many traditional Privacy-Enhancing Technologies are relatively mature and benefit mostly from actions to support adoption whereas others (e.g. multi-party computation) would benefit most from continuing the strengthening of the technology next to activities to support demonstration of its potential and enable early adoption".[55] This connects to the call made by ENISA to (self-)assess Privacy-Preserving and Privacy-Enhancing Technologies via a maturity model in order to develop a better overview of the different stages of development of the different technologies.

4   Recommendations for Privacy-Preserving Technologies

From the three trends mentioned above we formulate the following recommendations.

Development of Secure Data Storage Spaces
The growing use of digital services is pressing technologists to find privacy engineering solutions to alleviate the general concerns on privacy. The GDPR, among others, aims at providing legal assurances concerning the protection of personal data,

---

[53]https://github.com/aicis/fresco

[54]This point has been acknowledged by ENISA, who have developed a "Privacy-Enhancing-Technology self-assessment" toolkit in order to self-assess the market-readiness, or maturity, of a particular technical solution – see https://www.enisa.europa.eu/publications/pets-maturity-tool/at_download/fullReport

[55]Based on an interview with SODA researcher Paul Koster, Senior Scientist, Digital Security, Data Science, Philips Research.

while an increasing number of frameworks, tools and applications demand personal data. On the one hand, laws and regulations for guaranteeing privacy, for protecting personal data and for ensuring usable digital identities have never been so rigorous, but on the other hand, compliance with the GDPR and other relevant data regulation remains challenging with today's threat landscape, making the risks of data breaches larger than ever. The GDPR imposes a number of onerous cybersecurity and data breach notification obligations on organisations across Europe, with strong enforcement power for data protection authorities, and this generates a frightening situation for companies when it comes to working with (big) data. Beyond engineering solutions, which already exist, another business opportunity is opening up: secure data storage environments (which may be part of personal, industrial or even hybrid data platforms). These are digital environments that are topic oriented, linked and certified by data protection authorities, offering the possibility to train algorithms that need to be trained on real data while offering guarantees of IPR protection and making sure that databases in these environments are accurate. Within experiments and testing phases, such secure environments would exempt the enterprises that need data from the responsibility to prove that they have all the necessary security measures in accordance with the legal precepts. Combined with such approaches, lessons learnt from cases and best practices should feed into the updating of current data policies according to the use cases in the different industrial sectors. This would allow Europe to move forward in making business from AI/ML taking into account Privacy-Preserving Technologies.

Continued Support for Research, Innovation and Deployment of Privacy-Preserving Technologies
As stated above, the E-SIDES project has performed an in-depth gap analysis concerning the uptake of Privacy-Preserving Technologies. One of the main challenges identified and broadly underlined by the BDV PPP stakeholders that participated in this chapter is that of scalability. The main argument here, as also posed earlier by the E-SIDES project, is that the uptake of Privacy-Preserving Technologies suffers from a bootstrapping problem: the more certain solutions are used, the better they become; but in order for companies and SMEs to start using them, they need to be good (i.e. robust, verified, standardised, known in the industry, etc.). Many types of solutions emerge from research and development communities in privacy engineering. Within privacy engineering, solutions can come from community-identified problems that emerge during the development of digital services; they can come from dedicated programmes in which solutions are pitched for known and existing problems in society; or they can originate from demands posed by regulation of a certain digital technology. Without active developer communities and without support to get solutions and ideas from these communities into the real world, many potential solutions will never come to fruition. As such, more efforts into community building and support is necessary, combined with strengthened research and innovation actions to develop solutions that address the communities' requirements. There are already many efforts to strengthen the connection between large enterprises, SMEs and R&D in privacy engineering and the

implementation of Privacy-Preserving Technologies.[56] However, this still requires significant knowledge and awareness about data processing, big data analytics and data protection issues. Already existing infrastructures such as Digital Innovation Hubs[57] and Big Data Centres of Excellence[58] could also act as knowledge transfer centres for education, implementation and expertise for Privacy-Preserving Technologies, although for now Privacy-Preserving Technologies are not their main focus. Continuous efforts should be provided to develop training material, tutorials and tool support (e.g. libraries, open-source components, testbeds) and to incorporate them into formal and non-formal education. Highlighting and following best practices of implementation of Privacy-Preserving Technologies per sector would be a good way to allow companies to learn from – and improve – Privacy-Preserving Technology uptake.

Support and Contribution to the Formation of Technical Standards for Preserving Privacy

Different applications of big data technologies lead to different types of potential harm that require different responses and technological measures. Whereas we have provided a high-level overview of privacy (and confidentiality) threats and corresponding technical solution areas, more work is needed to capture, understand and communicate which type of solution fits a particular problem. This would benefit data-driven companies, start-ups and SMEs tremendously. The work done by ISO standardisation bodies and others that tackle the challenge of classification of technologies is crucial in understanding, shaping and prioritising challenges and solutions in the field of privacy engineering. The sanitisation efforts by projects mentioned earlier also push forward the creation of a common privacy language and semantics between machine and human language. This is a necessary step for automating compliance and for preparing good data for AI.[59] We need to continue work on maturity modelling and to support an EU-driven marketplace for Privacy-Preserving Technologies. Moreover, we need to keep supporting efforts to increase the development and implementation of technological standards around Privacy-Preserving Technologies. In terms of privacy regulation, despite the complexities and difficulties regarding its implementation, the GDPR can still be seen as a major step to strengthen protection of personal data for individuals. However, there is still uncertainty about the practical implications of the GDPR, also in combination with other data-related regulation (as such, the GDPR is merely one piece in the data-regulation puzzle). If risks to Europe's technology industry and big data strategy materialise in a significant way and aspects of the GDPR weaken competition and

---

[56]See, for instance, the SMOOTH project: https://smoothplatform.eu/

[57]https://ec.europa.eu/digital-single-market/en/digital-innovation-hubs

[58]http://www.bdva.eu/node/544

[59]See https://www.mckinsey.com/featured-insights/europe/ten-imperatives-for-europe-in-the-age-of-ai-and-automation

competitiveness, lawmakers should not hesitate to make necessary adjustments, wherever possible.[60]

Acknowledgements We thank the following contributors: Rosa Araujo (Eurecat, Spain), Alberto Crespo Garcia (Atos Spain S.A., Spain), Ariel Farkash (IBM), Antoine Garnier (IDSA, Germany), Akrivi Vivian Kiousi (INTRASOFT Intl, Greece), Paul Koster (Philips, the Netherlands), Antonio Kung (Trialog, France), Giovanni Livraga (Università degli Studi di Milano, Italy), Roberto Díaz Morales (Tree Technology S.A., Spain), Melek Önen (EURECOM, France), Ángel Palomares (Atos Spain S.A., Spain), Angel Navia Vázquez (Univ. Carlos III de Madrid, Spain) and Andreas Metzger (paluno, Germany). Research leading to these results received funding from the European Union's Horizon 2020 research and innovation programme under grant agreement no. 732630 (BDVe).

References

Bayamlıoğlu, E., & Leenes, R. (2018). The 'rule of law' implications of data-driven decision-making: a techno-regulatory perspective. *Law, Innovation and Technology*, 1–19. doi:https://doi.org/10.1080/17579961.2018.1527475

Cavoukian, A. (2008). Staying one step ahead of the GDPR: Embed privacy and security by design. *A Peer-Reviewed Journal, 2*(2).

Curry, E. (2016). The big data value chain: Definitions, concepts, and theoretical approaches. In J. M. Cavanillas, E. Curry, & W. Wahlster (Eds.), *New horizons for a data-driven economy: A roadmap for usage and exploitation of big data in Europe*. New York: Springer. https://doi.org/10.1007/978-3-319-21569-3_3

De Hert, P., Papakonstantinou, V., Malgieri, G., Beslay, L., & Sanchez, I. (2018). The right to data portability in the GDPR: Towards user-centric interoperability of digital services. *Computer Law & Security Review, 34*(2), 193–203.

Gardner, K. C., Broda, T., Jackson, T. C., Solnit, M., Sharma, M., Bubenheim, B., & Cosby, K. (2017). *Patent No. 9,720,569*. USA: U.S. Patent and Trademark Office.

Gellert, R. (n.d.). For an explanation and discussion on the risk-based approach in the GDPR, see Gellert, R. (2018). Understanding the notion of risk in the General Data Protection Regulation. *Computer Law & Security Review, 34*(2), 279–288.

Gürses, S., Berendt, B., & Santen, T. (2006). Multilateral security requirements analysis for preserving privacy in ubiquitous environments. *Proceedings of the UKDU Workshop*, 51–64.

Heurix, J., Zimmermann, P., Neubauer, T., & Fenz, S. (2015). A taxonomy for privacy enhancing technologies. *Computers & Security, 53*. https://doi.org/10.1016/j.cose.2015.05.002

Hildebrandt, M. (2015). *Smart environments and the end(s) of law. Novel entanglements of law and technology*.

Hoepman, J.-H. (2020). *Privacy design strategies*.

Hoepman, J.-H., Jensen, M., & Schiffner, S. (2016). *Readiness analysis for the adoption and evolution of privacy enhancing technologies*.

Kerr, O. S. (2012). *The Mosaic Theory of the Fourth Amendment. Rev. 311*.

Koops, B.-J., & Leenes, R. (2013). Privacy regulation cannot be hardcoded. A critical comment on the 'privacy by design' provision in data-protection law. *International Review of Law, Computers & Technology, 28*, 1–13. https://doi.org/10.1080/13600869.2013.801589

Leenes, R., & Kosta, E. (2015). Taming the cookie monster with Dutch law – A tale of regulatory failure. *Computer Law & Security Review, 31*. https://doi.org/10.1016/j.clsr.2015.01.004

---

[60]See also the recent policy briefs by the Transforming Transport project mentioned earlier.

Mann, Z. Á., Salant, E., Surridge, M., Ayed, D., Boyle, J., Heisel, M., et al. (2018). *Secure data processing in the cloud. Advances in service-oriented and cloud computing: Workshops of ESOCC 2017* (pp. 149–153). New York: Springer.

Newman, N. Fletcher, R. Kalogeropoulos, A., Levy, D., & Nielsen, R. K. (2017). *Reuters Institute Digital News Report 2017.*

Purtova, N. (2018). The law of everything. Broad concept of personal data and future of EU data protection law. *Law, Innovation and Technology, 10*(1), 40–81.

Schaub, F., Balebako, R., & Cranor, L. (2017). Designing effective privacy notices and controls. *IEEE Internet Computing, 21*, 70–77. https://doi.org/10.1109/MIC.2017.75

Stojmenovic, I., Wen, S., Huang, X., & Luan, H. (2016). An overview of Fog computing and its security issues. *Concurrency and Computation: Practice and Experience, 28*(10), 2991–3005.

Zarsky, T. Z. (n.d.). *See the personal data definition in the GDPR and its incompatibility as described in, for example: Zarsky, T. Z. (2016). Incompatible: The GDPR in the age of big data. Seton Hall L. Rev., 47, 995.*

Zillner, S., Curry, E., Metzger, A., Auer, S., & Seidl, R. (Eds.). (2017). *European big data value strategic research & innovation agenda.* Retrieved from Big Data Value Association website www.bdva.eu

A Best Practice Framework for Centres of Excellence in Big Data and Artificial Intelligence

Edward Curry, Edo Osagie, Niki Pavlopoulou, Dhaval Salwala, and Adegboyega Ojo

Abstract This chapter presents a best practice framework for the operation of Big Data and Artificial Intelligence Centres of Excellence (BDAI CoE). The goal of the framework is to foster collaboration and share best practices among existing centres and support the establishment of new Centres of Excellence (CoEs) within Europe. The framework was developed following a phased design science process, starting from a literature review to create an initial framework which was enhanced with the findings of a multi-case study of existing successful CoEs. Each case study involved an in-depth analysis and a series of in-depth interviews with leadership personnel of existing CoEs.

The resulting best practice framework models a CoE using open systems theory that comprises input (environment), transformation (CoE) and output (impact). The framework conceptualises the internal operation of the CoE as a set of high-level capabilities including strategy, governance, structure, funding, and people and culture. The core capabilities of the CoE include business development, collaboration, research support services, technical infrastructure, experimentation/demonstration platforms, Intellectual Property (IP) and data protection, education and public engagement, policy outreach, technology and knowledge transfer, and performance and impact assessment. In this chapter we describe the best practice framework for CoEs in big data and AI, including objectives, environment, strategic and operational capabilities, and impact. The chapter outlines how the framework can be used by a CoE to support its strategic direction and operational decisions over time, and how a new CoE can use it in the start-up phase. Based on the analysis of the case studies, the chapter explores the critical success factors of a CoE as defined by a survey of CoE managers. Finally, the chapter concludes with a summary.

Keywords Centres of Excellence · Research management · Research organisational design · Research capabilities

E. Curry (✉) · E. Osagie · N. Pavlopoulou · D. Salwala · A. Ojo
Insight SFI Research Centre for Data Analytics, NUI Galway, Galway, Ireland
e-mail: edward.curry@nuigalway.ie

© The Author(s) 2021
E. Curry et al. (eds.), *The Elements of Big Data Value*,
https://doi.org/10.1007/978-3-030-68176-0_8

1   Introduction

This chapter presents a best practice guide for the operation of Big Data and Artificial Intelligence Centres of Excellence (BDAI CoE). The goal of the guide is to foster collaboration and share best practices among existing Centres of Excellence (CoEs) and support the establishment of new CoEs within Europe.

The best practice guide is conceptualised as a framework to capture appropriate practices for operating a BDAI CoE. The framework was developed following a phased design science process, starting from a literature review to create an initial framework which was enhanced with the findings of a multi-case study of existing successful CoEs. Each case study involved an in-depth analysis and a series of in-depth interviews with CoE leadership.

The resulting best practice framework models a CoE using open systems theory (Von Bertalanffy 1968) that comprises input (environment), transformation (CoE) and output (impact). The framework conceptualises the internal operation of the CoE as a set of high-level capabilities including strategy, governance, structure, funding, and people and culture. The core capabilities of the CoE include business development, collaboration, research support services, technical infrastructure, experimentation/demonstration platforms, Intellectual Property (IP) and data protection, education and public engagement, policy outreach, technology and knowledge transfer, performance and impact assessment.

Initial insight from this work indicates that there is a wide range of practices that are needed to operate a BDAI CoE successfully. Some practices (governance, financial management, human resources) in the BDAI CoE environment are arguably the same as found in traditional businesses. However, other practices are unique to a BDAI CoE and are substantially different from conventional business practices. In particular, collaboration is a crucial practice between the CoE and industry players, balancing the need for scientific advancement and the transfer of technology to the industry.

The rest of the chapter is structured as follows: the Sect. 2 details what a CoE is and how it plays a fundamental role in the creation and sharing of research and innovation within the local and national innovation ecosystems. Section 3 sets out the methodology used in the design and refinement of the framework. In the Sect. 4, we describe the best practice framework for CoEs in big data and AI, including objectives, environment, strategic and operational capabilities, and impact. Section 5 outlines how the framework can be used by a CoE to support its strategic direction and operational decisions over time, and how a new CoE can use it in the start-up phase. Section 6 explores the critical success factors of CoE as defined by a survey of CoE managers. Finally, the chapter concludes with a summary.

2   Innovation Ecosystems and Centres of Excellence

To understand the essence and nature of a CoE, it is essential to understand the wider setting and context in which the CoE is situated. To this end, we introduce the key elements of the innovation ecosystem in which CoEs exist to understand their role within the national innovation ecosystem and the broader technological ecosystem.

National Innovation Ecosystems constitute networks of public and private sector institutions that generate value from the development and applications of new technologies. They play a crucial role in the socio-economic development of countries (Mowery et al. 1993; Fagerberg and Srholec 2008).

In this chapter, we focus on the networks around big data and AI technologies and their roles in the creation and sustainability of CoEs. In particular, our interest lies in the national or pan-European innovation systems that have a significant investment regarding funding and workforce directed towards addressing the challenges and leveraging the opportunities of big data and AI. We focus on the concept of the CoE to identify the characteristics of the thriving organisations, mainly public sector and universities that are leading the technological developments around big data and AI in Europe.

In natural ecosystems, smart organisms control their energy. In business ecosystems, smart companies manage their information and information flows (Kim et al. 2010). Regarding data, the ecosystem metaphor is used to describe the data environment supported by a community of interacting organisations and individuals. Data Ecosystems are formed in different ways around an organisation and community technology platforms, or within or across sectors. Data Ecosystems exist within many industrial sectors where a vast amount of data moves between actors within complex information supply (Cavanillas et al. 2016).

Beyond data, the AI Innovation Ecosystem (Zillner et al. 2020) is complex and diverse. It contains multiple types of stakeholders, and, to be effective, there needs to be alignment and collaboration between them. It is central for the sharing of assets, technology skills and knowledge. It provides a scale to achieve consensus and critical mass around the generation of value through innovation that no single partner alone could achieve. It expresses the collaborative purpose that binds organisations and individuals together in achieving success in deploying AI. A functional data and AI ecosystem must bring together the key stakeholders with clear benefits for all. The key actors in a big data and AI ecosystem (Zillner et al. 2017), as illustrated in Fig. 1, are as follows:

- **End-User:** a person or organisation from the public or private sector that leverages AI technology and services to their advantage
- **Application Provider:** an organisation that uses AI technology for developing a vertical AI application (e.g. to be offered as AI service)
- **User:** a person who either knowingly or unknowingly uses or is impacted by a system product or service that uses AI
- **Data Supplier:** a person or any organisation (public or private) that creates, collects, aggregates and transforms data from both public and private sources

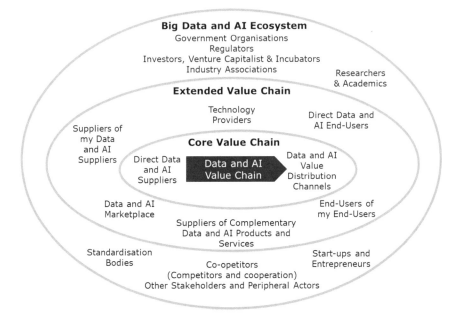

Fig. 1 The micro, meso and macro levels of a big data and AI ecosystem (Curry 2016)

- **Technology Creator:** typically, an organisation (of any size) that creates tools, platforms, services, hardware and technical knowledge
- **Broker:** an organisation that connects the supply and demand for AI assets (such as skills, data, algorithms, and infrastructure) needed for developing AI applications by providing a channel for exchanging AI assets
- **Innovator and Entrepreneur:** drives the development of innovative AI technology, products and services
- **Researcher and Academic:** researches and investigates new algorithms, hardware, technologies, methodologies and business models; provides skills and training in AI and assesses the societal aspects of AI
- **Regulator:** assesses AI systems for compliance with regulation, privacy and legal norms
- **Standardisation Body:** defines technology standards (consensus-based, de facto and formalised) to promote the global adoption of AI technology
- **Investor, Venture Capital and Incubator:** provides resources and services to develop the commercial potential of the ecosystem
- **Citizen:** a person who will or will not develop trust in AI technologies

An effective European AI Innovation Ecosystem facilitates the cross-fertilisation and exchange between participants that lead to new AI-powered value chains that can improve business and society and deliver benefits to people. A productive European AI Innovation Ecosystem is an essential component to overcome key adoption challenges. Within the ecosystem model, researchers and academics play research and innovation roles. Traditionally, within universities, academic departments and schools often work towards the establishment of a specific-purpose CoE to drive a research and innovation mission for big data and AI.

2.1   What Are Centres of Excellence?

Excellence as a concept has many varying definitions depending on the area of focus, that is, whether it is research, development, education or management. It is a complex concept that is difficult to define and operationalise due to its dynamic and multidimensional nature (Schmidt and Krogh Graversen 2017). Hellström states that "excellence is a term for the political and the scientific community: this is because its evaluative dimensions vary within a common theme which most researchers can relate to, and it is often tangible enough for external interests to partake and discuss its implications" (Hellström 2011). According to the OECD (2014), a CoE relates to promoting high-quality scientific research, facilitating basic research through funding, promoting the internationalisation of national research, raising the profile of the host institution through the establishment of a CoE, formulating influential research groups and collaborations, and attracting experts and highly skilled researchers. Another view by Ohno-Machado (2014) relates CoEs to data science skills, technical and policy infrastructure for data acquisition, efficient storage and management, knowledge generation, data security, and privacy protection and sector-wide collaboration. Aksnes et al. (2012) have identified three basic schemes for CoEs in Nordic countries, and these include programmes that focus on scientific excellence, schemes that aim for innovation excellence and programmes that address societal challenges.

Similarly, Hellström (2011, 2012) have developed an analytical framework for analysing CoE schemes according to their strategic orientation, institutional and operational conditions, and impact and capacity building attributes. In this regard, they classify CoE schemes according to the following strategic directions: basic and strategic research, innovation and advanced technological development, and social and economic development. In this context, we define a BDAI CoE as follows:

> "A Big Data and Artificial Intelligence Centre of Excellence is an organisation or organisational unit within a national system of research and education that provides leadership in research, innovation and training for Big Data and AI technologies."

The defining characteristic is its focus on enabling technologies and societal impacts of big data and AI. Within this broad scope, a CoE can serve as a common place for accumulating and creating knowledge that addresses challenges of big data and AI, opens new avenues of knowledge-based economies, guides policy instruments in the era of digital life, and informs the public about the externalities of technological advances based on information processing. Based on context consideration, we use the above-listed classification to categorise BDAI CoEs according to their primary strategic orientations.

3 Methodology

The framework was developed following a phased process, starting from a literature review to create an initial version which was enhanced with the findings of a multi-case study of existing successful BDAI CoEs. The CoEs were selected based on a mix of size, posture (from basic to applied research) and geographical balance.

The production of the framework followed two types of information-gathering exercises carried out on three BDAI CoEs that were selected as case studies. First, literature review (done as desktop research) provided secondary data on each case and, second, a series of interviews with senior managers (12 in total) of the selected CoEs produced primary data, also on each case. The elicited information was reviewed to cross-check correctness with the various sources and to fine-tune it for the best quality, including readability, understandability, navigability, organisation and presentation.

The methodology follows design science principles within a rigorous design process that facilitates the engagement of scholars, as well as ensuring consistency by providing a meta-model for structuring the methodology. The design science approach used is closely aligned with the three design science research cycles (Relevance Cycle, Rigor Cycle and Design Cycle) proposed by Hevner (2007).

In this approach, we had step-by-step activities that began with recognising the problem at hand, followed by statements of objectives to be actualised in the tasks. We engaged in the design and development of the framework. Next, we evaluated the framework, which was followed by the demonstration of how it could be used. Finally, we communicated the framework to users. The steps in the methodology are:

- Identification and motivation of the problem
- Definition of objectives for the framework
- Design and development of the BDAI CoE framework
- Evaluation of the framework
- Demonstration of the use of the framework
- Communication of the framework

A research methodology based on the Delphi method is employed for capturing the best practices and guidelines for CoEs (Linstone and Turoff 1975). The Delphi method is primarily used for forecasting with the help of a panel of experts over multiple iterations. Our methodology uses a two-round approach for capturing and refining best practices and guidelines with the help of a panel of seven CoE managers. The objective of the interviews was to capture the collective intelligence and experience of the interviewees within a framework for BDAI CoEs. The experts on BDAI CoEs were interviewed with participation from several CoEs across Europe.

4   Best Practice Framework for Big Data and Artificial Intelligence Centre of Excellence

The objective of the framework is to develop a best practice guide for use in promoting value generation and sharing of ideas within the big data and AI innovation ecosystem.

> **The goals are to:**
> – Foster collaboration and promote sharing of best practices and know-how among CoEs and national initiatives
> – Provide expert guidance and (non-financial) support to member states looking to establish a new national CoE for big data and AI.

Within the framework as illustrated in Fig. 2, there is a process flow in the form of a value chain starting from the environment (which supplies input) through the core BDAI CoE capabilities (which process the input) to the output represented by the impact of the output received by the society under various categories: economic,

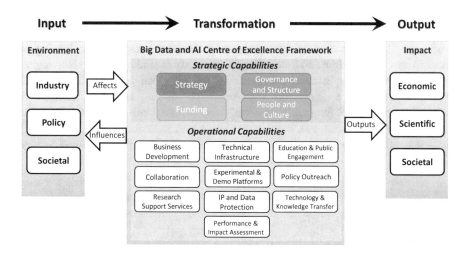

Fig. 2 Framework for Big Data and AI Centre of Excellence

scientific and societal. There is a backward flow (feedback) from the impact of a CoE back to the CoE and to the environment in which the CoE operates. For example, a CoE may hire personnel it trained as postgraduates or receive income from services rendered to a partner, which can return value to the CoE. Similarly, the impact created can influence the environment in which it operates, particularly regarding policymaking and funding decisions. The quality of output from a CoE is often the most significant determinant of funding decisions by funding agencies.

4.1   Environment

As described in the literature on organisational science, the "environment means forces difficult to control from inside that demanded a response" (Weisbord 1976). The external environment comprises forces that initiate organisational change (Burke and Litwin 1992). In the context of a BDAI CoE, the environment is defined as three forces: industry, policy and citizens.

4.1.1   Industry

The term industry refers to companies, start-ups and businesses that are carrying out economic activities related to big data and AI. While the big data and AI industry would directly affect the strategy and performance of a BDAI CoE, the relative strengths or weaknesses of other industrial sectors may be reflected in the core elements of the BDAI CoE framework. A recent Norwegian study indicated that the industry provides increasingly significant financial support (more than doubled since the 1980s) for academic research while the proportion of basic funding is decreasing (Gulbrandsen and Smeby 2005). In a study carried out among Norwegian university professors, a clear relationship exists between industry funding and research performance. Professors with industrial funding are often engaged in applied research and frequently produce entrepreneurial results, they collaborate more with other researchers both in academia and in industry, and they report more scientific publications (Perkmann and Walsh 2007; Gulbrandsen and Smeby 2005).

> **The industry in the context of the BDAI CoE framework is defined as follows:**
> "The ecosystem of companies surrounding a big data and AI Centre of Excellence that is associated with the creation of economic value, at both national and European levels."

Establishing and maintaining strategic industry—research collaborations should be a priority for BDAI CoEs. Inter-organisational network relationships in the context of "open innovation", the role of practices such as collaborative research, university-industry CoEs, contract research, and academic consulting are the basic needs of existing CoEs (Perkmann and Walsh 2007).

The industry demands for big data and AI tools and services drive research focus on the development of these innovative technologies through collaborative research, contract research and consultation services with industry participants. Industry-focused CoEs are highly user-centric in the design of their technologies, and, as such, they work very closely with their end-users to co-design functional solutions to the users' respective challenges.

In the field of big data and AI, CoEs within the EU focus on different domains and trends, while the industries mainly drive decisions within a country. However, international development in science and technology also has an important impact on local trends and decision-making by the management of organisations. For example, within Ireland, the IT, medical and pharmaceutical industries are significant parts of the economy; therefore, data analytics research CoEs focus on providing cutting-edge technology tools and services for these sectors. Centres within economies dominated by petrochemicals focus on the development of data analytics for the digitalisation of oil and gas exploration and related developments in geology domains.

New or emerging CoEs should focus on the areas of interest of the country of operation to align the CoEs' strategic interest with the national strategic interest. This enables the country to provide better funding and policy support for a CoE. As seen from the case studies, where interests diverge, a CoE could run into problems in balancing its priorities. Evidence from the survey indicates that internal capabilities, such as supportive governance, exemplary strategy implementation, the existing units for business development, a simplified IP arrangement process and advanced outreach programmes, are needed to promote university—industry collaborations.

4.1.2 Policy

Policies and regulations can be divided into two broad categories: research and innovation policy and data protection policy. The first policy defines the goals of funding available to CoEs and influences the alignment of the elements within a CoE with those goals. The second policy primarily focuses on clarifying rules about data usage, data ownership, data localisation and data portability (Ron 2016), which are critical to the operation of a CoE.

Policy in the context of the BDAI CoE framework is defined as follows:
"The policy is defined as the set of public laws, regulations and principles that govern research and innovation activities at the national and European level, as well as dictating the access, manipulation and distribution of data."

A dedicated agency or agencies in each country support research activities and provide funding support when needed. The reason for the use of dedicated agencies to fund and support research institutions is that these agencies are specialised in designing arrangements and policies that help to align the research institutions' strategic interests with the country's overall educational system, particularly STEM subjects, research and development, and development of expertise. The agencies help to prioritise areas of research, not just for the country but also among existing CoEs in the country to avoid unnecessary duplication of research effort and funding. The agencies also monitor the performances of CoEs to ensure impacts are up to expectations considering the investment funding provided to them. For example, the Department of Business, Enterprise and Innovation (DBEI) in Ireland has the responsibility of enacting research-related policies and helps in setting national strategic directions regarding stem disciplines, including Science and Technology and Innovation (STI). In addition to the DBEI, the Science Foundation of Ireland (SFI), Enterprise Ireland (EI) and the Industrial Development Authority (IDA) are Irish Government agencies that not only fund Research and Innovation (R&I) development initiatives but also play crucial roles in planning and deciding the direction of the country's technological development, including the development of expertise. Generally, policy formulation fosters academic-industry collaboration as a way to facilitate technology transfer from the academic/research institutions to the industry where research results are applied in practice. Successful CoEs have developed strong working relationships with these agencies to implement policy, but also to shape it.

It is essential for new and existing CoEs to ensure close alignment with funding agencies and national research and innovation agendas. For example, one CoE was aligned with a national digital transformation agenda. As part of the transformation process, the CoE was charged with the research and development initiatives for the CoE of a specific sector of national importance. There could arise a considerable number of funding issues, where a CoE interest fails to align well with the national research agenda that is pursued by the funding agencies.

4.1.3 Societal

Citizens or civil society communities play an important role within the external environment of a BDAI CoE. Social, political and cultural values influence the progress of scientific research and technological innovation in society (Bijker et al. 1987). The state of a societal environment around a BDAI CoE can be assessed using frameworks produced by organisations such as the Organisation for Economic Co-operation and Development (OECD) or the United Nations (UN). In this regard, we use the following three indices: the Human Development Index (HDI), the Global Competitiveness Index (GCI) and the Global Innovation Index (GII).

> **Societal in the context of the BDAI CoE framework is defined as follows:**
> "The societal environment of a BDAI CoE comprises the state of human development as measured by composite statistics and indexes, and the national priorities for human development regarding the UN Sustainable Development Goals and H2020 Societal Challenges."

There is a feedback loop between the societal influence on a CoE and the impact of the CoE's output on the society. Society influences a CoE through various policies and research agenda directives. The societal influences on a CoE include the existing science and technology goodwill of a country, the ability to attract high-level research expertise and industry, and the ability to harness the available expertise and research output. The presence of more expertise and companies enables research institutions to produce higher-quality outputs that are driven by the demand for the output and the availability of quality skills. The identified interdependence between society and research institutions works systematically to sustain the research environment as well as the industrial environment. In this sense, the industrial or corporate entities serve as the user entities for research output, as well as research collaborators providing the problems and challenges for which solutions need to be designed.

Thriving research organisations prioritise the publication of research output, attend international science and technology conferences, and get involved in collaborative research contracts or projects. These are avenues that publicise the inventions of a CoE and add to its popularity, helping it to stand out from the crowd. The CoEs within our study had an excellent national and international record of performances in science and technology development initiatives. The CoEs support the countries' rise in the Global Indicators, which creates a positive feedback loop by attracting an inflow of personnel and companies which further drive quality output.

4.2 Strategic Capabilities

The strategic capabilities of the framework include strategy, governance, structure, funding, people and culture.

4.2.1 Strategy

> **Strategy in the context of the BDAI CoE framework is defined as follows:**
> "The means by which a CoE intends to achieve its overall mission and goals."

A dynamic and innovative research environment has a clear and visible strategy which has been formulated by a senior research and management group (Schmidt and Krogh Graversen 2017). Successful CoEs have well-defined, distinct, narrow-ranged research areas which are unique in their region (or country) (Schmidt and Krogh Graversen 2017). The strategy of a CoE is not limited to corporate body management activity. Unlike companies which define their future goals and can independently plan how they achieve them, CoEs often have research agendas handed down to them by funding agencies in a top-down approach. This commonly results in a situation where CoEs force severe performance challenges, which can create occasional conflicts of interest between a CoE and its funding partners and host university or affiliated educational institutions. The management act of strategising is needed to define goals to be pursued by the CoE and to plan ways to achieve them. Prioritising strategic goals is critical to make the best use of available resources and create a focus on the mission of the CoE.

The BDAI CoE study discovered that the strategy design processes in the studied CoE cases were similar. For example, in all cases studied, the management

- clearly defines and lists its strategic goals and objectives
- tries to align the CoE's strategic goals and mission with national (and European) research goals and objectives
- is market-focused and directed towards industry challenges and demands when designing its future goals and objectives
- is working hard to attain knowledge development and technology transfer to the industry through collaboration with industry partners

On the other hand, there are specific approaches that are different in the case studies.

For example, some CoEs carry out widespread consultations to gather information to formulate strategies. Such consultations included dialogue with stakeholders in the research ecosystem and with their staff, and research and funding organisations and affiliated educational institutions.

Some CoEs break down strategic goals into manageable objectives or activities and use Key Performance Indicators (KPIs) to measure performances towards objectives, goals, mission and vision. These KPIs cover impact areas including economic, commercialisation and academic, and they are operationalised.

Applied CoEs focus on developing a robust interface with industry partners. This approach helps the CoEs to:

- Identify technical, social and cognitive barriers to use of technology
- Define, reinforce and maintain mutual understanding and a shared vision with industry partners
- Track evolving technologies and challenges
- Establish new industrial collaborations

Through this approach, CoEs can identify constraints in existing tools, identify opportunities for changes to transform end-user work practices, and transfer knowledge and expertise via a feedback loop in the innovation cycle. This end-user knowledge allows them to engage in industrial projects and to justify continued basic funding from funding agencies.

Finally, decision-making through consensus of all members at the CoEs on major matters requires holding several meetings and using procedures to prepare and anchor decision-making and to run processes that enable achievement of a consensus.

The BDAI CoE study reveals that in the strategy design process, CoEs consider the following factors in the definition and design of future goals, objectives and priorities:

- Design strategic goals to align with the national research agenda
- Design strategic goals to align with market demands and trends, bearing in mind future needs and developments
- Break down overloaded research agenda from funding institutions into strategic goals and objectives
- Break down goals into manageable activities and measure each with KPIs
- Operationalise the KPIs into daily activities

Strategy Formulation A broad dialogue is necessary to design robust strategies for a CoE. The formulation of the strategy needs to go beyond the senior management group and be inclusive of all stakeholders, including researchers and students. The process of soliciting contributions to strategy design needs to be all-inclusive. For example, one CoE holds an annual general strategy meeting to gather ideas from everyone on how to advance the CoE. It is also crucial that the strategy formulation opens a dialogue with industry stakeholders, host university(s) and researchers from the broader ecosystem. This dialogue with stakeholders is regarded as very important for a CoE's future success as it offers the stakeholders an opportunity to articulate their priorities. For instance, some stakeholders may prefer the development of an international profile, while others suggest the development of national and local priorities.

Alignment of KPIs with Strategy As part of the strategic initiatives of a CoE, the management should strive to design KPIs to measure the performances of their organisation towards the set goals.

In this sense, the CoE's management should operationalise some clearly defined strategies by formulating them into objectives that are measurable using properly designed KPIs. The measurement of those KPIs should be on a regular periodic basis, for example quarterly, bi-annually or annually.

4.2.2 Governance

> **Governance in the context of the BDAI CoE framework is defined as follows:**
> "The means by which a CoE achieves decision-making and operations."

Joynson and Leyser (2015) propose a set of good research practices for high-quality science regarding research governance and integrity, which include training in good research practice, openness about the consequences of misconduct, and adoption of appropriate ethical review processes.

Core to the effective governance of a CoE is a strong governance body and management team. The governance body of a CoE can go by a range of names, which include Governing Council (GC), Centre Steering Committee (CSC) or General Assembly (GA). The composition of the governing body usually consists of both internal and external members. Internal members typically include the CoE's Director or Chief Executive Officer (CEO) and a few top-level officials which could be both academic and non-academic staff. External members can be drawn from industry partners. Despite the similarity in the composition of the governing body, differences exist to some extent. For example, some CoEs include an independent observer, an official from the Technology Transfer Office (TTO), or members of governmental departments.

The governing body of a CoE holds regular meetings, about twice a year in some CoEs and up to three or four times a year in other CoEs. Some CoEs use a Strategy Board to complement the activities of the governing body. The Strategy Board is charged with the responsibility of drafting the strategic goals as well as overseeing the day-to-day operations of the CoE. These boards are composed of the top leadership personnel of the CoE. Often CoEs maintain an Executive Team and together with the CEO of the CoE report the CoE's operations to the GC. In reverse, the GC disseminates its information through the Executive Team to the general members of the CoE. This approach is bottom-up and top-down information dissemination.

The management team of a CoE needs to plan and coordinate research activities, define and prioritise research target areas, and emphasise research productivity and quality (Schmidt and Krogh Graversen 2017). The management team should lead by example by supporting high ethical standards and paying attention to the responsible conduct of research. They should ensure policies that promote being the "best" within the scientific enterprise, and within a context that encourages responsible conduct (Schmidt and Krogh Graversen 2017).

In general, the governing body has the role of making the top-level decisions and approving the strategic goals, objectives and priorities of the CoE. Whatever the composition is, there is a significant value that each member brings to the governing board. For example, an independent observer assumes the role of suppressing biases in judgements or dealing with areas of conflict of interest during decision-making processes. Similarly, the role of the Principal Scientific Investigator in the governing body is to introduce ideas from an in-depth research point of view, which is necessary for delivering research targets.

The bottom-up and top-down information dissemination approach is useful in ensuring accountability, contribution to the decision-making process and an allowance for general inclusivity. It also enables the governing body to monitor the CoE's performances through KPIs.

4.2.3 Structure

Structure in the context of the BDAI CoE framework is defined as follows:
"How a CoE is designed in terms of levels, roles, units, decisions, and accountability."

An appropriate CoE structure depends on the type of institutions and the level of decision-making, as defined by Bleiklie and Kogan (2007)

- CoE as a "republic of scholars": Leadership and decision-making is at the collegial level by independent scholars.
- CoE as a "stakeholder organisation": (1) Institutional autonomy is considered a basis for strategic decision-making by leaders who are assumed to see it as their primary task to satisfy the interests of major stakeholders. (2) The interests of other stakeholders therefore circumscribe academic freedom.

Schmidt and Krogh Graversen (2017) identified that dynamic research environments have flexible organisational structure which may consist of a core researcher group and some attached members or affiliates. Successful CoEs have an organisational structure with high adaptability to internal and external changes.

One of the most critical findings in the case studies is that the structure of a CoE is designed to ensure representation of stakeholders, including host institutions (or affiliate educational institution), industry partners, funding agencies and key staff of the CoE. The structure is designed to facilitate operations and support decision-making and governance that enables coordination and integration of the activities of the CoE for consistency and synergy.

In the design of the structure of a CoE or in guiding the evolving features of the structure, it is important to consider the size of the CoE and the scope of activities. It is also essential to consider the interdependency of the various roles that must work together to optimise resource utilisation to maximise outcomes. Structures enable the efficient running of an entity – the roles, the reporting lines and the accountability for the respective responsibilities. The structure facilitates information dissemination and enforcement of rules and regulations, and thus can also play a key role in the development of suitable cultural practices.

4.2.4 Funding

Funding in the context of the BDAI CoE framework is defined as follows:
"The availability, diversity and sustainability of the monetary support for carrying out research and educational activities in the CoE."

Funding practices for a CoE need to ensure that it is provided with sufficient funding and that it has diverse external funding sources to supplement basic research funding. Funding practices in CoEs with a focus on applied research look to secure funding in the form of collaborative or contract research, with industry partners facilitating technology transfer. Joynson and Leyser (2015) highlight two good research practices for high-quality science through the adoption of diverse funding approaches and the clear communication of funding opportunities and assessment criteria funding that are critical to the recruitment of new researchers, which is a key success factor of a CoE.

From the BDAI CoE case study, the result shows that funding models are provided in a cycle with a fixed period to address specific long-term objectives (e.g. 4, 6 or 8 years). Funding schemes come in mixed models comprising diverse funding sources. A mixed funding model pushes a CoE to explore multiple funding sources such as national, industry (local and international) and European funding sources (e.g. H2020). The industry funding sources could further be broken down into contract research with large multinational companies or with small and medium enterprises (SMEs), as well as with start-up companies. However, there are challenges involved in dealing with SMEs and start-up organisations because of their income level and undefined strategies and goals. Extra funding sources beyond the basic sources usually supplied by funding agencies can also be in the form of services delivered as consulting services by CoEs to other corporate entities or organisations in the not-for-profit sector or even educational sector. The extra funding could also come from national funders that facilitate organisations to sponsor projects financially for a CoE to execute them. In the European Commission (EC), most international funding sources come from EC H2020 and FP7 projects. Participation in projects sponsored by these funding sources in addition to collaborative research with industry partners helps CoEs to obtain extra income to augment their basic funding requirements.

A CoE's sources of funding can be listed as follows:

- Engagement or collaboration with industry partners in collaborative research and consultation services to industry members.
- Participation in EU projects under mainly Horizon 2020 and also FP7 Research and Innovation projects.

- A business development role can be used to develop better engagement capability with the industry partners, which helps a CoE to negotiate more contracts for more income.
- Some CoEs pursue commercialisation and spinouts as well as scientific inventions and publications. However, some are more specialised in the scope of research and innovation development.

Additional funding is often needed to enable a CoE to finance specific interests that the funding agencies may not want to fund. However, funding policy requirements may pose some challenges for a CoE in that it may be required to provide a given amount of its funding needs to become eligible for funding supply from its financiers. For example, one CoE studied needs to provide up to 25% of its funding needs to be eligible for continued funding from funders. This places the management under pressure to collaborate with industrial partners even when it is not a priority to enter into such a contract.

4.2.5   People

People in the context of the BDAI CoE framework are defined as follows:
"The people required to carry out specific tasks towards the goals of the organisation."

CoEs are affiliated to educational institutions, which appear, in most cases, to be the primary sources of personnel supply, particularly CoEs that run academic courses such as master's, PhD and postdoctoral training. In the case of all CoEs, the host universities provide the human resources policies that guide the personnel practices in the CoE.

To gain a broader scope of expertise to bring into their CoEs, the management of research institutions advertise vacancies in both local and international fora, and this enables them to build a range of options into the selection process. CoEs also use some cultural practices:

- To keep their people in a good state of mental health and social well-being (e.g. community volunteering programmes, excursions, walking and cycling activities). If the CoE is not hosted at a single physical site, these can be online activities (e.g. online mindfulness, virtual coffee sessions and online game tournaments, especially during the Covid-19 pandemic).
- To help build their skill and careers through various activities—for example, lunch seminars with invited speakers and on-the-job training of partners' workers on internship programmes. A programme at the CoE combines researchers and partners representatives for cross-fertilisation of skills, weekly meetings featuring occasional invited speakers and thesis programmes involving public speaking.

- To help in the integration of new in-takes through mentoring programmes and to get them up to speed with others.
- To eliminate preferential treatment (e.g. unconscious bias programme) and ensure gender equality and gender mix, programmes like staff diversity, gender equality and women's networking, are organised. In addition, local language training programmes for non-native speakers take place.

To make people feel at home, CoEs use programmes to bring about a feeling of togetherness in a common purpose. For example, one CoE organises an International Cultural Day, which is an event where the different cultures of the various represented nationalities are displayed and celebrated, including the provision of food from various nationalities. A feeling of togetherness can also be achieved through the creation of a friendly environment, where individuals voice their concerns. This helps achieve collaboration and teamwork necessary for productivity in the CoE. Joynson and Leyser (2015) propose a set of good theoretical research practices for high-quality science. These practices include providing adequate training programmes for researchers, being open and clear about consequences of misconduct, and the adoption of appropriate ethical review processes.

4.2.6 Culture

Culture in the context of the BDAI CoE framework is defined as follows:
"The underlying values, beliefs and norms that drive the teams and the CoE as a whole."

Culture is a critical part of the CoE. Schmidt and Krogh Graversen (2017) identified that a successful CoE has a working climate based on internalised norms grounded in a research tradition. The working environment should be open to new ideas, methods and approaches. Staff within the CoE have research autonomy during the research process. The working climate is based on teamwork with close cooperation among research staff. Finally, they identify that culture encourages internal professional and social dialogues.

The case results point to the common fact that most CoEs have a mix of local and international culture. A key question is how CoEs use cultural practices to achieve a spirit of togetherness and inclusivity that reduces conflicts, eliminates preferential treatment and maximises productivity.

The effective use of cultural practices in CoEs helps the management to mitigate problems and helps staff to attain high levels of productivity:

- Integration of new in-take
- Collaboration and teamwork
- Welfare programmes
- Researchers/staff personal skills development

- Inclusivity and voice
- Support for outreach
- Elimination of preferential treatment and achieving gender equality

Culture plays a vital role in the level of interrelationship and interaction existing between people in an organisation. Culture is connected to the degree of collaboration that is possible in an organisation and has a direct impact on the success of the organisation. Culture is developed or guided to evolve into practices that support healthy sharing, caring and support of one another, a situation that enables people in an organisation to feel a sense of togetherness, giving them an opportunity to voice their concerns and contribute to decision-making processes and general shared goals. Like corporate organisations, research institutions also recognise the strong need for good cultural practices in a workplace and how to use their impact to direct success.

The BDAI CoE study reveals that CoEs use various programmes to enhance cultural practices and to make things happen in the way they are desired. For example, in the case of integration of new in-takes, some CoEs use mentoring and orientation programmes to familiarise recruits with their operation and culture. Welfare programmes cater for students and staff to make them feel valued and to get the best out of them for their success and that of the CoE. As they cooperate and collaborate to deliver for the success of their institutions, researchers in research institutions, particularly student researchers, often have personal career development needs. To compensate for their individual needs, leading research institutions provide career and personal development programmes for their workers.

4.3   Operational Capabilities

Operational capabilities in the context of the BDAI CoE framework are defined as follows:
"The operational capability is the ability of a CoE to perform a coordinated set of tasks, utilising organisational resources for the achievement of its mission and goals."

The BDAI CoE framework identifies a set of operational capabilities needed to operate a CoE. These capabilities are detailed in Table 1.

Capabilities maintained by a CoE are partly dependent on their areas of focus and partly conditioned by their need to meet stakeholder demands. There is a wide range of capabilities within the studied CoEs. Some of the highlights from the case studies are as follows:

One CoE exercises an elaborate plan of outreach in the form of Education and Public Outreach (EPE) programmes for which a Subject Matter Expert is employed. The elaborate EPE process is informed by the importance attached to it by the

Table 1 Core operational capabilities of the BDAI CoE framework

Operational capability	Definition
Business development	How the CoE develops new business opportunities and manages its partnerships
Collaboration	How the CoE enhances academic-academic and academic-industrial interactions
Research support services	The local research support services implemented by the CoE
Technical infrastructure	Computing resources used to support the research and innovation activities of the CoE
Experimentation & demonstration platforms	The platforms that support the scientific and innovation activities of the CoE
IP and data protection	How the CoE approaches intellectual property management and data protection
Education and public engagement	How the CoE's dissemination activities inform the public of the science and technology developments
Policy outreach	How the CoE tries to influence future policy
Technology and knowledge transfer	How the CoE drives the transfer of know-how and adoption of its technology
Performance and impact assessment	How the CoE identifies and tracks its performance and impact

government's interest in making the public aware and also taking advantage of science and innovation outcomes.

A CoE with an applied focus to bring the best of services and products to their industry partners adopts a practical process of demonstrating their prototypes contained in a catalogue of demonstrators, IPs and the state-of-the-art analytics and visualisation technology reviews to their partners. This capability brings research outcomes to its network of industry members to which it also delivers services such as seminars, conferences and consultation to create awareness and disseminate information to the end-users of its technologies. The process of garnering collaboration with partners uses two calls for demonstrator proposal. Later, a team filters the proposals received and rates the accepted ones. Finally, the rated proposals are decided upon by the senior management of the CoE, which makes final proposal choices.

Another CoE developed an iterative three-stage process of innovation cycle methodology called Scalable Innovation Cycle (SIC), in which the CoE carries out a user-led generation of ideas and validation of results. The CoE's processes are highly user-centred, and hence it aligns them closely with the end-user-centric methodologies. The goal of this methodology is to combine research with real-world deployment to meet real business problems. Being iterative, SIC requires the use of a series of feedback among pilots, prototypes and experiments to identify new challenges and gaps to perfect results.

The results of these case studies show that there are various capabilities, and these capabilities tend to differ from CoE to CoE depending mostly on their strategic research domain and end-user needs. With this in mind, it is hard to pinpoint one capability as the best approach to research as there are reasons that support the use of individual capabilities in each CoE.

However, whatever capability is in use in a CoE, there is a need for it to be regularly well operationalised and measured for the desired outcome. KPIs should be designed by breaking down a capability into stages of work, and metrics should be put in place to measure performances at each stage over a given time interval or periodically.

4.4   Impact

Impact in the context of the BDAI CoE framework is defined as follows (Harland and O'Connor 2015):
"The direct and indirect 'influence' of research or its 'effect on' an individual, a community or society as a whole, including benefits to the economic, social, human and natural capital."

The definition of the impact metrics and their measurement methods are a significant part of the impact assessment methodology. The following subsections provide guidelines from the literature on how to measure the economic, scientific and societal impact of research output. The impact on the environment and society would be seen in reports of innovation activities derived from field research about impact areas such as economic, scientific and societal. The parameters to understand impacts could be measured through the KPIs being monitored by the BDAI CoE and those monitored by the country government agencies in which the BDAI CoE is located. For example, the economic impact could be how a CoE and industry partnership or collaboration in research and technology is bringing about a measurable increase in commercial activities, companies created through commercialisation, spinouts and jobs creation, and skills development. There are reports which provide a narration of these measures for the government and government agencies to use in support of policymaking for performance review and educational purposes.

4.4.1   Economic Impact

Economic impact in the context of the BDAI CoE framework is defined as follows:
"The economic impact is the effect on commerce, employment, or incomes generated from Big Data and AI research in general and by the CoE in particular."

As described in Adams (2016), the examples of best practices for the assessment of economic impact are:

- Funders need to be sure that job creation is reported consistently across multiple organisations so researchers need an agreed standard such as "full-time equivalent jobs created" to avoid counting part-time roles.
- Claims of impact remain assertions unless there is an independent validation of impact evidence.
- Evaluators require an audit trail to use impact data for evaluation purposes.
- Impact evidence must be collected over time, attributing each impact to original research or expertise and tracing the developing sequences of activities.
- Evidence types can vary widely depending on the discipline, the stakeholders and the changes that have occurred.
- Impact evidence can include quantitative reports of increased sales for a commercial stakeholder or quality of life improvements.
- Qualitative testimonials can directly attribute changes to the research, or the contributions made by researchers because of their expertise.
- Impact information needs a standard structure and categorisation.

A digital research report by Digital Science & Research Ltd. that was released in March 2016 suggests the following best practices for a Research Excellence Framework to improve both the quality and value of future CoEs (Adams 2016):

- To ensure that the full range of meaningful impacts can be recognised, consider extending eligible periods both for impacts and for the research on which they were based.
- Require listing of funders and grant references in the case study template.
- To aid assessment and further use, consider developing guidance on certain types of evidence where appropriate, e.g. sales, staff numbers, company investment.
- Where possible, re-use information from other systems, e.g. ORCID.

4.4.2 Scientific Impact

Scientific impact in the context of the BDAI CoE framework is defined as follows:
"The scientific impact of a CoE is the returns on research investment assessed qualitatively or quantitatively within the academic sphere."

The assessment of the scientific impact of a CoE helps funding agencies to evaluate returns on research investment from a research impact perspective. The scientific result can be assessed qualitatively or quantitatively. An analysis carried out by Sutherland et al. (2011) identifies the following practices for quantifying the impact and relevance of scientific research:

- **Qualitative approaches:** This approach involves expert panels evaluating impact, for example as high, medium or low, based on written descriptions of impact.
- **Quantitative approaches:** This approach involves numerical indicators derived from scoring systems or questionnaires focused on the various possible impacts of a research programme or project.

4.4.3 Societal Impact

Societal impact in the context of the BDAI CoE framework is defined as follows:
"The societal impact of a CoE is its impact on human lives and health, organisational capacities, societal behaviours and the environment."

A variety of frameworks and models are proposed to quantify and measure societal impact (Penfield et al. 2014; Bornmann 2013; Sutherland et al. 2011). Such a variety of frameworks might also be reflected by the impact assessment methods adopted by national funding agencies across Europe. Regardless of the specifics of assessment tools or methods, the underlying objective of assessing societal impact is to understand the social externalities of research and innovation activities undertaken in a BDAI CoE.

Impact on the environment and society can be captured by reporting activities which are conducted by several agencies such as the United Nations Human Development Index (UNHDI), GCI, GII, Knowledge Impact (KI) and Knowledge Fusion (KF) rankings agencies or organisations. These rankings are measurements that also categorise measures into impact areas such as economic, scientific and societal. The parameters to understand impacts could also be measured through some KPIs being monitored by the individual BDAI CoE, on the one hand, and those monitored by the research-funding agencies and other government agencies of the country in which the BDAI CoE is located, on the other hand.

Societal impact can be reached through various practices that CoEs can adopt to influence the relationship between research and society (non-academic community). Societal impacts, as defined by Molas et al. (2002), are part of a conceptual framework for analysing third-stream activities and categorised as follows:

- Research CoEs have capabilities in two main areas: (a) knowledge capabilities and (b) physical facilities. These capabilities are developed as CoEs that carry out their core functions of teaching and research.
- Using the means at their disposal, CoEs carry out three main sets of activities; they (c) do research, (d) teach, and (e) communicate the results of their work.

The type of economic impact a CoE has on the economy in which it exists is dependent on the research areas it specialises in and how that drives economic output. For example, a large-scale CoE may have broad research areas which cut across data analytics applicable in many domains such as media analytics, optimisation and decision analytics. It also participates in other domains such as personal sensing, sustainable IT, e-government, machine learning and Semantic Web. On the other hand, a CoE may have a narrower domain focus with industry-centric capability for producing various data analytics and visualisation tools. Centres may also focus on a single industrial domain. The visible outcome of a CoE does not depend entirely on its output because it also depends on the amount of publicity the CoE has provided on its scientific outcome. Publicity on a CoE's research result is essential in that it helps to create public awareness (locally and internationally) and attract partners for collaboration, creating an avenue for technology transfer.

Conversely, collaboration opportunities previously involved have the potential to bring more opportunities to the CoE because previous engagements serve as an opening for further engagements. This is the cyclical aspect which calls for adequate investment in various ways by which research output can be publicised, and it should include the national agenda of the country in which the CoE is located, as well as the funding agencies' contribution towards publicity and exposure to opportunities. Many countries have put in place policies to drive outreach activities from CoEs to the public, while individual CoEs also make an effort to get involved in presentations at conferences as well as sending entries to scientific publications. Another important consideration for impact is the quality of research output. Good-quality and innovative research output sells itself while bad results fail. This would therefore be a good reason to invest in world-class researchers and infrastructure, in addition to a continuous study of the trends in the markets both in the local and international environment.

Scientific impact is constituted by additions to the state of the art in science and technology which are made known to the public through publications in scientific journals and conferences, as mentioned above. A culture of documentation of research processes and findings on a regular basis can help provide information necessary for preparing articles on the outcome of research endeavours. Documentation should be given priority in research exercises not only for project purposes but also for article writing and presentation at scientific conferences. Societal impact is linked to economic impact with the use of research outcomes in the industry, thereby creating new companies, jobs and economic values which benefit the entire society. Also, societal impact refers to the direct benefit derived by people when they use technology items and when technology helps to create better conditions around them, e.g. reduction in poverty levels and crime and disease control and prevention, as well as helping humanity sustain a greener environment in any way possible.

4.4.4 Impact Measured Through KPIs

Whichever category an impact belongs to, it can be measured through specific indicators that can capture perceivable improvements due to the outcomes of a CoE. KPIs (as described in Table 2) are basic indicators that can be measured with

Table 2 Sample impact KPIs

Economic KPIs	Scientific KPIs	Societal KPIs
• Participation in major EU initiatives • Coordination of major EU initiatives • European Research Council awards • Amount of research income from non-Exchequer, non-commercial sources • Amount of research income from commercial sources • Spinout companies formed • Commercialisation awards • License agreements	• Number of journal publications • Number of conference publications • Impact factors of venues • Number of publication downloads • Number of publications views • Publication citations • Number of European Research Council awards	• Number of master of science (MSc)/master of engineering (MEng) graduates • Number of PhD graduates • Percentage trainee departures with industry as the first destination • Research impacts on UN Sustainable Development Goals • Number of contributions to EPE (e.g. school visits, public seminars, citizen science experiments, dialogue with policymakers, etc.)

defined metrics designed to provide measures of benefits produced regarding economic, scientific and societal advantages. For example, in Ireland, the principal research financing agencies, such as SFI, EI and IDA, have together developed a set of KPIs to measure research performances and their impacts on the nation's goals based on their research outcomes. SFI demands that a research centre's targets be ambitious and achievable and reflect the strategic and commercial positioning of the centre. The centre's targets will therefore be part of the basis for evaluation of the centre's proposal. Also, funded centres' metrics will be reported against defined KPIs and evaluated against the targets on an annual basis (Roche et al. 2013). SFI selected 13 KPIs and used these to score each centre under relevant performance indicators and targets broken down into four categories: academic outputs, human capital outputs, funding diversification and commercialisation. All of these must be aligned with the objectives of the research centres' programmes as well as the overall SFI objectives per Agenda 2020.

SFI evaluates a research centre's performance periodically using evaluation instruments such as the Metrics Governance report and balanced score card, the annual report of the centre, the annual census report (including financial reporting) and site visitations with the external panel (Roche et al. 2013).

5 How to Use the Framework

An assessment of how the capabilities are contributing to the CoE's overall goals and objectives has taken place. This gap analysis between what the CoE wants and what it is achieving positions the framework as a management tool for aligning the operational capabilities of the CoE with its objectives.

The framework focuses on the execution of two key actions:

- Define the goal and posture of the CoE.
- Develop and manage the CoE's strategic and operational capability over time.

Here we outline these actions in more detail and discuss their implementation.

Defining the Scope and Goal First, the CoE must define the scope of its efforts. Agreeing on the desired posture (from basic to applied research) has a significant impact on the CoE and thus on its goals and priorities. Second, the organisation must define the goals of its effort. It is important to be clear on the CoE's objectives and the role of its capabilities in enabling those objectives. Having a transparent agreement between the internal and external stakeholders of the CoE can tangibly help achieve those objectives.

Develop and Manage Strategic and Operational Capabilities Once the scope and goals of the CoE are clear, the CoE must identify its current capabilities by examining across its different operational and strategic functions. This helps the CoE to have a clear view of its current capabilities. Comparisons with the best practices identified within the framework can help identify key areas for action and improvement. To develop capability over time, the CoE should:

- Develop a roadmap and action plan
- Add a yearly follow-up review of capabilities to ensure their fitness for purpose and alignment with the CoE's objectives

Agreeing on stakeholder ownership for each priority area is critical to developing both short-term and long-term action plans for developing and improving capabilities.

The decision to use the BDAI CoE framework to improve operations of a CoE should not involve re-inventing the wheel. The concepts it contains have been theorised and applied extensively by many successful CoE organisations. These concepts and the manner of implementing them can be harnessed to support the development and growth of big data and AI-oriented research entities. The plan to use the BDAI CoE framework may need to incorporate an enhancement of the operations of existing big data and AI CoEs, including the manner of drafting the strategic direction, seeking funding, collaboration, information dissemination and outreach practices. By considering the elements of the BDAI CoE framework, which include strategy, governance and structure, funding, culture and capabilities, it is clear that appropriate practices under each of these elements may need to be (re-) designed into the activities and the general operations of a CoE may need to be performed in achieving the strategic goals of the CoE.

The management team needs to evaluate all factors in the framework, such as environmental, industry and societal, which have a significant influence on the way a CoE may be run. They should consider the needs of its "customers" to know what is currently in demand as well as industry trends. Within such a competitive research and innovation landscape, the management team must decide on the specific value direction the CoE must explore so they can guide the process of resource allocation and talent development or recruitment.

5.1   Framework in Action

The framework is being used by a number of CoEs which contributed to its creation. The CoEs use the framework in different ways, from the training and onboarding of staff to planning the design of new or enhanced organisational capabilities. The framework has also been used to guide the creation of a new CoE. The GATE project[1] was a Horizon 2020 WIDESPREAD-2016-2017 TEAMING Phase 1 programme that aspired to create a sustainable business plan for the creation of the first CoE in big data in Bulgaria. The purpose of this big data centre is to produce excellent science by seamlessly integrating related fields and associating complementary skills. GATE aspired to add value to knowledge, to strengthen the capacity of researchers, to educate and train early-stage researchers, to disseminate and promote projects, and to achieve international visibility and scientific as well as industrial connectivity. With innovation pillars like data-driven government (public services based on open data), data-driven industry (manufacturing and production), data-driven society (smart and sustainable cities) and data-driven science (big data technology stack in the scientific community), GATE had set its aim high to fulfil its goal.

  The framework was used in an advisory capacity in the GATE project by sharing best practices at several meetings and workshops. The framework also supported the CoE to determine their research strategy and business plan. Overall, very positive feedback has been received by the GATE as they built the first big data CoE in Bulgaria, paving the way for more CoEs to start spreading in Eastern Europe in the future. In the words of Professor Sylvia Ilieva (Director GATE CoE) the framework helped the centre "in difficult very first steps of structuring and organisation [and] guided the building of GATE sustainable model on the collective experience and best practices". She continued that it "helped at specialising [the] GATE mission and focus to be complementary, but competitive to the other 55 Centres in Western Europe".

6   Critical Success Factors for Centres of Excellence

Critical success factors are a range of key enablers that CoEs, like corporate bodies, employ to achieve success in their operations. While some are very easily identifiable, e.g. funding availability and a mix of employees' capabilities and cooperation,

---

[1] https://www.gate-coe.eu/

other success factors are not quite salient, e.g. the role of culture in the success of a CoE. Similarly, some success factors are common to a majority of CoEs, e.g. the importance of enough funding towards success, possession of world-class researchers, collaboration with important partners and output publicity. Other factors are peculiar to individual CoEs because certain factors apply to the research focus of a CoE. However, whatever the key success factor is, it is the responsibility of the management team to identify it early enough and to harness it to drive success in the required direction.

This section reports the findings of the BDAI CoE case studies as success factor recommendations for existing CoEs and potential ones for their research operations. These factors are gathered from interviews with the CoEs' senior management using a series of open-ended questions:

1. What are the common difficulties faced by the CoE in achieving its objectives?
2. What factors contribute to/enable the success of the CoE?
3. What are the typical mechanisms deployed to address success factors and challenges in the CoE?
4. What would you need to do to be more successful?

Challenges are the drawbacks to the progress of any organisation, while the success factors facilitate progress. Therefore, the management team of an organisation, according to its mandate, has to devise strategies and practices to eliminate or at least mitigate challenges and other risks to success. Success factors can be leveraged to drive the development of capabilities to meet the CoE's goal.

6.1 Challenges

The key challenges identified in our interviews are detailed in Table 3. They are aligned to the related strategic or operational capability. The list does not have an order of priority.

6.2 Success Factors

The factors with which the CoEs' leadership contribute to their success are detailed in Table 4. They are aligned to the related strategic or operational capability.

6.3 Mechanisms to Address Challenges

The mechanisms deployed by the CoE's leadership to address their challenges are detailed in Table 5. They are aligned to the related strategic or operational capability.

Table 3 Summary of challenges

Challenge	Related capability
To remain a going concern – sustainability in the research industry.	Funding
Ensure essential funding to pursue basic research	Funding
Satisfying high-performance targets for the CoE	Strategy
Encourage more collaboration and partnership arrangements achieved	Collaboration
Lack of autonomy. The lack of a separate legal entity status	Governance
The need to ensure that governance adds value to the CoE's operations creates some concerns	Governance
Competing interests – funders' objectives versus researchers' objectives	Strategy
Human resource availability and retention, e.g. recruitment of PhD-level graduates with significant industry experience, can be a challenge	People and culture
Working with SMEs is challenging due to their resource availability problems, lack of clearly defined objectives and the fact that they often have short-term plans	Collaboration
Physical separation from important partners limits interaction and knowledge of themselves	Collaboration
Facilitation of a flowing, open discussion of technology and solutions between the CoE and industrial partners	Technology and knowledge transfer
Capability and capacity to assure partners that the CoE will help them to solve their challenges	Collaboration
Bridging the knowledge gaps between academic IT, commercial IT and the associated research and business problems	Business development
The need to bridge the gap between people with knowledge of the business problem and those with knowledge of theory	Knowledge transfer
Maintaining a flow of new project contracts and adequacy in project management expertise	Research support services
Industry funding policy demands up to 25%–50% of its funding needs from industry. It creates a challenge of how to balance the interests of researchers with partners	Funding
Aligning portfolio with the strategy to meet partners' demands. This also creates project selection and investment challenges, which often lead to frustration in researchers and industry partners	Structure
Work overload arises from too many activities at the CoE, which is, perhaps, contributed by the funding policy	People and culture
The trade-off between expediency and consensus in making decisions and at the same time gaining staff commitment to achieve the CoE's goals	Structure
Leading knowledge workers who are not driven by ordinary incentives like salaries because they have their own career agendas	People and culture
There is a need for the "cross-pollination" of cultures between research and industry environments	Knowledge transfer

Table 4 Summary of success factors

Success factors	Related capability
Ability to attract grant funding is based on reputation (of both the CoE and individuals) for excellent research outputs	Funding
Local presence of big industry players in tech, medical, pharma, etc. offers opportunities for collaboration and industry funding	Funding
The stock of a talented team of people: the capability to assemble world-class academic talents attracts and satisfies stakeholders	People and culture
Ensuring that the people in the CoE can develop themselves and their careers	People and culture
Effective public outreach that translates science into something easy to understand for non-scientists	Education and public engagement
Maximises outputs by providing (i) *space* (infrastructure and labs) that attract academics, (ii) *money* and (iii) *reputation* of the individual members and the team	Structure
Research turnover ensures that the CoE is fresh and relevant to the industry	People and culture
The more the CoE collaborates and works together, the more successful it will be	Collaboration
Focus on projects that are proposed by industry members. This ensures that what is produced will have an immediate and beneficial impact	Strategy
The produce-for-immediate-impact dynamic is highly motivating for the CoE to get to work on a huge variety of projects across many industries every 6 months	Experimentation & demonstration platforms
Deep collaboration with industry partners provides the CoE with a huge opportunity for success, as it is involved in industry-focused research	Collaboration
The support of the funding agencies is received in two ways – in the form of funding supply and help in the prioritisation of the research agenda	Funding
The CoE is structured to support balancing scientific excellence and supporting business partners	Structure
The CoE supports academic researchers in their career development and the goal of the CoE through operationalisation of both agendas in daily activities. This decision enables a robust structure that allows people to be focused both on their personal needs and the needs of the CoE	People and culture
The committed and hardworking young scientists of international combinations make significant contributions	People and culture
The industrial experience of the management team, which possesses a unique skillset in communication and industry-research collaboration and capability to speak/understand the languages of both the academics and industry	Business development

Table 5 Summary of mechanisms to address challenges

Practice	Related capability
Planning and measuring process: – Development of a strategic plan, an annual appraisal plan and KPIs plan to align with the CoE's goals – Measured and reviewed monthly – Iterative planning process: over time, a plan may need to be reviewed and adjusted because initial factors affecting the plan have changed	Strategy
Communicate the progress of the CoE regularly to all members of the CoE to promote unity and focus on the common goal	Collaboration
Publish a strategic plan and allow people at all levels to engage. This allows people to engage with the vision	People and culture
To help attain very high targets: break the KPIs down into manageable pieces that people can handle	Strategy
Align research agenda with the National Government's science and technology agenda and the goals of industry partners and domain trends	Strategy
Using media publicity on current trends and using the media to create awareness about its research output	Education and public engagement
The CoE maintains a market-focused approach by engaging with industry and other CoE representatives at different events	Business development
Enables funding agencies to help prioritise their research agenda	Funding
Meet with industrial stakeholders twice yearly to deliberate and to set research agenda as well as help in decision-making processes	Collaboration
Arrangement for obtaining IP is quick and straightforward. This attracts industry partners to sign a contract for a collaborative project	IP and data protection
A one-on-one mentorship programme with industry to enrich the CoE's experience in the development of researcher talent	People and culture
A monthly meeting with industry partners' representatives to monitor and discuss the progress of the CoE. Meetings ensure regular engagement of industry partners and increase awareness of industry role in making the CoE a success	Collaboration
Internal meetings (weekly and monthly) enable the management team to get constant visibility of the CoE's internal operations	Collaboration

6.4   Ideal Situation

According to the CoEs' leadership, the ideal conditions for the operation of their CoEs are detailed in Table 6. They are aligned to the related strategic or operational capability.

Table 6 Summary of ideal situations

Ideal situation	Related capability
Separate legal entity status may allow the CoE to evolve into a larger entity to deal with SMEs, become self-sustaining and be able to deliver all its mandates	Governance
The right balance of resources to deal with all challenges involved to meet increasing knowledge and demands for data analytics outputs	Strategy
Academic service-level agreement between the CoE and academics who are working for the CoE	Structure
Meritocracy – a basis for decision-making on funding, performance and rewards	Governance
A Strategic Investment Fund can provide flexibility, particularly in a situation where a merit-based funding policy is lacking	Funding
Collaboration-seeking techniques to attract people to collaborate across non-traditional boundaries, both internally and externally	Collaboration
Having a less divided funding framework, an increased funding level and an aligning funding interest with stakeholders' interest and CoE's ambition	Funding
Increase the cash contribution from industry partners	Funding
Division of labour in a more balanced way among the CoE's employees	Structure
Need to develop international networks and collaborations	Collaboration

7 Summary

This chapter presented a best practice framework for the operation of BDAI CoE. The goal of the framework is to foster collaboration and share best practices among existing centres and support the establishment of new CoEs within Europe. The framework was developed following a phased design science process, starting from a literature review to create an initial framework which was enhanced with the findings of a multi-case study of existing successful CoEs. The chapter outlined how the framework can be used by a CoE to support its strategic direction and operational decisions over time, and how a new CoE can use it in the start-up phase. Based on the analysis of the case studies, the chapter explored the critical success factors of CoEs as defined by a survey of CoE managers.

Acknowledgements Research leading to these results received funding from the European Union's Horizon 2020 research and innovation programme under grant agreement no. 732630 (BDVe). This publication has emanated from research supported in part by a research grant from Science Foundation Ireland (SFI) under grant no. SFI/12/RC/2289_P2, co-funded by the European Regional Development Fund.

References

Adams, J. (2016). The societal and economic impacts of academic research. *Digital Science* (March).

Aksnes, D., Mats, B., Borlaug, S., Hansen, H., Kallerud, E., Kristiansen, E., et al., (2012). *Excellence in the Nordic countries. A comparative study of research excellence policy and excellence centre schemes in Denmark, Finland, Norway and Sweden.*

Bijker, W. E., Hughes, T. P., & Pinch, T. (1987). The social construction of technological systems: New directions in the sociology and history of technology. *The Social Construction of Technological Systems, 60*(3), 428.

Bleiklie, I., & Kogan, M. (2007). Organization and governance of universities. *Higher Education Policy, 20*(4), 447–493.

Bornmann, L. (2013). What is societal impact of research and how can it be assessed? A literature review. *Journal of the American Society for Information Science and Technology, 64*(2), 217–233.

Burke, W. W., & Litwin, G. H. (1992). A causal model of organizational performance and change. *Journal of Management, 18*(3), 523–545. https://doi.org/10.1177/014920639201800306

Cavanillas, J. M., Curry, E., & Wahlster, W. (Eds.). (2016). *New horizons for a data-driven economy: A roadmap for usage and exploitation of big data in Europe.* New York: Springer. https://doi.org/10.1007/978-3-319-21569-3

Curry, E. (2016). The big data value chain: definitions, concepts, and theoretical approaches. In J. M. Cavanillas, E. Curry, & W. Wahlster (Eds.), *New horizons for a data-driven economy: A roadmap for usage and exploitation of big data in Europe.* New York: Springer. https://doi.org/10.1007/978-3-319-21569-3_3

Fagerberg, J., & Srholec, M. (2008). National innovation systems, capabilities and economic development. *Research Policy.* https://doi.org/10.1016/j.respol.2008.06.003

Gulbrandsen, M., & Smeby, J.-C. (2005). Industry funding and university professors' research performance. *Research Policy, 34*, 932–950. https://doi.org/10.1016/j.respol.2005.05.004

Harland, K., & O'Connor, H. (2015). *Broadening the scope of impact: Defining, assessing and measuring impact of major public research programmes, with lessons from 6 small advanced economies* (Vol. 2).

Hellström, T. (2011). Homing in on excellence: Dimensions of appraisal in Center of Excellence program evaluations. *Evaluation, 17*(2), 117–131. https://doi.org/10.1177/1356389011400891

Hellström, T. (2012). *Centres of excellence as a tool for capacity building – Sweden Case Study Programme on Innovation, Higher Education and Research for Development (IHERD).*

Hevner, A. R. (2007). A three cycle view of design science research. *Scandinavian Journal of Information Systems, 19*(2), 87–92. Retrieved from https://doi.org/http://aisel.aisnet.org/sjis/vol19/iss2/4

Joynson, C., & Leyser, O. (2015). The culture of scientific research. In *F1000Research.* https://doi.org/10.12688/f1000research.6163.1

Kim, H., Lee, J.-N., & Han, J. (2010). The role of IT in business ecosystems. *Communications of the ACM.* https://doi.org/10.1145/1735223.1735260

Linstone, H. A., & Turoff, M. (1975). The Delphi method: Techniques and applications. *The Delphi Method - Techniques and Applications, 29*, 1–616. https://doi.org/10.2307/1268751

Molas, J., Salter, A., Patel, P., Scott, A., & Duran, X. (2002). *Measuring third stream activities.*

Mowery, D. C., et al. (1993). National innovation systems: A comparative analysis. In *National innovation systems: A comparative analysis.*

OECD. (2014). *Promoting research excellence: A new approach to funding.* OECED Publishing.

Ohno-Machado, L. (2014). NIH's big data to knowledge initiative and the advancement of biomedical informatics. *Journal of the American Medical Informatics Association.* https://doi.org/10.1136/amiajnl-2014-002666

Penfield, T., Baker, M. J., Scoble, R., & Wykes, M. C. (2014). Assessment, evaluations, and definitions of research impact: A review. *Research Evaluation, 23*(1), 21–32. https://doi.org/10.1093/reseval/rvt021

Perkmann, M., & Walsh, K. (2007). University–industry relationships and open innovation: Towards a research agenda. *International Journal of Management Reviews, 9*(4), 259–280. https://doi.org/10.1111/j.1468-2370.2007.00225.x

Roche, S., Driscoll, S. O., Higgins, L., & Neill, O.O. (2013). *Science Foundation of Ireland Research Centres Programme: Overview of Research Centres Programmes.*

Ron, D. (2016). *Big data and data analytics: The potential for innovation and growth.*

Schmidt, E. K., & Krogh Graversen, E. (2017). *Persistent factors facilitating excellence in research environments.* New York: Springer. https://doi.org/10.1007/s10734-017-0142-0

Sutherland, W. J., Goulson, D., Potts, S. G., & Dicks, L. V. (2011). Quantifying the impact and relevance of scientific research. *PLoS One, 6*(11). https://doi.org/10.1371/journal.pone.0027537

Von Bertalanffy, L. (1968). *General system theory.*

Weisbord, M. R. (1976). Organizational diagnosis: Six places to look for trouble with or without a theory. *Group & Organization Management, 1*(4), 430–447.

Zillner, S., Curry, E., Metzger, A., Auer, S., & Seidl, R. (Eds.). (2017). *European big data value strategic research & innovation agenda.* Retrieved from Big Data Value Association website www.bdva.eu

Zillner, S., Bisset, D., Milano, M., Curry, E., Hahn, T., Lafrenz, R., et al. (2020). *Strategic research, innovation and deployment agenda - AI, data and robotics partnership. Third Release (3rd).* Brussels: BDVA, euRobotics, ELLIS, EurAI and CLAIRE.

Data Innovation Spaces

Daniel Alonso

Abstract Within the European Big Data Ecosystem, cross-organisational and cross-sectorial experimentation and innovation environments play a central role. European Innovation Spaces (or i-Spaces for short) are the main elements to ensure that research on big data value technologies and novel applications can be quickly tested, piloted and exploited for the benefit of all stakeholders. In particular, i-Spaces enable stakeholders to develop new businesses facilitated by advanced Big Data Value (BDV) technologies, applications and business models, bringing together all blocks, actors and functionalities expected to provide IT infrastructure, support and assistance, data protection, privacy and governance, community building and linkages with other innovation spaces, as well as incubation and accelerator services. Thereby, i-Spaces contribute to building a community, providing a catalyst for engagement and acting as incubators and accelerators of data-driven innovation, with cross-border collaborations as a key aspect to fully unleash the potential of data to support the uptake of European AI and related technologies.

Keywords Data-driven innovation (DDI) · Data experimentation environment · Data space · Data platform · Data sharing · Data ecosystems and community · Digital Innovation Hub (DIH) · Federation of DIHs

1 Introduction

The term Data Innovation Space (in short i-Space) was initially coined by the Big Data Value Association (BDVA) and included in the first version of its Strategic Research and Innovation Agenda (SRIA) (Zillner et al. 2017) as one of the mechanisms identified to implement its research and innovation strategy, together with (i) lighthouse projects (large-scale demonstrators aimed to showcase the applications

D. Alonso (✉)
ITI, Valencia, Spain
e-mail: dalonso@iti.es

E. Curry et al. (eds.), *The Elements of Big Data Value*,
https://doi.org/10.1007/978-3-030-68176-0_9

of data-driven solutions to different sectors), (ii) technical projects (addressing specific data issues and technical aspects) and (iii) cooperation and coordination projects (to enable international cooperation for efficient information exchange and coordination of activities).

This chapter presents Data Innovation Spaces as environments to test, experiment and deploy new data-driven innovations. More specifically, Sect. 2 introduces the concept of Data Innovation Spaces and their main characteristics. The key elements of Data Innovation Spaces, as well as basic expected services, are presented in Sect. 3. Section 4 presents the role of i-Spaces in the European landscape and their alignment with other initiatives. Section 5 explains the specific certification process implemented by the Big Data Value Association (BDVA) to recognise relevant initiatives in Europe. The impact of the BDVA-recognised i-Spaces in their respective ecosystems is presented in Sect. 6. General collaboration between Data Innovation Spaces and a specific example of creating a European federation are explained in Sect. 7. Finally, the chapter ends with learnt stories and success stories as part of Sect. 8.

2 Introduction to the European Data Innovation Spaces

European Data Innovation Spaces are the main elements to ensure that research on BDV technologies and novel BDV applications can be quickly tested, piloted and thus exploited in a context with the maximum involvement of all the stakeholders of BDV ecosystems. The objective is to facilitate large and small companies, public administration, and European and national projects and society, in general, in easily accessing economic opportunities offered by the BDV and developing working prototypes to test the viability of actual business deployments. As such, i-Spaces enable stakeholders to develop new businesses facilitated by advanced BDV technologies, applications and business models. i-Spaces bring together not only technical and application developments but also all aspects needed to foster skills, competencies and best practices. i-Spaces usually rely on national and regional initiatives, federating, complementing and leveraging activities of similar national incubators/environments, existing Public—Private Partnerships and other national or European initiatives.

The main characteristics of a Data Innovation Space are as follows (as shown in Fig. 1)**:**

- Forming **hubs** to bring technology and application developments together while catering for the development of skills, competencies and best practices. These environments provide new and existing technologies and tools from industry and open-source software initiatives as a basic service to tackle the big data value challenges.
- Ensuring that **data** is at the centre of big data value activities. i-Spaces make data assets based on industrial, private and open data sources accessible. They are

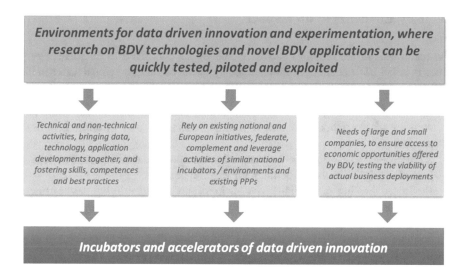

Fig. 1 Data Innovation Spaces concept

secure and safe environments that ensure the availability, integrity and confidentiality of data sources.

- Serving as **incubators** for the testing and benchmarking of technologies, applications and business models. This provides early insights into potential issues and helps to avoid failure in the later stages of commercial deployments. In addition, it is expected that this activity will provide input for standardisation and regulation.
- Developing **skills** and sharing best practices is an important task of i-Spaces and their federation. They will also link with other existing initiatives at both the European and national level.
- New **business models** and ecosystems will emerge from exposing new technologies and tools to industrial and open data. i-Spaces are a playground for testing new business model concepts and the emerging ecosystems of existing and new BDV "players".
- Gaining early insights into the **social impact** of new technologies and data-driven applications and how they will change the behaviour of individuals and the characteristics of data ecosystems.
- Acting as a **catalyst** to foster data-driven communities in the ecosystem and accelerate value creation.

The establishment of European Data Innovation Spaces and their evolution is reflected in the roadmap of the implementation of the Big Data Value Public-Private Partnership (BDV PPP), as detailed in Chap. "A Roadmap to Drive Adoption of Data Ecosystems". Phase 1 of this roadmap (2016–2017) is devoted to the establishment of the ecosystem (including i-Spaces and their collaboration towards a federation or network of i-Spaces), phase 2 (2018–2019) proposed disruptive forms of big data solutions, and phase 3 (2020) considers the sustainability and the benefits of the carried-out actions.

3   Key Elements of an i-Space

As mentioned in the previous section, i-Spaces are conceived as interdisciplinary hubs to target BDV challenges encountered by SMEs and small regional actors in the following different dimensions (see Fig. 2).

- **Technical**, providing infrastructure for testing, giving advice on architecture and security of the workspace and tool implementation, and offering help-desk support
- **Application**, supporting the building of precompetitive application, developing (visual) analytics tools and settings for specific domains
- **Business**, creating new data-driven business models, identifying new business opportunities with already existing data, and providing proof of impact and ROI
- **Social**, supporting SME uptake in digitisation, offering services for cultural heritage and local governments, and providing digital solutions for policy development

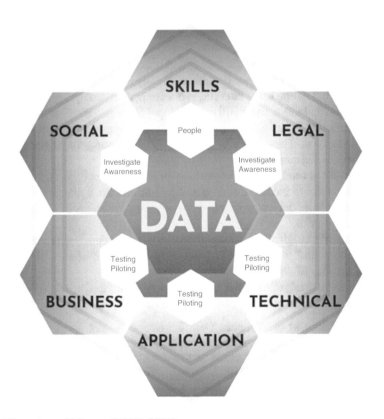

Fig. 2 Dimensions of i-Spaces (BDVA SRIA)

- **Skills**, training and educating employees to make use of big data technologies and build on data expertise, and providing master's-level students with industrial problems and specific data
- **Legal**, providing templates for data provision contracts and consulting on internal secure data management process and architecture

In terms of services, i-Spaces are supposed to provide to SMEs and industry, society and other European initiatives (including projects) a set of basic tools to allow the demonstration, experimentation and training, testing, showcasing and benchmarking of their data-driven solutions and products, before going to the market. This set of basic services includes:

- **Community Building**: Contributing to the identification and management of stakeholder ecosystem communities along thematic and/or regional dimensions.
- **Asset Support:** Supporting data providers in integrating datasets in a quality-secured way while maintaining a catalogue of available data assets.
- **ICT Support:** Providing basic ICT assistance as well as focused support from big data scientists and data specialists, and business development during research and innovation projects. This includes assistance in benchmarking datasets, technologies, applications, services and business models.
- **On-boarding:** Running an induction process for new project teams.
- **Resourcing:** Allocating the resources (computing, storage, networking, tools and applications) to individual research and innovation projects and scheduling these resources among different projects.
- **Data Protection and Privacy:** Data protection, including ensuring compliance with laws and regulations such as the EU GDPR (General Data Protection Regulation), and the deployment of cutting-edge, state-of-the-art security technologies in protecting data and controlling data access, privacy and anonymisation in terms of handling and deleting personally identifiable information (PII).
- **Data Governance:** Taking into account privacy and protection issues, defining the rules for accessing and sharing data. This includes the standardisation of procedures for sharing metadata, defining the (smart) contract between stakeholders, assessing technologies such as encryption and blockchain, and formulating the necessary solutions to orchestrate the agreed governance.
- **Federation:** Supporting linkages to other innovation spaces and facilitating experiments across multiple innovation spaces. An effective federation will help to support research and innovation activities through accessing and processing data assets across national borders (data spaces).
- **Business Support:** Facilitating start-ups and SME inclusion in the value creation process by leveraging community engagement.
- **Incubation and Acceleration:** Delivering all forms of suitable support to data-driven value creation projects by liaising with existing thematic, national or regional initiatives.

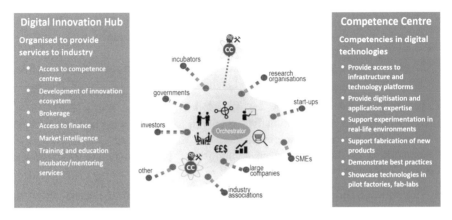

Fig. 3 Competence Centres and Digital Innovation Hubs (Source: European Commission) (by European Commission licensed under CC BY 4.0)

4 Role of an i-Space and its Alignment with Other Initiatives

As mentioned above, the concept of Data Innovation Space was initially coined in 2014 by the BDVA and identified as a key instrument to foster data-driven innovation based on experimentation, testing and benchmarking. Since then, many other instruments have appeared in Europe, aimed at bringing innovation closer to industry and society, and more specifically to those actors with no capacity to benefit from the latest European digital innovations.

In this way, and considering that only about 1 out of 5 companies across the EU is highly digitalised, and around 60% of large industries and more than 90% of SMEs lag in digital innovation, the European Commission introduced in 2017 the concept of the Digital Innovation Hub (DIH),[1] to ensure that every company, small or large, high-tech or not, can take advantage of digital opportunities. DIHs are one-stop shops that help companies become more competitive with regard to their business/ production processes, products or services using digital technologies. DIHs provide access to technical expertise and experimentation so that companies can "test before invest". They also provide innovation services, such as financing advice, training and skills development, that are needed for a successful digital transformation.

A Digital Innovation Hub brings many actors together, to develop a coherent and coordinated set of services that are needed to help companies (especially SMEs or enterprises from low-tech sectors) that have difficulties with their digitisation through a one-stop shop. However, the core of a DIH is the Competence Centre, which provides technical expertise and access to advanced facilities (see Fig. 3).

---

[1] https://ec.europa.eu/futurium/en/system/files/ged/dei_working_group1_report_june2017_0.pdf

Fig. 4 Data Innovation Space vs. DIH and Competence Centre

The European Commission has developed an online catalogue[2] to provide a comprehensive picture of DIHs in the EU across varying competences structures and service offerings. It is a repository with more than 400 DIHs, over 200 of which are fully operational, including information on the technology and application specialisation, geographical coverage, markets addressed and general digitisation support available. According to this catalogue, there are around 190 DIHs in Europe specialised in data mining, big data and database management, meaning that these data-driven DIHs are ready, based on the expertise provided by their Competence Centres, to support companies in their respective ecosystems in the development, adoption and testing of data-driven solutions.

In this way, the concept of Data Innovation Space is aligned with that of a Competence Centre on Big Data, in the sense that it provides access to infrastructure, expertise, support to experimentation and production of new services, and best practices regarding data-driven solutions and products. On the other hand, it can also offer advanced services such as brokerage, access to finance, training, and incubation and acceleration. In this case, it would act as a Data-Driven Innovation Hub (actually, all BDVA i-Spaces are recognised DIHs on big data), bringing together not only technical competencies but all tools and aspects needed to allow SMEs to put their data-driven services and products into the market. Taking all of the above into consideration, and depending on the offered services, a Data Innovation Space would range between a Competence Centre on Big Data and a Data-Driven Innovation Hub (see Fig. 4).

Other important instruments developed to mobilise data and foster data sharing and reuse are data platforms and data spaces. According to a BDVA position paper on data sharing and data spaces,[3] a data space is an ecosystem of data models, datasets, ontologies, data sharing contracts and specialised management services (e.g. as often provided by data centres, stores and repositories, individually or within "data lakes"), together with soft competencies around it (i.e. governance, social interactions, business processes). These competencies follow a data engineering approach to optimise data storage and exchange mechanisms, in this way preserving, generating and sharing new knowledge. On the other hand, data platforms refer to architectures and repositories of interoperable hardware/software components, which follow a software engineering approach to enable the creation, transformation, evolution, curation and exploitation of both static and dynamic data in data spaces.

---

[2]https://s3platform.jrc.ec.europa.eu/digital-innovation-hubs-catalogue

[3]https://www.bdva.eu/node/1277

Specific examples of data space and data platforms are mentioned in this BDVA paper, and it is also worth mentioning the nine innovation actions funded by the European Commission under the topic "Supporting the emergence of data markets and the data economy", especially aimed to address the necessary technical, organisational, legal and commercial aspects of data sharing/brokerage/trading, both for personal and industrial data.

These instruments incorporate in Data Innovation Spaces (and Data-Driven Innovation Hubs) the dimension of data sharing, data trading and data reuse, allowing Data Innovation Spaces to share datasets and data sources with other Data Innovation Spaces, and providing interoperability and scalability in terms of data.

The new Digital Europe Programme will reinforce the role of Digital Innovation Hubs and European Data Spaces as the main instruments to increase the competencies and bring innovation to the European industry and society in terms of data. This programme also includes technology infrastructures with specific expertise and experience of testing mature technology in a given sector, under real or close to real conditions (e.g. smart hospital, smart city, experimental farm, corridor for connected and automated driving), which are the Testing and Experimentation Facilities (TEFs) on AI.

These TEFs will exploit, test and validate data spaces to test AI-powered solutions, also enriching them by providing user feedback. TEFs will contribute to data spaces by collecting and providing data from experimentation. On the other hand, the Digital Innovation Hubs will act as a distribution channel for AI to empower all local companies and users.

Figure 5 shows the different dimensions provided by different European instruments.

According to the European Commission, a Digital Innovation Hub relies on four pillars to increase the competitiveness of companies with regard to their business/production processes, products or services using digital technologies. These pillars are: (i) access to an innovation ecosystem with connection and networking with multiple stakeholders, (ii) test before invest, with access to technical expertise and experimentation, (iii) support to find investments and (iv) skills and trainings. With respect to this last aspect, to find alignments and synergies with the so-called centres of excellence, organisational units within a national system of research and education that provides leadership in research, innovation and training in digital technologies are of utmost importance, given the regional/national scope of both types of initiatives and their complementarities. In the case of big data, the connection between Data-Driven Innovation Hubs and the network of Big Data Centres of Excellence is valuable in identifying gaps in the industry demand side (workforce) at regional level and jointly planning a training programme to fill those gaps. Further details on big data and AI Centres of Excellence are available in Chap. "A Best Practice Framework for Centres of Excellence in Big Data and Artificial Intelligence".

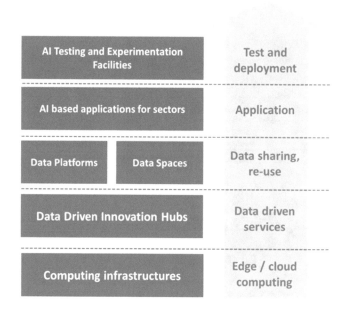

Fig. 5 European instruments to foster data-driven innovation and experimentation

5   BDVA i-Spaces Certification Process

With the objective of identifying relevant and qualified initiatives in Europe aligned with the concept of i-Spaces, the BDVA launches yearly public calls that are open to any innovation hub on big data[4] in Europe. The candidates are evaluated in terms of infrastructure and technologies provided, the services that are offered, projects and applications where the DIH is involved, the impact on the local/regional and national/European ecosystem, and the business strategy and sustainability. After the review process, those initiatives that meet specific criteria are qualified as BDVA i-Spaces. This call has been launched over the last 5 years, and during the several editions, new i-Spaces have been incorporated, composing the current group of 15 BDVA i-Spaces (see Fig. 6).

The different steps of the labelling process are as follows:

- Launch of the open call, aimed at any data-driven competence centre, DIH on big data and AI, etc. in Europe, interested in having the recognition of BDVA as a qualified Data Innovation Space. This recognition guarantees that the innovation environments provided meet the requirements to boost data-driven and AI-based innovation at a local level, and the collaboration with similar initiatives to foster adoption at European level.

---

[4]http://www.bdva.eu/node/1173

Fig. 6 Map of recognised BDVA i-Spaces 2019

- Online survey/questionnaire. Candidates are invited to fill in an online questionnaire, to collect information from their initiatives in the following domains:

  - Infrastructure, including computing, storage and communication capacities, allocation of resources, data access methods and tools, policies, standards and certificates
  - Services, including technical support, data management, analysis and visualisation, data governance, privacy and protection, incubation and acceleration, business support, skills and training
  - Projects and sectors, including most relevant projects and aggregated number of experiments per year
  - Ecosystem and collaborations, including actors engaged in the ecosystem, involvement in regional clusters, outreach and collaborations
  - Business strategy, including growth, impact and sustainability models

- Review (including review committee meeting). Received applications are reviewed by a review committee composed of external experts also recruited through an open call. Each of the five domains of the applications is scored

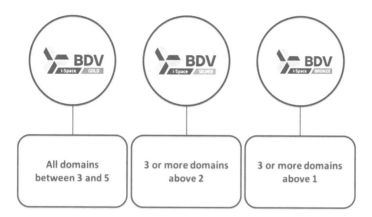

Fig. 7 i-Spaces labelling criteria

between 1 and 5. Final results are agreed in a review committee meeting. Applications are granted either a gold, silver or bronze label according to the criteria shown in Fig. 7.

- Proposal to the BDVA Board of Directors and announcement to i-Spaces. The results from the review committee are submitted to the BDVA Board of Directors for approval and communicated to the candidates.
- Trophy hand-out ceremony, usually co-located with the European Big Data Value Forum (www.ebdvf.eu) and where trophies are handed out to i-Spaces on stage by the BDVA president.

6 Impact of i-Spaces in Their Local Innovation Ecosystems

Digital Innovation Hubs, in general, and BDVA i-Spaces, in particular, are expected to contribute to the digital transformation and development of their respective ecosystems. They should be deeply rooted in innovation ecosystems and offer digital transformation services to companies in their proximity. They are also expected to contribute to the development of the RIS3 (Research and Innovation Strategies for Smart Specialisation) strategy.[5] To illustrate this, below we sketch several specific actions carried out by the BDVA i-Spaces supporting the emergence of their respective ecosystems.

CeADAR: Ireland's Centre for Applied Artificial Intelligence
The CeADAR centre is a main plank in Ireland's Smart Specialisation Strategy, particularly in applied AI and data analytics. The centre is directly funded by the

---

[5]https://s3platform.jrc.ec.europa.eu/s3-guide

Department of Business, Enterprise and Innovation through its two main industry agencies, Enterprise Ireland (EI) and the Industrial Development Authority (IDA), which are in charge of the S3 R&I strategies and priorities for Ireland. In 2018, CeADAR went through an international review process where it was referred to as a key contributor to the digital transformation of Ireland's industry. As part of this review, the centre has received funding from the State Agencies of €12 million to drive its data analytics and artificial intelligence agenda. CeADAR as the National Technology Centre for Applied Data Analytics and Artificial Intelligence has developed links with some of the other technology centres to combine their domain knowledge in specific areas with their expertise in different fields of AI.

CINECA

Embedded in the Italian national HPC centre, CINECA i-Space operates at the intersection of big data, HPC and deep learning technologies to support research and innovation with the most advanced infrastructure, tools, services and skills. The RIS3 Emilia-Romagna strategy is based on four strategic priorities: (i) to increase Emilia-Romagna enterprise competitiveness, (ii) to sustain the emerging specialisation areas, (iii) to provide orientation to the digital transformation and (iv) to develop services of excellence, in four specialisation areas: (a) building and construction, (b) mechatronics and motoring, (c) health and wellness industries and (d) cultural and creative industries. CINECA developed dozens of projects involving large companies and SMEs of all specialisation areas, providing value-added services rooted in advanced simulation, big data and AI technologies.

EURECAT/Big Data CoE Barcelona

The Barcelona Big Data Centre of Excellence (Big Data CoE) is an initiative led by EURECAT, which was launched in February 2015 with the support of the Barcelona City Council, the Government of Catalonia and Oracle. Its impact in the regional ecosystem includes actions as being:

- A pillar of the SmartCat Strategy led by the Catalan Government to promote key enabling digital technologies which include big data and data analytics.
- An evolution phase to embrace not only data-related technologies but also AI technologies is a core element for the deployment of the Catalan AI strategy.
- Developing projects with local companies aligned with the RIS3CAT strategy in Catalonia, notably in sectors like Digital Health, Industry 4.0 and Tourism.

ITAINNOVA/Aragon DIH

DIH on "HPC-Cloud and Cognitive Systems for Smart Manufacturing processes, Robotics and Logistics" is the Aragonese initiative that, within a framework of European cooperation (DIH), extends the strategy of economic and industrial promotion of Aragon and the intelligent regional strategy of Aragon, forming the technological and innovative action of the Aragonese Innovation System. Within the National Strategy for Industry 4.0, it has developed an advisory action that will identify the degree of digitisation of the Spanish Industry. Only 15 entities have been selected to carry out this advisory task throughout Spain. ITAINNOVA has been selected as a qualified consultancy entity for the development of these actions in its areas of influence. This will allow Aragon DIH the ability to offer its services fully integrated into the national strategy of digitisation of the industry.

ITI/Data Cycle Hub

The Data Cycle Hub, coordinated by ITI, is a Digital Innovation Hub composed of a consortium of organisations with complementary experience that supports companies and the public sector in the Valencia region in their digital transformation. The Valencian Institute of Business Competitiveness (IVACE) is the coordinator of the RIS3CV (development of the RIS3 strategy specifically for the Valencia region). ITI has been working with IVACE in the RIS3CV strategy since the beginning, carrying out the ICT secretariat and working with all the ICT ecosystems. ITI also developed the Industry 4.0 agenda in the Valencia region. Activities of the Data Cycle Hub are aligned with almost all of the RIS3CV areas, including industry (working directly with the Industry 4.0 Lab with IVACE), Health, Tourism, Agrifood, Habitat and Cities, Transport and Energy (also working in Smart Grid Lab with IVACE) – all of them included in the RIS3CV priorities.

Know-Center

Know-Center Graz was founded in 2000 within the framework of the COMET K1 program, and became Austria's leading research centre for data-driven business innovative information and communication technologies. It actively integrates into national cooperation and networks including Green Tech Cluster, AC Styria, Human. Technology Styria, Styrian Service Cluster, Silicon Alps Cluster and IT Clusters. It has close ties with competence centres such as Pro2Future, Virtual Vehicle, Materials Center Leoben and Large Engines Competence Center.

RISE/ICE by RISE

ICE, the Infrastructure and Cloud datacenter test Environment, is a research data centre inaugurated in January 2016. The facility is open to use primarily for European projects, universities and companies. However, customers and partners from all over the world are welcome to use ICE for their testing and experiments. ICE's mission is to contribute to Sweden being at the absolute forefront regarding competence in sustainable and efficient data centre solutions, cloud applications and data analysis, including links with other regional DIHs such as EIT RawMaterials CLC North, Luleå EIT InnoEnergy. ICE is fully aligned with the regional development plan and is running an S3 pilot for an AI and big data ecosystem in the region.

Smart Data Innovation Lab (SDIL)

The SDIL supports pre-commercial research between academia and industries, especially SMEs, in the areas of smart infrastructure, medicine and Industry 4.0. Its potential analysis service under the programme Smart Data Solution Center Baden-Württemberg (SDSC-BW) aims to facilitate entry into smart data analytics application and Industry 4.0 for SMEs. All of these correspond to the digitisation strategy of Germany as well as the RIS3.

TeraLab

TeraLab provides AI and big data "one-stop shop" support to research organisations, web innovators, start-ups, midcaps and large groups, as well as governmental and educational organisations. TeraLab is actively involved in France's regional and national initiatives around AI and big data:

- It is a consortium partner of the regional initiative PACK IA.
- It contributed to the national AI mission led by Cédric Villani.
- It participated in the projects ADMIRR, EXPRESSO, GeoLytics, M4P, PULSE and Data&Musée.

Universidad Politécnica de Madrid/Madrid's i-Space for Sustainability/ AIR4S DIH

This DIH/i-Space, aligned with the RIS3-Madrid priorities, supports the digitisation of industry, especially SMEs but also midcaps, big companies and public administrations, to improve their products, services and processes, by introducing the great advantages of artificial intelligence and robotics into their business. AIR4S provides companies in all disciplines with a multidisciplinary and personalised approach and consequently addresses multisector domains in a confident way. It brings together world-class technological expertise and infrastructure on AI and robotics but also deep knowledge on how to apply these technologies on different market domains, while being aligned with the Sustainable Development Goals and being respectful of the social, legal and ethical aspects of these technologies.

In the context of data spaces and data communities, AIR4S supports the creation of links between different local initiatives related to access to open data and facilitates cooperation among different data holders at the local level. These links can be created and maintained thanks to the permanent collaboration among European DIHs and the connection to local public systems.

7 Cross-Border Collaboration: Towards a European Federation of i-Spaces

To fully exploit the benefits that the different Digital Innovation Hubs (DIHs) are bringing to the industry, one step beyond in the collaboration among those initiatives and towards a network of DIHs is necessary. In the report "Digital Innovation Hubs: Mainstreaming Digital Innovation across All Sectors",[6] the creation of a Europe-wide network of DIHs supporting any business at a "working distance" is seen as an ambitious but achievable objective. In this way, the EC has invested EUR 500 million in the Horizon 2020 programme in initiatives for:

---

[6]https://ec.europa.eu/futurium/en/system/files/ged/dei_working_group1_report_june2017_0.pdf

- Networking and collaboration of digital competence centres and cluster partnerships
- Supporting cross-border collaboration of innovative experimentation activities
- Sharing of best practices and developing a catalogue of competencies
- Wide use of public procurement of innovations to improve efficiency and quality of the public sector

As a result, there exist some running initiatives whose objectives are to break silos, find synergies and foster collaboration among DIHs in different technologies and domains (as relevant examples, the AI DIH network (https://ai-dih-network.eu) aims at establishing a framework for continuous collaboration and networking between DIHs focusing on artificial intelligence, MIDIH project (https://www.midih.eu) aims to create a network of manufacturing DIHs in the area of IoT/Cyber-physical systems (CPS), DIHNET (https://dihnet.eu) supports collaboration among DIH networks across Europe, and DIHelp (https://dihelp.eu) is a mentoring and coaching programme supporting 30 DIHs to develop and/or scale up their activities).

This role of DIHs is reinforced in the envisioned Digital Europe Programme[7] (see Fig. 8), as a means to ensure the digital transformation of all businesses as well as public administrations, in a broad roll-out of digital technologies and digital skills to the entire economy. DIHs are supposed to work closely with the relevant specialised centres and make sure that companies and public administrations can experiment with those technologies (test before investing) and develop skills to meet their needs. As part of this programme, the European Commission also envisages the creation of a network of European DIHs including all regions of Europe, to cover activities with a clear European added value and promote the transfer of expertise.

Regarding big data, the creation of a European federation of Data-Driven Innovation Hubs was included as part of the H2020 programme in 2020, under the topic DT-ICT-05,[8] with the main challenge of breaking "data silos" and stimulating sharing, reusing and trading of data assets, federating data sources and fostering collaborative initiatives with relevant digital innovation hubs, with the ultimate objective of contributing to the creation of the European Common Data Space. The call explicitly mentioned the BDVA i-Spaces among those initiatives to coalesce towards this federation of Data-Driven Innovation Hubs.

The concept is completely aligned with the strategy of the BDVA i-Spaces group, as is reflected in the BDVA SRIA, where supporting linkages to other innovation spaces and facilitating experiments across multiple innovation spaces is seen as a crucial point towards an effective federation that will help to support research and innovation activities through accessing and processing data assets across national borders. The i-Spaces group has been working in recent years with that objective in mind, to foster collaborations and define the processes towards the creation of a

---

[7]https://ec.europa.eu/futurium/en/digital-innovation-hubs/digital-innovation-hubs-digital-europe-programme

[8]https://ec.europa.eu/info/funding-tenders/opportunities/portal/screen/opportunities/topic-details/dt-ict-05-2020

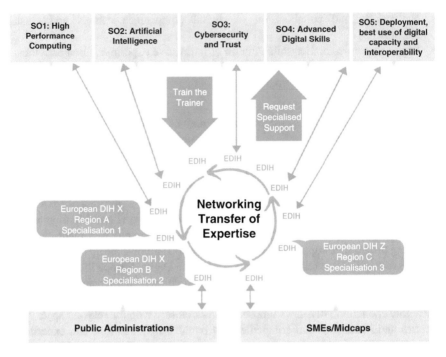

Fig. 8 Schematic overview of the role of EDIHs in Digital Europe Programme (European Commission) (by European Commission licensed under CC BY 4.0)

network of i-Spaces. Among those activities, it is worth mentioning the organisation of the workshops "Towards a Federation of European Data Spaces" (BDV PPP Meetup, Sofia, May 2018), "Shaping the European Ecosystem: From i-Spaces and Centres of Excellence to Big Data DIHs" (European Big Data Value Forum 2018, Vienna, October 2018) and "Federation of data services to foster the adoption of data-driven AI in Europe" (BDV PPP Summit, Riga, June 2019), and the joint participation in the 5th meeting of the Working Group on DIHs: Big Data and AI,[9] organised by the EC in Brussels (November 2018), where i-Spaces shared knowledge, experiences, best practices and their views towards a federation of DIHs on big data.

This collaboration crystallised in a successful project proposal under the call DT-ICT-05. This EUHubs4Data project started in September 2020 and will run for 3 years, with the overarching objective of creating the reference federation in Europe for big data cross-border experimentation and innovation, providing a complete pan-European catalogue of data sources and services to foster data-driven innovation at local and regional level. The project also aims to:

---

[9]https://ec.europa.eu/digital-single-market/en/news/fifth-meeting-working-group-digital-innovation-hubs-big-data-and-artificial-intelligence

- Contribute to the creation of the European Common Data Space, by mobilising, sharing and making available all types of data (close/open, personal/industrial, private/public, research, etc.), with the objective of generating data value from them and fostering data-driven innovation in Europe.
- Lay the foundations for the creation of a pan-European federation of initiatives focused on data-driven innovation and experimentation (DIHs on big data) that, based on strong collaboration and value co-creation, will support European business in their development, and launch data-driven products and solutions to the market, assisting them in their whole journey along the data value chain.
- Add value to the ecosystem of existing initiatives in Europe, positioning a one-stop shop for data-driven innovation and experimentation; building community around the data economy; establishing a liaison among data-driven research and innovation, regulatory bodies and policy makers, industry and data service providers; and bringing together and aligning all actors necessary to boost data-driven innovation.

To accomplish its objectives, the EUHubs4Data project will rely on the following pillars:

- A starting point, with an initial ecosystem composed of the BDVA i-Spaces and some relevant players to link with data-driven initiatives in Europe
- The expansion of the ecosystem during and after the lifecycle of the project, defining a model to incorporate new DIHs into the federation of DIHs; access to local, national and European data incubators; and the involvement of more SMEs
- The offer of the federation, with a global catalogue of data-related services, which will configure the offer of the federation of DIHs to end users. This global catalogue will rely on the individual catalogues of the DIHs, will be enriched with outcomes and assets coming from past and existing European actions (projects) and will be accessible at the local level through the regional DIH or local access point.
- The attraction of the demand side, by a cross-border data innovation programme, with the three-fold objective of (i) attracting the demand side to use the federated services in a cross-border basis, (ii) testing the model of service provisioning and (iii) defining the model to be applied once the project is finished.
- The community around the federation, with whom links will be established with the objective of bringing together all European initiatives working around the data economy and data technologies.
- Business and sustainability, to define a model that includes all aspects that guarantee the continuity of all activities of the federation once the project is finished.

The main outcome of the project will be a federated catalogue that will be made available to companies in the different European regions through their respective DIHs, which will provide access to specific federated services following the paradigm "European catalogue, regional offer" (as reflected in Fig. 9). Specificities about the federated catalogue and how the local offer is instantiated by the regional DIH

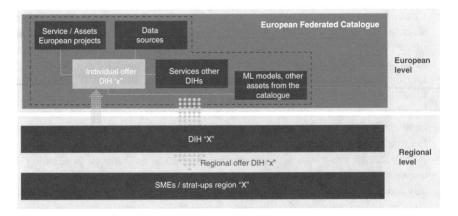

Fig. 9 EUHubs4Data European catalogue and regional offer

based on the catalogue will remain transparent for local companies, which will have access to an improved offer through its regular point of sale. Hence, DIHs of the federation will act as bridges for European SMEs to a unique catalogue that will include European data-driven innovations coming from multiple stakeholders.

Another important aspect of the EUHubs4Data project will be to actively contribute to the alignment of existing European initiatives towards the common objective of mobilising, sharing and making available all types of data (close/ open, personal/industrial, private/public, research, etc.), in order to get value from them, foster data-driven innovation in Europe, and contribute to the creation of a Common European Data Space. To achieve this, a specific task of the project will be devoted to (i) identifying relevant existing European initiatives on big data and related technologies, (ii) defining a clear value proposition in order to define the guidelines of collaboration with the mentioned objectives in mind, (iii) establishing the necessary links with those initiatives and (iv) specifying a roadmap that defines the work to be done (Fig. 10).

8   Success Stories

Below, we report on the success stories for each of the different BDVA i-Spaces, particularly highlighting their contribution and use in key actions and projects.

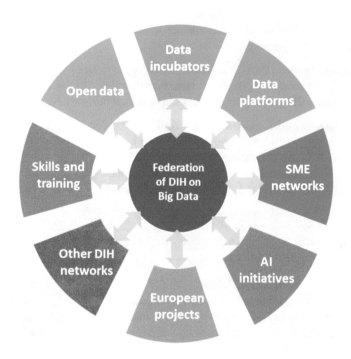

Fig. 10 EUHubs4Data community

8.1 CeADAR: Ireland's Centre for Applied Artificial Intelligence

Bespoke Innovation and Collaborative Projects CeADAR provides translational research projects to companies for integration in their operational/production systems. As part of this service, companies benefit by (i) starting their data and artificial intelligence journey, (ii) outsourcing key problems to explore new technological avenues, (iii) developing their own in-house data science team and (iv) participating in consortiums to tackle big challenges.

65 Market-Oriented Demonstrators CeADAR delivers approximately eight demonstrator projects per year in two cycles of 6 months, each in collaboration with industry partners. Each project is proposed by the industry members and is focused on a close-to-market challenge. Project development costs are met from the Centre's core budget. The Centre aims to deliver the following for each project: (i) state-of-the-art review, (ii) technical specification, (iii) demonstrators and (iv) assistance with member on-premise demonstrator evaluation. The extensive catalogue of over

65 technology demonstrators from previous platform research is available to all member companies (https://www.ceadar.ie/outputs/our-demos). These demonstrators have proven very useful for companies to start tapping into the benefits of data analytics in their organisations.

Data Science Awards CeADAR is a co-founder of the DatSci Awards, the National Data Science Awards (https://www.datsciawards.com). This is the major annual event in Ireland showcasing and celebrating data analytics and AI talent.

Industry Impact and Economic Value Add CeADAR has been in existence for over 7 years and in 2018 went through its 5-year term review, achieving the highest marks on each of the evaluation criteria with an international panel of experts from industry and academia. Due to this success, associated government agencies have increased (by 2.5 times) the funding to the centre for the next 5 years.

8.2   CINECA

Anomaly Detection in an HPC System Inside the project "Deriving and Validating Models for the Infrastructure Monitoring", the anomaly detection project, carried out by the Multithermal Lab of the University of Bologna on CINECA monitoring data, identified a deep learning model able to achieve high accuracy (90–97%) with a semi-supervised learning approach. This use case is peculiar as CINECA's role is that of data provider and, of course, of data user, and the automation of the anomaly detection would improve its services. These monitoring data are in the orders of TBytes, are currently used for different purposes (deriving thermal models for each core in the system, predicting a specific algorithm computation time, predictive maintenance, etc.) and are undergoing a process of anonymisation in order to be shared with a larger community of researchers.

Risk Management Code Optimisation for a Large Insurance Company The risk assessment in the life insurance field may require considerable computing power. The algorithm that the large insurance company was previously using took many hours and would not allow for calculating the risk measurement with a nested Monte Carlo approach. In fact, nested Monte Carlo involves two stages, scenario generation (outer stage) and portfolio re-valuation (inner stage), that produce millions of Monte Carlo trajectories to be executed for each of the millions of life policies. The simulation becomes an immediate computational challenge. The insurance company asked CINECA to develop a Proof of Concept (PoC) to demonstrate the improved efficiency that could be obtained with efficient code parallelisation and optimisation. The nested Monte Carlo with parameters 100000×100 for all of the 12M policies was achieved. The insurance company then decided to establish a commercial contract with CINECA for the provision of the service.

Sequential patterns of errors from on-board diagnostic devices for TEXA, a European leader company on electronic diagnostic. In the PRESERVE project, which has been funded within the Fortissimo EU project, sensor data from TEXA on-board diagnostic tools have been analysed in order to identify the driving habits on the one hand and patterns of operating parameters that are predictive of failures and damages on the other. The result is a portfolio of prototypes of services that can predict failures, mechanical problems or damage at the component level, and offer the manufacturer detailed information to better re-design or upgrade their spare parts or vehicle. The return on innovation investment (ROI2) for TEXA from this project has been estimated as 2,72.

LIGA: A Platform for the Game-Content Market LIGA is a project funded within the Fortissimo EU project in partnership with CNR (Consiglio Nazionale delle Ricerche) and Kumo (an SME in the field of 3D technologies and digital asset creation and management).

The current advantage of Kumo is that it is a platform for collecting, sharing, managing and collaborating on 3D content, where consumers of 3D content can access leading museums, gaming and other brands' data. At the end of July 2018, LIGA stored 25 million entries in its database, describing the popularity of game entities among players. Assuming no new game entities will be created in the future, LIGA will add 12 million of entries per month to its database, resulting in 720 million database rows by mid-2023.

Tax Fraud Detection for SOGEI, the Italian Revenue Agency Computing Centre CINECA, with its IOP4HPDA data scientists, developed predictive models of the fraudulent behaviour of companies in the entailment of tax credit and provided methodological solutions for impact and compliance assessment, in particular relating to training sample bias and model estimation and evaluation. The fraudulent behaviour model increased the auditing success rate from 39% to 65% (precision).

Managing Scientific Data for Various Scientific Communities Among the scientific research projects that the HPC department of CINECA supports, many can be reported as being both very successful and data-intensive projects, e.g. EMODnet (European Marine Observation and Data Network; http://www.emodnet-chemistry. eu/) and SPHINX (Data Storage and Preservation of High-resolution climate experiments; http://sansone.to.isac.cnr.it/sphinx/).

8.3   EGI

EOSC-Hub (www.eosc-hub.eu) EOSC-hub brings together multiple service providers to create the hub: a single contact point for European researchers and innovators to discover, access, use and reuse a broad spectrum of resources for advanced data-driven research. The project mobilises providers from the EGI Federation, EUDAT CDI, INDIGO-DataCloud and other major European research infrastructures to deliver a common catalogue of research data, services and software for research.

EOSC-hub collaborates closely with GÉANT and the EOSCpilot and OpenAIRE-Advance projects to deliver a consistent service offer for research communities across Europe:

- Start: January 2018
- End: December 2020
- 100 partners from 53 countries, including 19 research communities
- 13 work packages
- 49 services ready for use

eXtreme DataCloud (http://www.extreme-datacloud.eu) The eXtreme DataCloud (XDC) is an EU H2020-funded project aimed at developing scalable technologies for federating storage resources and managing data in highly distributed computing environments. The services provided will be capable of operating at the unprecedented scale required by the most demanding, data-intensive research experiments in Europe and worldwide. XDC will be based on existing tools, whose technical maturity is proved, and the project will be enriched with new functionalities and plugins already available as prototypes (TRL6+) that will be brought at the production level (TRL8+) at the end of XDC. The targeted platforms are the current and next-generation e-Infrastructures deployed in Europe, such as the European Open Science Cloud (EOSC), EGI and the Worldwide LHC Computing Grid (WLCG), and the computing infrastructures funded by other public and academic initiatives.

8.4 EURECAT/Big Data CoE Barcelona

Big Data and IoT to Improve Tourism Management in Barcelona With the goal of improving real-time decision making of tourism management in Barcelona as well as in policy definition, Barcelona Big Data CoE conceptualised and executed a big data and IoT-based project in partnership with the Barcelona City Council, the GSM Association Mobile World Capital and Orange. The target was the Sagrada Familia district, the city's hottest tourist attraction which causes severe mobility disruption in this area. We studied the macro-mobility (at district level) using call data records from Orange as well as micro-mobility (at street level) using the dedicated infrastructure of 10 Wi-Fi and GSM sensors around the Sagrada Familia streets as well as 3D cameras at the exits of the closest Metro extensions. We made use of the DATURA platform to perform the analysis of more than 50 TB of data accounting for more than 20 million users (aggregating all sources with national and international tourists) over a year. The main results of the project include seasonal macro- and micro-mobility patterns as well as visitors' profiles (segmented into tourists, excursionists and nightlife visitors) (https://www.bigdatabcn.com/portfolio-item/bcn-tourism-management-big-data-iot-in-action/).

Leading eCommerce Company The objectives of the project were to design and develop a new data platform as a critical technology component for a large e-commerce organisation to become a data-driven company, better support existing core business, and provide new capabilities aimed at a more personalised interaction with the customers. The deployed big data analytics platform scales to support 28 million users' daily interactions around the world, both with batch and real-time use cases.

Advanced Analytics for Cruïlla Cruïlla is a very popular and crowded music festival that takes place every year in Barcelona. Today it is one of the most successful music festivals in Europe. The goal of the project, commissioned to the Big Data CoE by the festival sponsors, was to apply data analytics to improve customer knowledge and develop strategies to boost customer engagement with and loyalty to the festival. User profiling was used to improve customer experience, make better marketing decisions and perform customised campaigns that were monitored through Google analytics and social network data.

Analysis of Wi-Fi Data Sources to Extract Origin—Destination Patterns in a Tram Network TRAM is a company that exploits Barcelona's tram network. The project consisted of analysing data from Wi-Fi sensors installed in trains of the tram lines operated by TRAM. The purpose was to compute O/D (origin and destination) matrices and other indicators and visualise them in a dashboard. In the use case, three trains of two tram lines were equipped with Wi-Fi sensors, which count the aggregated information of MAC id corresponding to passengers' mobile phones with active Wi-Fi. These data are analysed to determine the position of the users and, later, to verify the first and last station of a trip, which is the basic information to compute the O/D matrix. The data are calibrated with IR data sensors (for presence detection), already installed in the trains. The use of accurate data filtering and validation techniques was fundamental to distinguish actual tram passengers from other pedestrians around the train, therefore obtaining realistic O/D matrices.

Data Analysis to Improve Mobility Decisions A Proof of Concept (PoC) was commissioned by AlphaNet Seguretat, an SME which provides a wide range of security services to municipalities. The PoC included the design and deployment of a data analysis solution whose data source was car license plate numbers provided by AlphaNet's infrastructure. The PoC also included the development of algorithms to achieve AlphaNet security objectives and the development of a control dashboard.

8.5   ITAINNOVA/Aragon DIH

The *Moriarty® platform* is the result of more than 15 years of research in the field of AI and cognitive systems. Moriarty® is a tool for the design and implementation of advanced artificial intelligence software solutions, developed by ITAINNOVA, that solves various business problems with large volumes of data (big data). With Moriarty® one will be able to understand and structure information, identify hidden patterns and correlations in data, induce knowledge as well as build learning systems. In an agile, precise and simple way, it will allow one to convert their data into valuable information, facilitating the making of strategic decisions.

A very recent success case using Moriarty is the *Aragon Tourism Smart Observatory*, which is a dashboard for the regional Tourism Authority in order to let them see the trends of users of social media networks (among other sources) talking about Aragon's tourist places.

This dashboard includes sentiment analysis, tourist places and products, Twitter trends, and semantic searches on relevant tourist websites. Information is updated and analysed in real time in order to provide the latest trends and comments by tourists in the region. This is a technological asset aimed to be used for controlling and developing the regional tourism strategy.

8.6   ITI/Data Cycle Hub

EUHubs4Data   The European Federation of Data-Driven Innovation Hubs (Coordinator) (1 September 2020–31 August 2023) (no website yet), with the ambition of becoming a reference instrument for data-driven cross-border experimentation and innovation, supports the growth of European SMEs and start-ups in a global data economy. ITI Data Space is the coordinator and leader of the project and one of the i-Spaces providing support to experiments (42 experiments).

REACH   REACH is a European incubator for trusted and secure data value chains (01 September 2020–31 August 2023) (no website yet). It is a second-generation incubator for data-fuelled start-ups and SMEs aiming to develop innovative experiments within data value chains. ITI Data Space is one of the nodes of REACH providing support to experiments and incubation.

TECH4CV   TECH4 CV is an alliance of Competence Centres in Enabling Technologies (https://tech4cv.com/data-hub/) (1 January 2018–31 December 2020). Especially those based on data, to solve the present and future problems of any company of the Valencian Community. ITI Data Space is leading the alliance and providing the Data Space infrastructure for experiments.

DATAPORTS   DATAPORTS is a Data Platform for the Cognitive Ports of the Future (https://dataports-project.eu/) (1 January 2020–31 December 2022). It provides a secure environment for the aggregation and integration of data coming from several data sources existing in the digital ports and owned by different stakeholders to improve processes, offer new services and devise new AI-based and data-driven business models. ITI Data Space provides knowledge, tools and methodologies related to big data and AI in digital infrastructures and data-driven business models.

TransformingTransport (TT)   TT (http://transformingtransport.eu/) (1 January 2017–30 June 2019) demonstrated transformations that big data will bring to the mobility and logistics market. ITI leads the Ports Domain and the Valencia Port Pilot, providing ITI Data Space for analysing data in ports.

8.7 Know-Center

"Mobile Phone Data Analysis" This is an extraordinary example of transferring research results into business. Geospatial data that is continuously generated by cell phones is used to analyse movements of groups of people, thus enabling innovative use cases in intelligent transportation systems (ITS) and in digital marketing. The usage of sensors and embedded technology in vehicles and transportation infrastructure yields new applications in the field of intelligent transportation systems (ITS), such as the prediction of traffic flows and critical transport situations, trip planning in multi-modal transportation and increased traffic. Yet such technology is not pervasively available. Therefore, the application of other location-aware services such as satellite tracking (GPS, Galileo) and cell phone networks is attractive. The latter is of high interest since the technology is available at low cost almost everywhere. Mobile phones regularly generate location-aware (geospatial) events. Other events are cell changes and whenever the user is taking a call or using the data connection. A first study looked at the feasibility of cell phone data in order to detect unusual events such as traffic congestions. The task was to identify congestions, especially on lower-ranking streets, by applying cell phone data without having access to the exact position of individuals, thus satisfying privacy concerns. An algorithmic challenge was how to deal with mobile phone events and their possibly inaccurate data in order to reconstruct trajectories. This resulted in a pool of knowledge, robust tools and scientific publications (Horn et al., 2014KC). Additionally, we addressed topics like transportation mode detection and map matching (Schulze et al. 2015KC). A further challenge was the processing of such data since it arrives as a stream of millions of users simultaneously.

Visual Multi-Perspective Optimisation of Logistic Processes A logistics dashboard is an interactive platform for optimising global logistics processes involving relevant stakeholders in the discussion of strategic alternatives. Logistical processes in production are characterised by a multitude of perspectives with orthogonal optimisation goals. This project addressed the problem of creating a global optimisation strategy for logistical processes through a data-driven visualisation which depicts key parameters and computes models to perspectives. In moderated discussions with the stakeholders different perspectives have been analysed, key parameters identified and interrelations between perspectives established. To inspect the logistical process from all perspectives, an optimum is devised from a dialogue between humans, machines and data. A crucial point that was successfully addressed is that in the optimisation process, human aspects and department interests play as much of a role as data and computational considerations. The interactive visual interface (dashboard) shows information for one or more selected parts. The parameters from various stakeholders are adjusted to view the impact on relevant key performance indicators. Green bars represent the optimum (i.e. corresponding to lowest costs). The key success factors of the resulting solution are both the model and the simulation, as well as the involvement of all stakeholders in discussing strategic alternatives based on real data.

"Participation in Global Scientific Challenges" Participating in global scientific challenges is our method of choice to benchmark ourselves with research teams worldwide, to test our skills and boost our motivation. We participate in global scientific challenges and compete with research teams worldwide to boost motivation and test our skills. Examples include SemEval, INEX, PAN, SciSumm and SemPub hosted at conference series like JCDL and ESWC, or at venues like CLEF. We won the Book Search shared task at INEX. We were awarded Most Innovative Approach at SemPub and we achieved Second Best Performance at SciSumm, with results having been presented at the SIGIR'17 and being an integral part of a master's thesis finished in 2018.

"Magna Painting Finishing Optimization" Based on the parameters of the paint job, MagnaPaint predicts the types of paint imperfections and informs the operator on which parameters have the strongest influence. Our industrial partner Magna is continuously trying to improve its processes and products via innovative technologies and methods. One focus area is the paint finishing process, where vehicles are coated with a protective lacquer. Due to external and internal influences, the coating may contain imperfections, which need to be manually removed, which is a costly process. By applying data science methods, we analysed the data and identified a number of root causes for various types of imperfections, which help the operator to increase the overall quality. The data consists of a large number of parameters, ranging from chemical measurements to process information. Together with the domain experts of our industrial partner, we developed a machine learning model, in order to forecast the expected quality of the processes. In cooperation with the Knowledge Visualisation Area, we developed a tool allowing the operator to visually interact with the learnt model. With this tool, the operator can experiment with different parameter sets and observe the predicted results, without the need to actually test these parameters in the production environment. This again saves time and costs and also avoids potential disruptions in the production process.

8.8 NCSR Demokritos/Attica Hub for the Economy of Data and Devices (ahedd)

National Network for Precision Medicine in Oncology Demokritos operates one of the four national units that are providing next-generation sequencing genetic diagnostics (solid tumours and peripheral blood) to the oncology clinics of Greece as well as management and big data analytics of the genetic archives.

National Network for the Environment and Climate Change: Demokritos operates a cluster of analytical laboratories evaluating toxic (particle, chemical and radioactive) pollution in the atmosphere, soil, water, the food chain and biological tissues.

NanoNOSE A recently initiated action with impact on both the agricultural and health sectors, NanoNOSE, will develop AI methodologies that will be used to combine expert input and advanced sensory data for identifying and predicting risk related to the existence of harmful microorganisms in crop silos.

Marie Curie fellowship for the design of material for gas separation membranes: The research will be based on the incorporation of machine learning techniques in a smart screening methodology that will illustrate the missing correlation between structural modification of the materials and their separation performance.

AI4EU The EU's landmark AI project (€20 million project, Jan. 2019–2022) seeks to develop an EU AI ecosystem, integrating the knowledge, algorithms, tools and resources available, and making it a compelling solution for users. Involving 80 partners across 21 countries, AI4EU will unify the EU's AI community.

IASIS Its aims are to seize the opportunity provided by a wave of data heading our way and turn this into actionable information that would match the right treatment with the right type of patient.

8.9   RISE/ICE by RISE

The aim of the D-ICE project is to establish an arena for data-driven innovation. The objective is to improve the conditions for value creation based on advanced data analytics in the industry and society.

The project is financed by national funding (Vinnova) over 21 months, and the partners are Ericsson, RISE SICS and the start-up Logical Clocks. The objective was to strengthen the Swedish competence in data handling, analysis and processing. The project built a collaboration (meeting and tools) platform for data owners and data analysis providers. The basis for the project is the national data centre initiative ICE with all server capacities; analytic tools, for example Flink and HOPS; and the data analytics and industry knowledge that exists within all parts of RISE.

The first pilot case in the project was done together with Scania, a supplier of heavy trucks to a global market. The number of connected Scania vehicles exhibits exponential growth, resulting in large amounts of streaming telematics data. In their own project FUMA, Scania's objective is to develop a big automotive data analytics framework that utilises its collected geolocation data to analyse the behaviour of vehicles from both an individual vehicle perspective and a fleet perspective.

When connecting FUMA to the D-ICE project, new possibilities were created for Scania, to be able to use our collaboration platform for testing new big data platforms and meet and work together with other organisations in our neutral third-party development environment.

The second pilot, for Mobilaris, was done to improve their product and service for positioning of mobiles and other connected equipment. Mobilaris's market is mobile operators, mining industries and public safety. The positioning system of users or equipment data has an operation user dashboard with analytics capabilities. The large dataset requires a distributed data management and analytics system to achieve low response times.

The services provided were Hadoop-as-a-service and analytics tools for the development of algorithms and queries, expert service for consultancy, and two racks of servers for comparison of different types of Hadoop distributions by different vendors.

The problem was solved with a Hadoop-based big data distributed file and analytics system. The i-Space provided a low-hurdle Hadoop as-a-service to get started with distributed data management and analytics, and an expert service as learning support and query analysis, as well as infrastructure in the form of two racks with 20 servers each for comparison operation of different types of Hadoop distributions for an understanding of product implementations.

The ICE i-Space can deliver a system and service not available in a smaller company that does not have the initial skills for operating a data centre and a Hadoop system and does not implement big data-based analytics. Smaller companies do not have the financial muscle to either do this by themselves to get started or to carry out a pre-study for decision making.

8.10   Smart Data Innovation Lab (SDIL)

Smarte Techniker-Einsatzplanung (STEP) The research project Smarte Techniker-Einsatzplanung, or "Smart Technician Mission Planning" (STEP), aims to simultaneously increase the efficiency of technician assignments and the availability of machinery. Information from and about machines generated by emerging technologies, such as predicted service demand, will be used. STEP is funded by the Federal Ministry for Economic Affairs and Energy (BMWi) in the context of the programme "Smart Service Welt I". Several project partners will work on the simulation model with real dispatching operation data. This requires a safe and cooperative setting which is offered by SDIL (http://www.sdil.de/en/projects/smart-technician-mission-planning-step/).

BigGIS: Fusion of Geospatially Distributed Heterogeneous Sensor Data BigGIS is a joint project between the regional office for environmental protection and various universities and firms in Baden-Württemberg. The project deals with big data and the fusion of uncertain geographic data. Increasing data volumes and increasingly complex calculation models require fast and robust procedures. Together with the SDIL, suitable algorithms are implemented, tested and further developed on the basis of temperature data. It aims at a scalable system that takes into account the peculiarities of spatial and temporal relationships. Therefore, the system must be able to merge the geospatial data as well as a model of its uncertainty, taking into account the heterogeneity of the data sources. The computing resources of the SDIL offer considerable added value for BigGIS, since data volumes in the gigabyte to terabyte range are processed (http://www.sdil.de/en/projects/biggis-fusion-of-geospatially-distributed-heterogeneous-sensor-data/).

Smart Data Solution Center Baden-Württemberg Project Networking Knowledge. Building a technology referral service is a complex venture. The demands on smart technologies and continuous evaluation are very high and require a well-established methodology. Coral Innovation, a young start-up of the University of Stuttgart, implemented just such a service and was supported by experts from SDSC-BW. The free-of-charge potential analysis with more than 8000 binary test classification questions was carried out on the SDIL platform and showed possible optimisation of the classification values (http://www.sdil.de/en/projects/sdsc-bw-networking-knowledge/).

TransformingTransport: Ports as Intelligent Logistics Hubs This project is part of the TransformingTransport EU lighthouse project that aims to demonstrate, in a realistic, measurable and replicable way, the transformative effects that big data will have on the mobility and logistics market. TransformingTransport brings together knowledge, solutions and impact potential of major European ICT and big data technology providers with the competence and experience of key European industry players and public bodies in the mobility and logistics domain. This project should demonstrate how solutions for objectives of a seaport pilot can be replicated and reused for the more challenging setting of an inland port. Compared to seaports, the added complexity in an inland port stems, for example, from the fact that the port is situated in the middle of a large city and at the centre of a large metropolitan area. This means that it has a multitude of roads, tracks and waterways that serve as entry and exit points for containers to and from the actual terminals and ports. In addition, roads need to be shared with many other cars within the metropolitan area. This task will extend the results of a large national innovation project on logistics control towers and enhance them with advanced big data analytics and visualisation capabilities that integrate the various relevant data sources from the port and terminals (http://www.sdil.de/en/projects/ports-as-intelligent-logistics-hubs).

8.11   TeraLab

MIDIH ("Manufacturing Industry Digital Innovation Hub", H2020, I4MS) (fully operational since October 2017): (www.midih.eu). MIDIH is a "one-stop shop" of services, providing industry with access to the most advanced digital solutions and industrial experiments, pools of human and industrial competencies, and access to "ICT for manufacturing" market and financial opportunities.

BOOST4.0, operational, started in January 2018 (www.boost.eu). *BOOST 4.0* "Big Data Value Spaces for Competitiveness of European Connected Smart Factories 4.0" will demonstrate, in a realistic, measurable and replicable way, an open, certifiable and highly standardised and transformative shared data-driven Factory 4.0. BOOST 4.0 will also demonstrate how European industry can build unique strategies and competitive advantages through big data across all phases of the product and process lifecycle (engineering, planning, operation, production and after-market services) building upon the BOOST 4.0 connected smart Factory 4.0 model to meet the Industry 4.0 challenges.

AI4EU will efficiently build a comprehensive European AI-on-demand platform to lower barriers to innovation, to boost technology transfer, and to catalyse the growth of start-ups and SMEs in all sectors through open calls and other actions. The platform will act as a broker, developer and one-stop shop providing and showcasing services, expertise, algorithms, software frameworks, development tools, components, modules, data, computing resources, prototyping functions and access to funding. Training will enable different user communities (engineers, civic leaders, etc.) to obtain skills and certifications.

Proof of ROI (Insurance) Client Profile: Mutual health insurance company (confidential). Client Needs: Early stage data experiment prototype scenario: A large French mutual health insurance company is considering an important strategic move towards novel big data techniques to improve knowledge of their subscriber behaviour. The business lines had identified several use cases, involving heavy machine learning algorithms. They requested support from the IT division, which evaluated the necessary investment. At this stage, the business lines were unable to provide ROI evaluation without concrete experimentation to allow authorisation of such an investment.

Access to research and technology (logistic).

Client Profile: La Poste, Mail Division.

Client Needs: Real value of the data collected by the mail sorting machines. Quality of data and then extraction of useful conclusions about the processes with the focus on two aspects: the fraud of the franking marks and data visualisation of the real process inside a sorting centre to be compared with the theoretical process.

Provided Solution to Meet the Needs: TeraLab provided a workspace and worked closely with La Poste to be able to get 15 Tbytes of data on TeraLab. A research team worked on anonymisation. An innovative company worked on the two use cases described previously. It was the first time La Poste was able to work on the entire dataset.

8.12 *Universidad Politécnica de Madrid/Madrid's i-Space for Sustainability/AIR4S DIH*

TransformingTransport (H2020-731932). This project demonstrated how big data can be used in the context of mobility and logistics (1 January 2017–31 July 2019) (https://transformingtransport.eu/). Our role in this project has been the creation of a data portal for all the open and closed data used by pilots in this project. The data portal is available at https://data.transformingtransport.eu/.

BigStorage (H2020-642963). This Marie Curie ITN project focused on training data scientists in order to enable them to apply holistic and interdisciplinary approaches, taking advantage of a data-overwhelmed world, which requires HPC and cloud infrastructures (1 January 2015–31 December 2018) (http://bigstorage-project.eu/). Our role in this project was in the development of efficient I/O techniques for big data management.

BigDataStack (H2020-779747). A project focused on delivering a completely open-source stack of high-performance technologies (1 January 2018–31 December 2020) (https://bigdatastack.eu/). Our role in this project is in the development of part of the open-source technology stack.

BigMedilytics (H2020-780495). A project focused on the application of big data technologies in the health sector (1 January 2018–28 February 2021) (https://www.bigmedilytics.eu/). In this project, our role is focused on the application of data mining and text mining techniques to health-related documents.

Ciudades Abiertas. This project is funded by the Spanish Government institution red.es, for the provision of Open Government solutions to cities in Spain, piloted in Madrid, Zaragoza, Santiago de Compostela and A Coruña (30 May 2018--31 December 2020 (https://ciudadesabiertas.es/). Our role in this project is the creation of ontologies to guide the publication of open data for these cities.

9 Summary

Despite the increasing relevance of the data economy in Europe, and the importance of data-driven innovation in fostering the digitalisation of companies and society, there are still many actors (small and medium) at national and regional level that do not have access to the benefits of data. There have been many efforts in recent years to solve this issue, from the European Commission, with the Digital Innovation Hubs as main instruments, and also from others, like the Big Data Value Association that is focused more on data, with the Data Innovation Spaces. This chapter presented these and other instruments, introducing their main aspects and characteristics and presenting alignments among them. It also focused on the certification process followed by the Big Data Value Association to recognise relevant initiatives in this field across Europe, and highlighted the importance of collaboration, with the project EUHubs4Data aimed at creating a European federation of Data-Driven

Innovation Hubs, as a meaningful practical example. Finally, the chapter presented some best practices and success stories that could be seen as experiences and lessons for the future.

Acknowledgements We are grateful for the contributions of the following individuals: Claudio Arlandini (Project Manager HPC for industry presso CINECA), Roberta Turra (Team Lead CINECA), Anne-Sophie Taillandier (Director of TeraLab), Natalie Cernecka (Head of Busines Development of TeraLab), Maria Eugenia Fuenmayor Garcia (Scientific Director of ICT and Media Areas Eurecat), Professor Stefanie Lindstaedt (CEO Know-Center and Director of Institute of Interactive Systems and Data Science at Graz University of Technology), Tor Björn Minde (Head of Lab at RISE ICE Datacenter), Dr. Ricardo Simon Carbajo (Head of Innovation and Development CeADAR – Ireland's Centre for Applied AI), Dr. Edward McDonnell (Centre Director, CeADAR – Ireland's Centre for Applied AI), Sergio Mayo Macías (Information Systems Project Manager ITAINNOVA), Michael Beigl (Head of national competence centre for Big Data AI, the Smart Data Innovation Lab SDIL), Sy Holsinger (Strategy and Innovation Lead/Business Development Manager at EGI Foundation), Periklis Terlixidis (Executive Officer of Attica Hub for the Economy of Data and Devices (ahedd) at NCSR "DEMOKRITOS"), Cristina Sandoval (International Projects Office at Universidad Politécnica de Madrid), Oscar Corcho (Professor at Ontology Engineering Group UPM). The research leading to these results received funding from the European Union's Horizon 2020 research and innovation programme under grant agreement no. 951771 (EUHubs4Data) and under grant agreement no. 732630 (BDVe).

Reference

Zillner, S., Curry, E., Metzger, A., Auer, S., & Seidl, R. (Eds.). (2017). *European big data value strategic research & innovation agenda*. Retrieved from Big Data Value Association website https://bdva.eu/sites/default/files/BDVA_SRIA_v4_Ed1.1.pdf

Part III
Business, Policy, and Societal Elements of Big Data Value

Big Data Value Creation by Example

Jean-Christophe Pazzaglia and Daniel Alonso

Abstract The Big Data Value contractual Public-Private Partnership between the European Commission and the Big Data Value Association (BDVA) was signed in October 2014. Since then, more than 50 projects and numerous BDVA members have explored how data can drive innovation across the data stack and how industries can transform business practices. Meanwhile, start-ups have been working at the confluence of new sources of data (e.g. IoT, DNA, HD pictures, satellite data) and new or revisited processing paradigms (e.g. Edge computing, blockchain, machine learning) to tackle new use cases and to provide disruptive solutions for known problems. This chapter details a collection of stories showing concrete examples of the value created thanks to a renewed usage of data.

Keywords Big data · Best practice · Data-driven innovation · Digital transformation · Success story

1 Introduction

Since the signing of the Big Data Value contractual Public-Private Partnership (PPP) in October 2014, more than 50 projects and numerous BDVA members have explored how data can drive innovation across the data stack and how industries can transform business practices. They are working at the confluence of new sources of data (e.g. IoT, DNA, HD pictures, satellite data) and new or revisited processing paradigms (e.g. Edge computing, blockchain, machine learning) to tackle new use cases and to provide disruptive solution for known problems (Zillner et al. 2017). The dimensions of big data value are multiple: they embrace data; skills; legal

J.-C. Pazzaglia (✉)
SAP, Mougins, France
e-mail: jean-christophe.pazzaglia@sap.com

D. Alonso
ITI, Valencia, Spain

and policy issues; technology leadership through research and innovation; transforming applications into new business opportunities; the acceleration of business ecosystems and business models, with a particular focus on SMEs; and successful solutions for the major societal challenges Europe is facing in areas such as health, energy, transport and the environment (Cavanillas et al. 2016).

With an initial indicative budget from the European Union of €534 million for the period 2016–2020 and €201 million allocated in total by the end of 2018, the BDV PPP has already mobilised €1570 million of private investments since the launch of the PPP (€467.47 million for 2018). Forty-two projects were running at the beginning of 2019 and the BDV PPP in only 2 years developed 132 innovations of exploitable value (106 delivered in 2018, 35% of which are significant innovations) including technologies, platforms, services, products, methods, systems, components and/or modules, frameworks/architectures, processes, tools/toolkits, spin-offs, datasets, ontologies, patents and knowledge. Ninety-three per cent of the innovations delivered in 2018 had an economic impact, and 48% had a societal impact. In 2018 alone, the BDV PPP organised 323 events (including own events, seminars and conferences) outreaching over 630,000 participants; and taking into account mass media, the Monitoring Report 2018 (*Big Data Value PPP Monitoring Report 2018* 2019) estimated the number of people outreached and engaged in dissemination activities as 7.8 million.

But how to make these numbers tangible? How to explain what the BDV PPP actors achieved? To answer these questions, in Spring 2019 the BDVA and the BDVe project launched the Best Success Story Award to identify and give visibility to success stories based on impact, developed in a way that can be easily explained to a broad audience. The first edition of the award enabled the five finalists to present their stories on stage at the BDV PPP Summit 2019 in Riga (Fig. 1).

The first edition, won by the TransformingTransport project with DataBio/Wuudis as runner-up, had the chance to have Mrs. Dace Melbārde, Member of the European Parliament and former Minister for Culture for the Republic of Latvia, award the prize to Rodrigo Castiñeira González, the project coordinator. The 2020 edition introduced a new category – SMEs and start-ups – and the awards ceremony took place during EBDVF 2020 with the Data Pitch project and the start-up Orbem as winners in their respective categories, while Ubiwhere was distinguished for the quality of its promotional video (Table 1).

In this chapter, we decided to present a set of success stories representative of the BDV PPP activities amongst the 2019 and 2020 participants. Each section shows the collateral provided by the contenders, a summary of the story and contact details to enable the reader to investigate further.

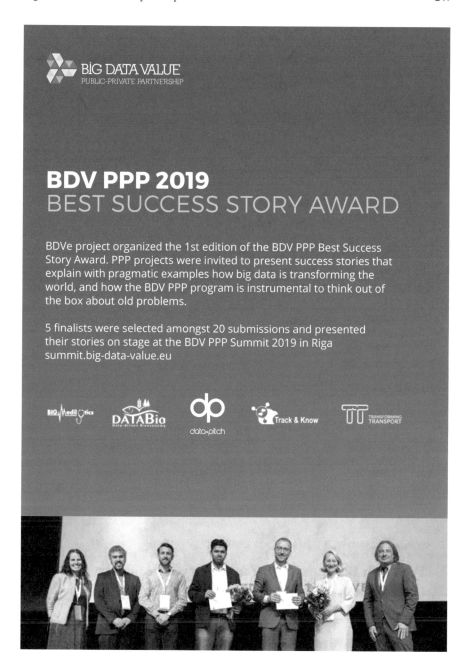

Fig. 1 BDV PPP 2019 Best Success Story Awards Ceremony

Table 1 Main characteristics of the stories

Title	Participant	Industries	Societal impact	SME enablement
How Can Big Data Transform Everyday Mobility and Logistics?	TransformingTransport	Transport	Transformative	
Digitalizing Forestry by Harnessing the Power of Big Data	DataBio/Wuudis	Agriculture	Environment	Yes
GATE: First Big Data Centre of Excellence in Bulgaria	GATE	Research and innovation	Digital divide	Yes
Beyond Privacy: Ethical and Societal Implications of Data Science	e-SIDES		EU policy	
A Three-Year Journey to Insights and Investment	Data Pitch	Incubator Healthcare	Personalised medicine	Yes
Scaling Up Data-Centric Start-Ups	Data Market Services	Incubator		Yes
Campaign Booster	EW-Shopp JOT	Retail		
AI Technology Meets Animal Welfare to Sustainably Feed the World	EDI Orbem	Food	Environment	Yes
Creating the Next Generation of Smart Manufacturing with Federated Learning	Musketeer Comau	Industry 4.0		
Towards Open and Agile Big Data Analytics in Financial Sector	I-BiDaaS CaixaBank	Financial	Cybersecurity	
Electric Vehicles for Humans	Track & Know	Transport	Environment	
Enabling 5G in Europe	Ubiwhere	Telecom		Yes

2 How Can Big Data Transform Everyday Mobility and Logistics?

TransformingTransport (TT) is one of the first two lighthouse projects of the EU Big Data Value Public-Private Partnership. The project, coordinated by Indra, has involved 49 partners. During its 31 months of execution, TT has been able to demonstrate the transformation that big data could bring to the mobility and logistics industries, which represent 15% of the global GDP and employ over 11 million people in the EU-28 zone. TransformingTransport leverages big data to reinvent and optimise mobility and the transport value chain. Significant results from pilots showed increased traffic observation of 70% in the city of Tampere (Finland), accurate traffic and accident predictions up to 2 h in advance on the AUSOL highway in Spain, reduced overall turnaround times and increased gate capacity of up to 10% at Malpensa Airport, reduced truck driving and handling process of 17% at a critical central EU Corridor (Amsterdam to Frankfurt), and reduced delivery vehicle usage at Valladolid (Spain) of 30% (Fig. 2).

3 Digitalizing Forestry by Harnessing the Power of Big Data

The importance of forests with carbon sink and wood as renewable materials to replace synthetic, oil-based materials is growing rapidly. For this, a digital forest management solution integrated with 'data to decisions' is essential as it makes the business value chain more efficient. The 'forestry pilot' implemented within the scope of the H2020 DataBio project introduced a new standard for a forest management plan to enable easy data sharing across the full range of forest stakeholders. Moving from the static paper-based forest management plan updated every 10 years, the Wuudis forest management platform was introduced to manage all of the forest business data in one place. The introduction of Laatumetsä ('quality forest' in English), a forestry-specific mobile solution for 'fieldwork quality monitoring' and 'forest threat data collection', enables both field workers and citizens to collect forest threat data leveraging AI for automatic image processing. This provides citizens with a unique e-tool to collect forest threat data, and it is the first ever tool in the EU where crowdsourced data has been utilised to control forest damage. Furthermore, the Wuudis platform standard interfaces are developed to integrate different forest data (e.g. data from drone monitoring, very high-resolution satellite data) to develop further services beneficial to the sector (Fig. 3).

Since March 2018, the available amount of open forest data has increased from 0.36 TB to 0.38 TB, the amount of downloaded data has exceeded 10.5 TB, and the service has been visited and data downloaded over 3.5 million times. It is worth noting that the innovations for better forestry developed in DataBio have been tested in the real business environment through customer pilots in Finland, Spain (Galicia), Belgium (Wallonia) and the Czech Republic. This confirms the industry's acceptance of the solutions (Fig. 3).

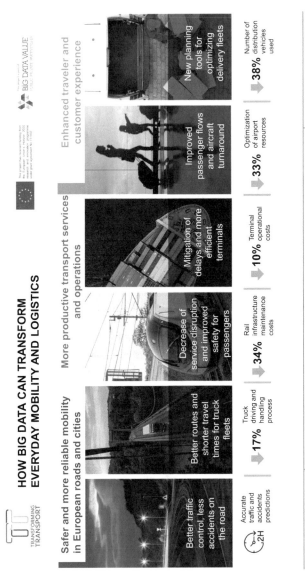

Fig. 2 "How can big data transform everyday mobility and logistics?" Entry

Fig. 3 "Digitalizing forestry by harnessing the power of big data" Entry

4 GATE: First Big Data Centre of Excellence in Bulgaria

The first Centre of Excellence (CoE) in Big Data and AI for Eastern Europe has been launched as 'Big Data for Smart Society' – GATE in Sofia, Bulgaria. The Centre is led by Sofia University 'St Kliment Ohridski', in partnership with Sweden's Chalmers University of Technologies and Chalmers Industrial Technologies (Fig. 4).

Catching the momentum within the booming data and AI-driven EU economy, and supported by the EU's Horizon 2020 Widespread programme, Regional Development Funds and industry, GATE creates a unique research environment and a globally competitive digital hub for big data and AI innovations in future cities, intelligent government, smart industry and digital health. The CoE also accumulates significant expertise and inspires and cultivates the next generation of AI and data scientists and professionals. Providing advanced infrastructure – platform, data, services, and testing and experimentation facilities – GATE City Living Lab, Digital Twin Lab and Visualisation Lab are the heart of a vibrant ecosystem where innovative ideas are generated, developed in projects and applied in effective collaboration with stakeholders. GATE pioneered the usage of the BDVe's best practice guide for big data CoEs, leveraging the collective experience of 31 EU centres on strategy, governance, structure, funding, culture, research-industry collaboration and outreach practice. GATE succeeded in a severe competition, created trust in EC and in the Bulgarian government and industry, and attracted more than €30 million in public and private funding for its operation in the next 7 years.

GATE boosts Bulgarian organisations in target sectors to become, and remain, competitive, thus increasing research capacity and reducing innovation gaps with other EU regions, and also creating confidence amongst citizens and businesses that Bulgaria can efficiently contribute to their needs for a data-driven society and economy (Fig. 4).

5 Beyond Privacy: Ethical and Societal Implications of Data Science

Everywhere we go, from our homes and workplaces to holiday destinations and shopping trips, we generate huge amounts of data which are stored, analysed and used by companies, authorities and organisations. Big data is a feature of our everyday lives (Fig. 5).

Data-driven innovation is deeply transforming society and the economy. Although there are potentially enormous economic and social benefits, this innovation also brings new challenges for individual and collective privacy, security, and democracy and participation. Within this framework, the EU-funded e-SIDES project has provided legal, ethical and economic guidance for big data and AI projects. e-SIDES has shown how these issues can be addressed through the use of privacy-preserving technologies leveraged and implemented in their research and

Fig. 4 "GATE: First Big Data Centre of Excellence in Bulgaria" Entry

Fig. 5 "Beyond privacy: ethical and societal implications of data science" Entry

architectures at design time. In 3 years, e-SIDES involved more than 3500 stakeholders in 25 events and was selected as the Success Story Innovation Highlight for DG Connect (Fig. 5).

6 A Three-Year Journey to Insights and Investment

At Data Pitch, we understand that data has the potential to create huge value for businesses, that start-ups and entrepreneurs have the initiative and ideas to create solutions to sector challenges, and that large organisations can unlock hidden potential in their businesses by sharing data and collaborating with start-ups. We set a range of Data Pitch challenges relating to the industries that are identified in the SRIA as having shown or predicted significant gains from data innovation. As an example, the aim of the 'Health and Wellness' Challenge – featured in the 2019 Best Success Story – was to identify and analyse patterns in patients' clinical pathways. This first cohort showed the importance for start-ups of working closely with medical data providers in order to manage the challenges surrounding sharing medical data. The result was an increase in client base and pilots' outreach, securing more than €7 million worth of new funding. By end 2019 – the official closure of Data Pitch – we supported 47 data-driven start-ups from 13 different EU countries. Collectively to date, the start-ups have amassed a total of €22.4 million worth of impact through further investment, sales and efficiencies. Not only have we seen great success in terms of impact, but the programme is also estimated to see just a mere 6% (3) death rate of companies over the same period (2022). Data Pitch has not only helped businesses and public sector organisations to unlock value from data, but the partners have also enabled early-stage companies to create viable long-term solutions. By working closely with the Big Data Value Public-Private Partnership (BDV PPP), we aim to share these insights and learnings to support other EU-funded programmes to achieve similar success in helping to drive a positive impact within the European data economy (Fig. 6).

7 Scaling Up Data-Centric Start-Ups

Data Market Services is a consortium of accelerators, investors, consultants, lawyers, universities and corporations created in 2019 under the European Union's Horizon 2020 research and innovation programme (Fig. 7).

Its objective is to serve as a gateway for data-centric SMEs and start-ups in Europe to overcome market barriers through the provision of free services. The list of services provided includes a data science academy, entrepreneurial training, IP and GDPR awareness, standardisation and data workshops, storytelling packages, trust-building, fund-raising packages, and mentoring and venture match-making activities that are tailor-made to the needs and characteristics of their product and

Data Pitch: A three year journey to insights and investment

At Data Pitch, we understand that data has the potential to create huge value for businesses, that startups and entrepreneurs have the initiative and ideas to create solutions to sector challenges, and that large organisations can unlock hidden potential in their businesses by sharing data and collaborating with startups.

As the three-year programme came to a close, we commissioned a study to explore the impact of the programme on its participants, interviewing startups and data providers about their experiences over the course of their participation.

The study, which used both qualitative and quantitative data, evaluates the short and long-term impacts of the Data Pitch programme and puts forward recommendations for future open innovation and data sharing programmes.

Lessons and insights

Below we highlight our key learnings and takeaways from the assessment, as well as sharing our insights on how other European accelerators might benefit from the findings:

- Data Pitch enabled data-driven innovation that would have not otherwise occurred. In providing a platform for data providers to try out open innovation in a low-risk setting, the programme addressed a gap in the innovation support landscape. Data Pitch also succeeded in laying the groundwork for a sustainable European data innovation ecosystem.

- A clearer focus on startups at 'acceleration-stage' maturity could increase overall performance. To support fairness and competition across the EU, the programme accepted applications from companies that were not considered an SME under the EU classification. The programme worked particularly well for startups who already had a product or service that could be developed further

- Data Pitch's pan-European approach was a huge bonus. Startups who were based outside of traditional innovation hubs, such as London and Berlin, found that the cross-national setup afforded them more opportunities than they would usually have access to. While it made 'serendipitous interactions' less likely, the virtual set-up gave them access to a much wider array of data.

- Access to data can affect startups' ability to attract funding. The study found that startups that could access data outside of the Data Pitch programme attracted, on average, €141,000 more in additional funding than startups that could not access other data. The data provided by the data providers was, therefore, an invaluable asset to many of the cohort's startups.

The full Assessment can be found here.

Investment and Long-term Impact

The Data Pitch programme officially ended in December 2019 following the Assessment detailed above. Over the duration of the three year programme, Data Pitch supported 47 data-driven startups from 13 different EU Countries. Collectively to date, the startups have amassed a total of €22.4M worth of impact through further investment, sales and efficiencies.

The Data Pitch partners received a total of €7.1M worth of EU funding to deliver the programme, and it is estimated within the Assessment that a total of 459% Return on Investment (ROI) will be achieved by the end of 2022. This assumption correlates with the current impact/value we have witnessed the programme participants attain. Some of the highlights can be seen below:

- Danish Startup Vital beats secured a total of €1.7M worth of funding from the European Commission's Eurostars Programme. They will use it to launch their AI alarm in the Netherlands, which supports remote care to prevent and treat cardiac disease by enabling early intervention.

- Radiobotics, a Copenhagen-based company working on an AI-based software solution that is challenging the status quo of musculoskeletal radiology, closed a €1.3M investment round.

- Berlin-based startup Phytics attempts to solve the problem of patent disputes with its market intelligence platform, which maps multiple data sources to allow companies to search and analyse technology, their competition, market landscapes, and patents. The startup has just raised several million euros in a financing round led by German investors eCAPITAL and Hightech-Gründerfonds (HTGF).

- French Fertility tech startup Mojo has announced a €1.7M seed round of funding led by Nordic seed fund Inventure. Mojo are aiming to make access to fertility treatment more affordable and accessible by using AI and robotics technology to assist in sperm and egg quality

Not only have we seen great success in terms of impact, but the programme is also estimated to see just a mere 6% (3) death rate of companies over the same period (2022). Data Pitch has not only helped businesses and public sector organisations to unlock value from data, but the partners have also enabled early stage companies to create long-term viable solutions.

By working closely with the Big Data Value Public Private Partnership (BDV/PPP), we aim to share these insights and learnings to support other EU funded programmes to achieve a similar success in helping to drive a positive impact within the European Data Economy.

---

data-pitch
INNOVATION PROGRAMME

Data Pitch
URL:
https://datapitch.eu/
Contact: Ryan Goodman
ryan.goodman@theodi.org

Industry:
Incubator/ healthcare
Award winner 2020

Fig. 6 "A three-year journey to insights and investment" Entry

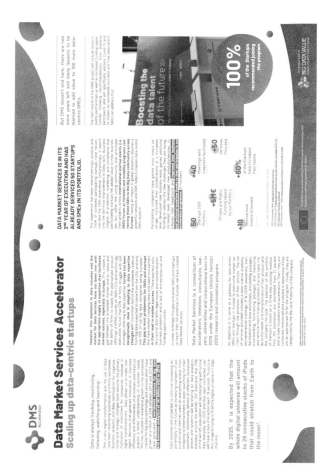

Fig. 7 "Scaling up data-centric start-ups" Entry

the company lifecycle. The selection of the portfolio of start-ups is based on a three-step scouting method. First, the businesses are shortlisted from EC-backed and private incubators and accelerators. Then, they are contacted, monitored and analysed to determine if they are an appropriate fit for the programme. Finally, they are categorised according to the lifecycle maturity of the company.

Over a year, Data Market Services recruited a portfolio of 50 start-ups, facilitated 40 meetings with investors and helped to secure €5 million in funding, with 60% of the start-ups increasing their teams (Fig. 7).

8 Campaign Booster

Digital marketing is evolving towards a content and message personalisation, adapting the services and products offered to the user's likes and needs. This trend is also influenced by external factors like weather and events, which strongly affect user digital behaviour (interests) (Fig. 8).

In this scenario, JOT has combined internal predictive tools and the EW-Shopp toolkit aimed at deploying and hosting a platform to easily integrate multilingual consumer-related data with weather and event data to support analytics on top of the enriched data. The toolkit has processed 2 years of marketing data statistics from Spanish and German campaigns, which represents 100 Gb of data on weather and events. More than 3000 models per region were generated.

This has enabled JOT to predict (1)when the campaign has to be launched, (2) which is the best location, (3) which will be the most relevant category and (4) the expected impact.

Thanks to this new analytical system, by activating campaigns activated relevant keywords, JOT is now able to generate relevant traffic data in 1 day with 30–50% of impressions (Fig. 8).

9 AI Technology Meets Animal Welfare to Sustainably Feed the World

Every year, the global poultry industry wastes 9 billion edible infertile eggs and kills 7 billion 1-day-old male layers. This is unethical, unsustainable and very expensive. Orbem – a start-up that made it to the final stage of the European Data Incubator (EDI) – is developing AI-powered imaging technology to address these problems (Fig. 9).

Orbem's AI technology combines non-invasive sensor technology with AI algorithms to automatically screen eggs. Specifically, we are developing the Genus: AI-powered magnetic resonance imaging (MRI) technology that predicts the fertility status of eggs before incubation and the sex of embryos in ovo. Throughout the EDI,

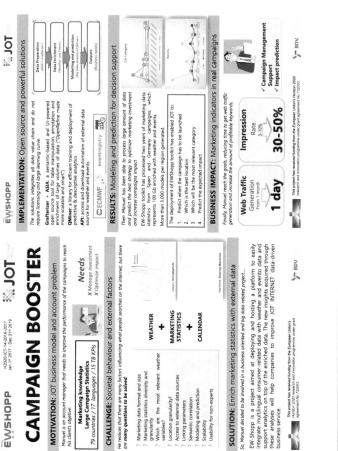

Fig. 8 "Campaign booster" Entry

ORBEM

Orbem: AI technology meets animal welfare to sustainably feed the world

The industrialization of poultry production resulted in the growth of two separate breeds: broilers for meat and layers for eggs. Broilers are optimized to gain weight as quickly as possible, but they suffer from lower fertility rates. Incubating infertile eggs results in a waste of time, energy, and resources, as these cannot be consumed once the incubation process has started. Layers are bred to lay as many eggs as possible, leaving males in an unfortunate situation: males don't lay eggs – obviously – but they are also too thin and have inferior meat quality in comparison to broiler meat, so they cannot compete with broilers in the meat market. Consequently, the modern poultry industry is driven to manually sex and then kill all newborn male layers.

Every year, the global poultry industry wastes *9 billion edible infertile eggs and kills 7 billion 1-day-old male layers*. This is unethical, unsustainable, and very expensive. Orbem – a startup that made it to the final stage of the European Data Incubator (EDI) – is developing AI-powered imaging technology to address these problems, making a difference to the triple bottom line: people, planet, and profit.

Orbem's AI technology combines non-invasive sensor technology with AI algorithms to automatically screen eggs. Specifically, we are developing the Genus AI-powered magnetic resonance imaging (MRI) technology that predicts the fertility status of eggs prior to incubation and the sex of embryos in ovo. Throughout the EDI, we adopted novel Big Data tools to improve AI model performance and to handle the large data streams demanded by the high-volume poultry industry. As a result, our technical solution evolved from proof of concept results to a minimal viable product operating on an industrial-scale computational unit. With these technical results at hand, we were able to confirm the impact of our technology across multiple dimensions.

By scanning eggs, we are creating a positive impact for our customers and the society with a triple bottom line strategy.

1. *People* – we're feeding the world by introducing billions of eggs into the food market.
2. *Planet* – we're preventing the unnecessary killing of billions of 1-day-old male chicks, reducing food waste and energy consumption along the way.
3. *Profit* – We're creating new revenue streams, increasing hatchery productivity and efficiency, and dramatically reducing incubation costs.

Orbem's impact is not only significant, it can be fully quantified. The introduction of 9 billion infertile eggs into the food market would be the equivalent of one egg per day for 24.6 million people. This represents 49.7% of 49.5 million children under 5 malnourished due to wasting according to UNICEF. The prevention of the purposeless killing of 7 billion newborn male chicks is also a measure of impact on its own. Moreover, impact towards this planet's species is also directly associated with impact to the planet. Specifically, we will enable sustainable production systems, reducing current hatchery waste by 94% (all infertile and males) and – accounting for the energy required to power our system– we can reduce energy consumption by up to 40%. Finally, in collaboration with the EDI incubator and poultry producers, we were able to quantify the commercial impact of our solution: we are creating a €2.3 billion yearly market opportunity.

Orbem's end users are enthusiastic by the prospect of installing our solution in their hatcheries. They say that "your technology is the future now" and "I can't see any reason why this machine wouldn't pay for itself". At Orbem, we share our customers' enthusiasm, and are looking forward to creating positive ripple effects across our society, our planet, and our economy.

Company Information
Web: www.orbem.ai
EDI media material: EDI Experiment Video
EDI presentation: Orbem EDI pitch
Orbem Genus: Genus High Resolution
Founders: Dr. Maria Laparidou, Dr. Pedro Gómez, Dr. Miguel Molina

This project has received funding from the European Union's Horizon 2020 research and innovation programme under grant agreement No 779790.

ORBEM

This project has received funding from the European Union's Horizon 2020 research and innovation programme under grant agreement No 779790.

Orbem by EDI
URL:
http://orbem.ai
Project:
http://edincubator.eu
Contact: Gomez Pedro
pedro.gomez@orbem.ai

Industry:
Food
SME Award Winner 2020

Fig. 9 "AI technology meets animal welfare to sustainably feed the world" Entry

Orbem adopted novel big data tools to improve AI model performance and to handle the large data streams demanded by the high-volume poultry industry. As a result, the technical solution evolved from proof of concept results to a minimal viable product operating on an industrial-scale computational unit. With these technical results at hand, they were able to confirm the impact of our technology across multiple dimensions, making a difference to the triple bottom line: people, planet and profit, creating a €2.3 billion yearly market opportunity and the introduction of 9 billion infertile eggs into the food market that would be the equivalent of one egg per day for 50% of 49.5 million children under 5 years of age who are malnourished (Fig. 9).

10 Creating the Next Generation of Smart Manufacturing with Federated Learning

The emerging data economy holds the promise of bringing innovation and huge efficiency gains to many established industries. However, confidentiality and the proprietary nature of data are often barriers as companies are simply not ready to give up their sovereignty. Musketeer offers the capacity to tackle these two dimensions by bringing efficiency while respecting the sovereignty of data providers in industrial assembly lines. Welding quality assessment can be improved using machine learning algorithms, but a single factory might offer too little data to create such algorithms. This requires accessing larger datasets from robots (Comau) located in different places to boost the robustness and quality of the machine learning model. Collecting manual ultrasound testing data and combining it with the welding data from the robot enables the algorithm to be trained locally. In parallel, this machine learning model is trained on different datasets from other factories. Trained models are eventually merged on the Musketeer platform (in a different location) to provide a robust model. Once the model is trained and has a satisfactory accuracy, thanks to this federated approach it becomes possible to provide the classification of the welding spot directly from the welding data. Massimo Ippolito, Head of Digital Innovation and Infrastructure at Comau, states that 'Using federated and collaborative Machine Learning techniques, Comau will be able to provide innovative maintenance services to their customers providing them more robust and more accurate predictive models, using data coming from different customers plants, while at the same time preserving privacy issues related to Company data' (Fig. 10).

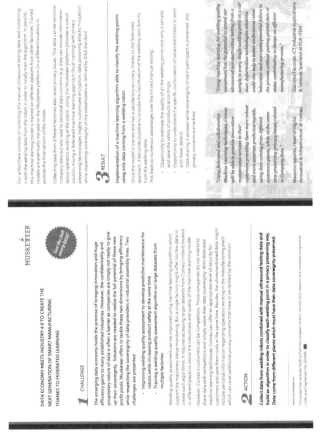

Fig. 10 "Creating the next generation of smart manufacturing with federated learning" Entry

11 Towards Open and Agile Big Data Analytics in Financial Sector

With more than 5000 branches, 40,000 employees and 14 million customers, CaixaBank is one of the largest financial institutions in Spain. Its consolidated big data models use more than 300 different data sources, and more than 700 internal and external active users are enriching its data every day, which is translated into a data warehouse with more than 4 petabytes that increases by 1 petabyte per year. Much of this information is already utilised by means of big data analytics techniques, for example to generate security alerts and prevent potential fraud. CaixaBank receives around 2000 attacks per month. Agility is key in this context, and CaixaBank needed to find ways to bypass rigid processes without compromising security or privacy. The GDPR limits the usage of customer data, even if used for fraud detection and prevention or for enhancing the security of customer accounts. The I-BiDaaS CaixaBank roadmap was a turning point for CaixaBank, and completely changed its approach from non-sharing real data at all positions to looking for the best possible way to share real data and perform big data analytics outside its facilities. I-BiDaaS helped to push for internal changes in policies and procedures and evaluate tokenisation processes as an enterprise standard to extract data outside their premises, breaking both internal and external data silos. This enabled a reduction of 75% of the time to access data by external stakeholders thanks to the use of synthetic data, breaking of data silos, external processing in a compliant way, and evaluation of external big data analytics tools in a much more agile manner (Fig. 11).

12 Electric Vehicles for Humans

Are electric vehicles (EVs) a viable solution for everybody? Within the Track & Know H2020 project, solutions are being developed and tested that, through a mix of mobility data analytics, trip planning and simulation, can analyse the current fuel-based mobility of a user and quantitatively describe the expected impact of switching to EVs on their mobility lifestyle. Electric mobility is frequently addressed as one of the future ways to make cities more sustainable and to improve the quality of life in urban environments.

However, when it comes to private vehicles, the switch has to face the practical difficulties that it might introduce in the lives of travellers, and this is currently a big deterrent for mass conversions to electric vehicles. Single users need to evaluate how their mobility lifestyle is going to change when their fuel-based vehicle is replaced by an electric one, given the various constraints it introduces – the foremost being less independence and (at present) lower availability of recharge points – and in most cases, their lack of means. Our approach includes two answers: 1) numerical Key Performance Indicator (KPI), in particular 'How often would I recharge?', 'How

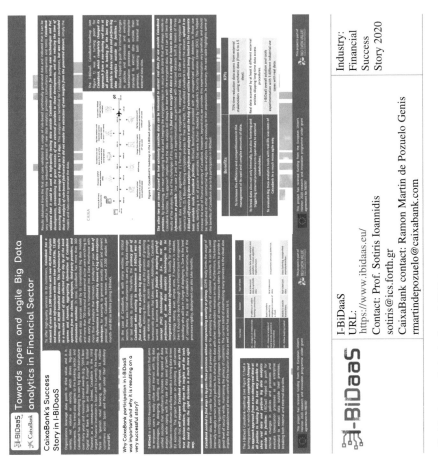

Fig. 11 "Towards open and agile big data analytics in financial" Entry

much time would I waste?', 'How much battery/how many euros would I spend?' and 'How much CO2 would I conserve?'; 2) impact on lifestyle, we place the (expected) recharge activities on the Individual Mobility Network (IMN), in order to understand which moments of a user's life will be affected: the home-to-work routine? Trips to occasional destinations?

A mass analysis of several users can help to identify those who easily convert to using EVs and those who have difficulties. Put on a map, this will help to shape market strategies that address different geographical areas in different ways (Fig. 12).

13   Enabling 5G in Europe

Rui Costa and Nuno Ribeiro were two young(er) researchers developing software for the telecom sector when they decided to take a chance and create their own business. The year was 2007, and Ubiwhere was born in the lovely city of Aveiro, on the sunny and windy coast of Portugal. With a team of three inspired and motivated people, the start-up was created to do precisely what the founders did best: research projects for the telecom sector. Building on its know-how, Ubiwhere focused on the research and development of innovative user-centred software solutions, with expertise in Internet-of-things (IoT) and machine-to-machine (M2M) solutions, data management and analysis, open data, and cloud-based services, targeting the future through innovation. In 2015, the company succeeded in taking the first steps into the next-generation network world. Having shown the SME's data analysis skills and ambition, Ubiwhere was invited to participate in two research projects funded by the European Commission, under the first phase of the 5G-PPP programme. This opened the doors to the creation of future-proof concepts and solution. All experts were present to propose an integrated approach for smart cities and city service providers and to combine multiple vertical domains into a unified ecosystem (mobility, environment and energy), allowing service providers to enhance their operational efficiency and cities to make better decisions based on data collected from diverse sources (Fig. 13).

Ubiwhere is now almost 13 years old, with around 70 employees, building solutions to connect people with everything and leveraging an infinite number of possibilities for services in several sectors that can have a real impact on people's lives. This motivation has led Ubiwhere to continually seek partners that can provide strategic value to both its research activities and commercial endeavours. Today, Ubiwhere is enhancing the future of 50 cities around the world (Fig. 13).

Electric Vehicles for humans

Mirco Nanni (ISTI-CNR, Pisa, Italy), Leonardo Longhi (Sistematica S.p.A., Rome, Italy)

EVs as a viable solution for everybody. Or not? Electric mobility is frequently addressed as one of the future ways to make cities more sustainable and to improve the quality of life in urban environments. However, when it comes to private vehicles, the switch has to face the practical difficulties that it might introduce in the life of travelers, and that are currently a big deterrent for mass conversions to Electric Vehicles (EVs). Single users need to evaluate how their mobility lifestyle is going to change when their fuel-based vehicle is replaced by and electric one, given the various constraints it introduces – the foremost being shorter autonomy and (at the present) smaller availability of recharge points – and in most cases they lack the means to do it.

Within the Track & Know H2020 project, solutions are being developed and tested that, through a mix of mobility data analytics, trip planning and simulation, is able to analyse the current fuel-based mobility of a user and quantitatively describe the expected impact of switching to EVs on her mobility life style.

First, know yourself, or at least your mobility. The fundamental requirement for understanding whether EVs are good for me, is to first understand my own mobility. While that might seem a trivial requirement, individual self-awareness is actually surprisingly overestimated. We all know the key components of our own daily life: going to work, coming back home, maybe shopping every Thursday afternoon, yet, getting a complete and detailed picture that includes occasional trips, deviations from normal routines, the distances covered everyday, which places we pass close by, etc. is well beyond the capability of most users. This precise awareness is what we want to achieve first, and we provide a tool for that, called Individual Mobility Networks (IMNs).

IMNs are concise graph representations of the mobility history of individuals (in our case, the mobility of their cars). From raw GPS traces the trajectories of a single mobility user are reconstructed and processed to refer the relevant locations that the user visited (the nodes of IMNs) and **aggregate** the trips between two locations (the edges of IMNs). Nodes and edges are enriched with several statistics of the associated trips, such as temporal distributions and distances. The following figure shows a pictorial representation of a IMN (left) and several examples extracted from different cities in Tuscany, which reveals a high variability of mobility behaviours, even between users in the same city.

Figure 1 IMN and examples from Tuscany. Thicker and red nodes/edges represent frequent ones

IMNs allow to automatically identify some of the key components of mobility (home, work, systematic movements, etc.) as well as provide detailed statistics about it (average distance of recurrent trips, stay durations, propensity to make occasional trips), allowing to make more precise evaluations and also to better understand the role of each trip in the overall context.

Simulating my new life with an EV. The approach developed starts from the trips performed by the user in her recent history, and estimates what would be the battery consumption needed to perform each trip, taking into consideration the vehicle type (and its battery capacity), the typical speeds of the roads traversed and the their slope. The battery level is therefore simulated along the whole history, and when it is not sufficient to reach the next destination, a stop at a recharge station is included. This simulation is repeated over various scenarios that consider different recharge options: indeed, some users might have a private garage to recharge during the night; others might work in an EV-recharge-equipped building; somebody might be so lucky to have both of them. The Figure below shows the different (and real) stories of a lucky user that recharges a little at home every day (green dots), and an unlucky one that has to rely on public recharge stations and therefore recharges a lot more each time, yet more rarely (red dots).

Although recharging rarely is appealing for a prospect user, the time spent in a recharge station is in most cases simply wasted, while home/work recharging goes more efficiently in parallel with her life.

An interpretable information for an informed choice. The analysis described above provides the means for measuring and characterizing the changes that EVs would have on my mobility. How to convey this information to the average user in an understandable and usable way? Our approach includes two answers: numerical KPIs, in particular "how often I would recharge", "how much time I would waste", "how much battery/euros would I spend", "how much CO2 I would save to the environment"; as numbers are not everything, we place the (expected) recharge activities on my IMN, in order to understand which moments of my life will be affected: the home-to-work routine? My trips to occasional destinations?

The business view. A potential EV user is a potential customer for EV providers and related services. A mass analysis on several users can help identifying those who are more easily convertible to EVs, and those that have more difficulties. Put on a map, this will help shaping market strategies that address different geographical areas in different ways.

Follow the updates on Track & Know H2020 project, G.A. n. 780754. https://trackandknowproject.eu/

Track & Know

URL:
https://trackandknowproject.eu/
Contact: Jennifer Rainbird
jenny.rainbird@inlecomsystems.com

Industry:
Transport
Success Story
2020

Fig. 12 "Electric vehicles for humans" Entry

ENABLING 5G IN EUROPE

WITH INNOVATION AND TALENT FROM PORTUGAL

Rui Costa and Nuno Ribeiro were two young(er) researchers developing software for the Telecom sector when they decided to take a chance and create their own business. The year was 2007, and Ubiwhere was born in the lovely city of Aveiro, on the Portuguese sunny and windy coast. With a team of three - inspired and motivated - people, the startup was created to do precisely what the founders did best: research projects for the Telecom sector.

Ubiwhere's first logo

While developing software solutions for the largest Portuguese telecommunications service provider's innovation lab, and working with disruptive technologies for mobile devices that no one remembers about nowadays, Ubiwhere engaged a telecom operator as its customer for the first time in 2008, having managed to join its first international project just one year afterwards. The team started to grow, and so did the ambition. With the technologies and know-how portfolio expanding, Ubiwhere was able to bring to market three new and innovative solutions:

- **UbiStudio**, a real-time (multi-user) collaborative whiteboard software suite, where users could import Powerpoint presentations or edit Office documents, browse websites directly from the slides and annotate the documents shown in the whiteboard from their desktop;

UbiStudio presented by Rui and Nuno at BETT (left), and CeBIT (right), in 2010, when cameras were not that good.

Smartlamppost online marketplace.

Guimarães (Portugal) was the first city to accommodate this pioneering solution, in 2019, to become smarter and more sustainable. It will soon be made available in Barcelona and Évora as well, with so many more to come. A highly strategic vision, solid partnerships (in projects and alliances like BDVA) and EU-funded research have been the building blocks to unlock the potential of novel technologies and successfully implement solutions that can change the world.

ubiwhere
SUITING THE FUTURE

Ubiwhere's most recent brand.

Ubiwhere is now almost 13 years old, with around 70 employees, building solutions to connect people with everything and leveraging an infinite number of possibilities of services in several sectors that can have a real impact on people's lives. This motivation has led Ubiwhere to continually seek for partners that can provide strategic value to both its research activities and commercial endeavours.

Today, Ubiwhere is suiting the future of 50 cities around the world.

ubiwhere
SUITING THE FUTURE

Ubiwhere	Industry:
URL:	Telecom
https://www.ubiwhere.com/	Success Story 2020
Contact: Ricardo Vitorino	Best Promotional Video
rvitorino@ubiwhere.com	

Fig. 13 "Enabling 5G in Europe" Entry

14 Summary

Ranging from industry transformation to promising start-ups, from agriculture to the retail industry, from the adoption of electric vehicles to ethical and societal policies, we hope that these brief descriptions of the stories give the reader the wish to know more about them. These 13 success stories are only the tip of the iceberg of all the work that is ongoing in the projects and companies from the BDV PPP ecosystem. Exploiting big data requires adding processing capabilities and smart algorithms: in addition to classical analytics tools, we have to highlight that AI technology, especially data-driven AI, is used in the majority of these success stories or the start-ups followed by our different incubators.

The know-how of our members is an extremely valuable asset for Europe, and it is no surprise that several BDV PPP members were instrumental in developing solutions to fight COVID-19 and that INRIA (FR), Orange (FR), INDRA (ES) and SAP (DE) were on the front line in the development of the tracing applications embedded in the privacy by design approach that conforms to the EU's fundamental values.

Choosing amongst all the stories was not an easy task, but we hope that this chapter encourages the reader to learn more about the featured stories and the other stories that we cannot feature due to space limitations. If the reader wants to know more details about these stories and all of the participants in the 2020 contest, they can visit the BDV PPP website at the following URL: https://www.big-data-value.eu/best-success-story-award-2020/.

References

Big Data Value cPPP Monitoring Report 2018. (2019). Retrieved from https://www.bdva.eu/MonitoringReport2018

Cavanillas, J. M., Curry, E., & Wahlster, W. (2016). The big data value opportunity. In J. M. Cavanillas, E. Curry, & W. Wahlster (Eds.), *New horizons for a data-driven economy: A roadmap for usage and exploitation of big data in Europe* (pp. 3–11). New York: Springer. https://doi.org/10.1007/978-3-319-21569-3_1

Zillner, S., Curry, E., Metzger, A., Auer, S., & Seidl, R. (Eds.). (2017). *European big data value strategic research & innovation agenda*. Retrieved from Big Data Value Association website www.bdva.eu

Business Models and Ecosystem for Big Data

Sonja Zillner

Abstract With the recent technical advances in digitalisation and big data, the real and the virtual worlds are continuously merging, which, again, leads to entire value-added chains being digitalised and integrated. The increase in industrial data combined with big data technologies triggers a wide range of new technical applications with new forms of value propositions that shift the logic of how business is done. To capture these new types of value, data-driven solutions for the industry will require new business models. The design of data-driven AI-based business models needs to incorporate various perspectives ranging from customer and user needs and their willingness to pay for new data-driven solutions to data access and the optimal use of technologies, while taking into account the currently established relationships with customers and partners. Successful data-driven business models are often based on strategic partnerships, with two or more players establishing the basis for sustainable win-win situations through transparent resource-, investment-, risk-, data- and value-sharing. This chapter will explore the different data-driven business approaches and highlight in this context the importance of functioning ecosystems on the various levels. The chapter will conclude with an introduction to the data-driven innovation framework, a proven methodology to guide the systematic investigation of data-driven business opportunities while incorporating the dynamics of the underlying ecosystems.

Keywords Big data · Business models · Data-driven innovation · Data ecosystems · Data economy · Innovation ecosystem

S. Zillner (✉)
Siemens AG, Munich, Germany
e-mail: sonja.zillner@siemens.com

© The Author(s) 2021
E. Curry et al. (eds.), *The Elements of Big Data Value*,
https://doi.org/10.1007/978-3-030-68176-0_11

1 Introduction

With the recent technical advances in digitalisation and big data, the real and the virtual worlds are continuously merging, which, again, leads to entire value-added chains being digitalised and integrated. For instance, in the manufacturing domain, all the way from the product design to on-site customer services, the entire value-added chain is digitalised. The increase in industrial data combined with big data technologies triggers a wide range of new technical applications with new forms of value propositions that shift the logic of how business is done.

Big data brings new value to existing and new businesses (Zillner et al. 2017). It enables the optimisation of established internal processes, such as the optimisation of logistics and operations, as well as the basis to monetise new offerings. In general, four different areas of value creation and business models can be distinguished. First, the optimisation and improvement of existing businesses mainly relies on the analysis of available data sources. Second, the upgrading and revaluation of businesses mostly relies on the integration of additional (often external) data sources. Third, monetising describes the realisation of new business opportunities that make use of available data sources. Finally, breakthrough business encompasses new ventures that rely on new data sources, which are often realised with new partners or even within new value networks.

To capture these new types of value, data-driven solutions for the industry will require new business models. The design of data-driven AI-based business models needs to incorporate various perspectives ranging from customer and user needs, their willingness to pay for new data-driven solutions to data access and the optimal use of technologies while taking into account the currently established relationships with customers and partners. In other words, the definition of promising data-driven business opportunities requires balancing the technical aspects on the supply side and the user perspective and market dynamics on the demand side.

In addition, successful data-driven business models are often based on strategic partnerships with two or more players establishing the basis for sustainable win-win situations through transparent resource-, investment-, risk-, data- and value-sharing. To connect all the partners and stakeholders, functioning ecosystems for data sharing, innovation and building value chains are needed.

In this chapter we describe the aforementioned challenges in further detail. To address these challenges, we sketch how the data-driven innovation (DDI) framework can be used to scope data-driven business opportunities by leveraging all needed partners and stakeholders, as well as by continuously aligning the needs on the demand side and the capabilities on the supply side.

This chapter starts by detailing central big data business approaches complemented by some analysis and examples.

In what follows, Sect. 2 gives insights into the different big data business approaches complemented by industrial usage stories. Section 3 elaborates on the nature of data-driven business opportunities, while Sect. 4 highlights the importance of and different levels of data ecosystems. Section 5 gives a short introduction to the

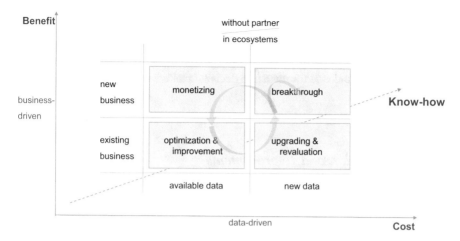

Fig. 1 Four variants of business patterns in the data economy (inspired by BITKOM 2013)

data-driven innovation framework as a possible way forward to scope data-driven business opportunities as well as adjacent ecosystems in a systematic manner. Section 6 concludes the chapter.

2 Big Data Business Approaches

The role of business models is to capture value from advancing technologies, such as big data. Business model decisions should not be driven by economic calculations only but should also consider the value opportunities for building up data asset and technology capability, as well as supporting ecosystems.

Within the data economy, we find various approaches to generating business value with big data technologies.[1] Four generic business patterns (see Fig. 1) can be distinguished: One can generate business value by using existing data sources or by integrating additional and new data sources. The offerings can be realised by a single organisation or within an ecosystem of partners. In addition, the added value might help to improve existing products and services within an established market or can even be used to generate new businesses and sometimes even new markets.

In the following, we elaborate these four patterns by highlighting the involved costs as well as benefits. To capture the cost of each business pattern, we analyse the underlying data complexity and business complexity, as these two factors will

---

[1] Our findings are based on a series of expert interviews we accomplished with project leads/ participants of industrial big data projects.

significantly drive the cost of implementation. To identify the benefits of each business pattern we refer to the value that is created. In addition, each business pattern will be illustrated with some industrial examples. We want to note that this simple classification clearly lacks scientific foundations. Its main objective is to provide strategic guidance for industrial decision makers when investing in big data projects.

2.1 Optimisation and Improvements

The business pattern "Optimisation and Improvements" relies on existing and already available data assets. These data assets require the typical efforts for data pre-processing and cleaning. Value is generated within the context of existing business processes.

Typical examples of this business pattern are as follows:

- Healthcare Domain: Administrative and financial data in hospital settings are analysed to increase the efficiency of the underlying administrative processes, such as scheduling of tasks or the utilisation of resources.
- Energy Domain: Sensor data of gas turbines are analysed to predict future damage and identify the cause of deviations in the process, in the material, etc.

For the above-mentioned examples as well as the business pattern in general, we can summarise the main characteristics of this business pattern.

Value Creation: The optimisation and improvement of existing process and businesses helps to reduce costs or to improve performance.

Data Complexity: In general, data assets are available, but their technical access needs to be ensured. Depending on the type of data source, e.g. sensor data or private data, the respective data governance challenges have to be addressed.

Business Complexity: The optimisation and improvement of established processes are in general a good starting point for data-enhanced offerings. By investigating available data sets, new insights regarding improvement potentials can be discovered while working with the data.

2.2 Upgrading and Revaluation

The business pattern "Upgrading and Revaluation" employs new data sources either by transforming internal raw data sources into a processable format (e.g. by semantic labelling of the content of medical images) or by integrating external data sources (e.g. weather forecast information) and developing new offerings.

Typical examples of this business pattern are as follows:

- **Energy Services Domain**: Data- and knowledge-based services enable the discovery of new insights about trends that help to increase the overall business performance. Analytics applications are applied to all kinds of data (e.g. product data, market data, competitor data, web data, customer data, financial data) to detect and respond to product, event, personnel, competitor, customer and market trends.
- **Healthcare Domain**: A radiologist's workflow can be improved significantly by establishing the means for seamless navigation between medical image and dictated radiology report data. Content information of medical images as well as dictated medical reports is semantically described and linked by metadata.
- **Industry Automation**: The data from Large Hadron Collider (LHC) automation and control components and systems from Siemens (WINCC OA) are collected (offline and online) for automated system health check and diagnostics in order to prevent future damage, which helps to significantly reduce the overall maintenance cost.
- **Global Production Chain**: Intelligent integration of global supply chain management information (e.g. via object tracking information) into the production planning processes increases the robustness and efficiency of the global value chain.
- **Smart Grid Systems**: Optimised energy production through the interactive planning and optimising of the top-level design of microgrids in collaboration with the user. The interaction relies on visual result analysis that enables the user to detect patterns in large and heterogeneous data sets, such as weather data, power demand data, time series and energy capacity data.

For all of the above-mentioned examples as well as the business pattern in general, we can summarise the main characteristics of this big data business pattern.

Value Creation: Here the underlying idea is to upgrade existing business processes and services by making use of additional data sources. By aggregating multiple data sources, insights about process performances and operational and financial measures as well as guidance for business decisions can be provided.

Data Complexity: The integration of new data sources implies efforts and investments for handling the various data governance challenges. In addition to the challenge of accessing external data sources, one might face the challenge of overcoming internal data silos (sometimes even organisational silos), as well as the challenge of pre-processing raw data that is only available in an unstructured format, such as images, videos or dictated text.

Business Complexity: The described data governance challenges might imply high investments such that a long-term business strategy is needed.

2.3 Monetising

The business pattern "Monetising" aims at generating new markets or revenue streams. By exploiting available data sources, completely new business scenarios, offerings and value streams are realised.

Typical examples of this big data business pattern are as follows:

- **Clinical Research**: Patient cohorts for clinical programmes can be identified more easily by using information extraction and advanced analytics on top of clinical data. In consequence, the feasibility of the clinical trials and thus the planning and designing of clinical programmes in the pharmaceutical domain can be significantly improved.
- **Information Service**: BLIDS[2] is a lightning information service, which offers energy providers, industry, insurance companies or event organisers precise information about the number of registered lightning activities. The service aggregates weather data from approximately 8 countries and more than 145 measuring stations in Europe, as well as enabling the user-adapted representation of content.
- **Smart Energy Profiles**: Increasingly, metering service providers, which service renewable decentralised energy resources and new types of demand, such as electric vehicles, can bundle the characteristic information of power feed-in and energy usage into smart energy profiles and sell these for profit. Currently the whole energy market operates with standard load profiles, which are inefficient.
- **Energy Automation**: Intelligent electronic devices deliver real-time high-resolution data on power network parameters. When transmission network operators install these data sources near bigger renewable energy resources, and utilise advanced analytics, they can resell the information gained on the characteristics of the wind park and the network area back to the operator of the park for operational efficiency increase on both sides.

Value Creation: Secondary usage of data, i.e. the user benefiting from the collected data, is outside the original context in which the data were produced and collected.

Data Complexity: It is important here to clarify whether the usage of the available data sources for other purposes is legally allowed.

Business Complexity: Bringing data-driven offerings to the market might trigger the challenges of building new markets and establishing new or redefining established business relationships.

---

[2]Offered by Siemens.

2.4   Breakthrough

Big data applications can lead to breakthrough scenarios that rely on collaborative ecosystems that establish new value networks by aggregating existing data sources with completely new data sources from various stakeholders.

Typical examples of the big data business pattern "Breakthrough" are as follows:

- **Healthcare:** Public health analytics application relies on the comprehensive disease management of chronic (e.g. diabetes, congestive heart failure) or severe (e.g. cancer) diseases that allow for the aggregation and analysis of treatment and outcome data, which again can be used to reduce complications, slow disease progression and improve outcomes.
- **Energy Efficiency:** Efficient energy use is highly dependent on energy automation – down to the device level in a private or commercial user. Learning systems are required which adapt to the preferences or business criteria of the energy user, along with efficient data exchange between retailers, energy markets and the network operators and the actual devices with the energy-efficiency service providers. Finally, smart meters and the metering service providers enable the billing of such complex but efficient energy usage.

Value Creation: Fundamental change of the established value generation logic.

Data Complexity: Heterogeneous data sets from various partners need to be exchanged and shared.

Business Complexity: The implementation of big data applications with break-through/disruptive potential is the most challenging business approach. Usually it relies on the interplay of various partners that have managed to establish an effective collaboration. New data sources are aggregated and used in order to develop new products and services. By addressing a new market (segment), breakthrough applications – as the name indicates – have the potential to revolutionise established market settings. It is likely that new players will emerge that are better suited to provide data-based service than the established player, and that the underlying business processes will change fundamentally.

3   Data-Driven Business Opportunities

In general, the concept of business opportunity is very broad, and is used to describe the chance to address a particular market need through the creative combination of resources that allows the delivery of advanced value propositions (Ardichvili et al. 2003). In this way, the definition of promising business opportunities relies on the balancing of – often mainly technical – capabilities on the supply side, with user needs and interests as well as market dynamics shaping the demand side. In addition, studies indicate that most successful entrepreneurs and investors continuously observe the demand side very carefully in order to understand what customers and marketplaces want, and never lose track of this information (Spinelli and Adams 2012). The knowledge reflecting the demand side is used to guide the scoping of offerings by combining own innovative technology components with reusable and available assets from others in a way that fosters competitiveness. In addition, the development of business opportunities is described as a continuous process that involves proactive efforts to explore all essential steps of a new business.

Any innovative technology that is not aligned with a concrete application triggering concrete demand is likely to fail. This is also true for big data solutions. Hence, the successful implementation of big data solutions requires transparency concerning the following four questions:

1. Who is using the new solution? (target user)
2. Who is providing the data? (data)
3. Who is paying for the solution? (revenue model)
4. Who needs to adopt the solution to bring it to the market? (ecosystem)

For instance, the implementation of health data analytics solutions for improved treatment effectiveness by aggregating longitudinal health data requires high investments and resources to collect and store patient data, for instance by means of a dedicated Electronic Health Record (EHR) solution (data). Although it seems to be quite obvious how the involved stakeholders, such as patients, payors, government or healthcare providers, could benefit from aggregated data sets (target user), it remains unclear whether they would be willing to pay (revenue model) or adopt such an implementation (ecosystem). In addition, as the sharing of personal health data is subject to high security and privacy constraints, one needs to clarify under which conditions the healthcare provider who produced and thus owns the data can and is willing to share the patient data (data).

The aforementioned responsibilities might be distributed across organisational boundaries. If the business approach is mainly targeting the optimisation and improvement of existing offerings, the identification of data-driven business opportunities is often within the scope of established partnerships and capabilities. However, if the business approach is aiming at a collaborative setting within new market and business domains, the scoping of business opportunities easily becomes a challenging task with many unknown variables that often cannot even be influenced by the organisation, as elaborated below:

- **Big data applications often rely on high investments to ensure data availability:** The collection and maintenance of comprehensive and high-quality data sets not only requires high investments but often takes some years until the data sets are comprehensive enough to produce good analytical results. For instance, in the medical domain, one would need to collect large-scale, high-quality and longitudinal data in order to gain reliable insight about the progress of diseases over time. As such high and long-term-based investments often can't be covered by one single party, the conjoint engagement of multiple stakeholders might be required.

- **Collaboration with partners with diverging interests:** As the impact of big data solutions increases and more data sources can be aggregated, an effective collaboration of multiple stakeholders with potentially diverging or even opposing interests needs to be established. In addition, the stakeholder's individual interests and constraints might even change over time. In the German electricity market liberalisation, a new market role of metering service providers was created in 2010. They are responsible for harvesting the energy usage data and could foster whole new branches of business. In addition, a range of stakeholders will require data on energy usage: retailers, network operators and new players that offer energy-related services like demand response. However, in order to establish the basis for an effective collaboration, the interests of the various stakeholders need to be reflected when developing the business case. Especially, the ambiguous regulatory framework on the rights and responsibilities of smart data usage prevents the existing and potential new players in utilities business to take on the new role of metering service provider.
- **Technological capabilities as well as its cost are a moving target:** Not only technological capabilities but also their cost factors are changing fast. Computing power and memory space per unit costs are still progressing exponentially according to Moore's Law. Additionally, an innovative and cost-effective form of information processing that is the main characteristic of all big data technologies decreases the cost and update cycles of technologies considerably. Thus, the cost factor of technological investments needs to be accounted for in the overall calculation.

Having explained why the development of data-driven business opportunities is very challenging, we need to emphasise that the lack of a business case should not hinder investments in big data projects. Instead, organisations should actively engage the emerging data ecosystems that will allow them to gain access to promising user groups and target customers, data assets and technologies, and stakeholders.

4   Leveraging the Data Ecosystems

As the impact of most big data applications increases exponentially, more data (scale) from different data sources (scope) can be integrated and analysed. In addition, the deployment of big data applications in industrial and public environments relies on incorporating the domain knowledge of underlying processes, as well as the alignment of many other horizontal technologies (e.g. cybersecurity, HPC, Internet of things, communication) and established systems. Therefore, the implementation of big data applications requires the collaboration of multiple – often competing – stakeholders on various levels: (a) for sharing the data assets; (b) for sharing technology, skills and knowledge with partners and stakeholders and (c) for establishing value networks generating new business.

Thus, the majority of big data business will take part in ecosystems. Successful ecosystems can help whole economic sectors as well as single players to prosper and develop. However, the governance of ecosystems relies on a balanced give and take. Looking at the various types of data, assets and actors in the data ecosystem will help to illustrate the underlying incentives and roles. The successful governance of big data ecosystems needs to reflect the interests and strategies of all players involved. We can distinguish ecosystems on three different levels.

4.1 Data-Sharing Ecosystem

The impact of big data applications increases if the multiple data sources from the various stakeholders of an industrial sector are integrated. For instance, in healthcare, by aggregating the administrative data and financial data with clinical data, it becomes possible to gain insights about the outcome of treatment bundles in terms of resource utilisation. Thus, cooperative settings for the sharing of data are needed. In order to establish sustainable data-sharing ecosystems, it is important to understand:

- Which data source(s) each actor can potentially provide
- What his or her sharing incentives are
- Which requirements (e.g. privacy standards, "opt out" ability, business models) need to be in place in order to enable/foster the sharing of data

For those who are providing data, a mechanism must be developed to ensure transparency and control of data usage, as well as some added value that is enough motivation to provide the data. Individuals might want to receive improved offerings and services with added value or better prices. Companies are interested in data to improve their knowledge about the consumer in order to customise their offerings, increase customer binding or optimise their pricing strategy.

4.2 Data Innovation Ecosystems

The data innovation ecosystem is complex and diverse. It contains multiple types of stakeholders, and, to be effective, there needs to be alignment and collaboration between them. It is the "agora" for the sharing of assets, technology, skills and knowledge. It provides scale to achieve consensus and critical mass around the development of AI value through innovation that no single partner alone could achieve (Zillner et al. 2020). It expresses the collaborative purpose that binds organisations and individuals together in achieving successful deployment of AI. The ecosystem is typically composed of the following roles:

- **End User:** Person or organisation from different sectors (private and public) that leverages AI technology and services to their advantage
- **Application Provider:** An organisation that uses AI technology for developing a vertical AI application (e.g. to be offered as AI service)
- **User:** A person who either knowingly or unknowingly uses or is impacted by a system product or service that uses AI
- **Data Supplier:** Person or any organisation (public or private) that creates, collects, aggregates and transforms data from both public and private sources
- **Technology Creator:** Typically, an organisation (of any size) that creates tools, platforms, services, hardware and technical knowledge
- **Broker:** An organisation that connects the supply and demand for AI assets (such as skills, data, algorithms and infrastructures) needed for developing AI applications by providing a channel for exchanging AI assets
- **Innovator/Entrepreneur:** Drives the development of innovative AI technology, products and services
- **Researcher/Academic:** Researches and investigates new algorithms, hardware, technologies, methodologies and business models; provides skills and training in AI and assesses the societal aspects of its impact
- **Regulator:** Assesses AI systems in compliance with regulation, privacy and legal norms
- **Standardisation Body:** Defines technology standards (consensus-based, de facto and formalised) to promote the global adoption of AI technology
- **Investor/Venture Capitalist:** Provides resources and services to develop the commercial potential of the ecosystem
- **Citizen:** A person who will or will not develop trust in AI technologies

An effective data innovation ecosystem facilitates the cross-fertilisation and exchange between stakeholders that leads to new data-powered value chains that can improve business and society and deliver benefits to citizens.

4.3 Value Networks in a Business Ecosystem

Business ecosystems can be defined as "a dynamic structure which consists of an interconnected population of organizations. These organizations can be small firms, large corporations, universities, research centres, public sector organizations, and other parties which influence the system" (Brynjolfsson and McAfee 2012; Peltoniemi and Vuori 2004). They allow organisations to access and exchange many different aspects of value, resources and benefits.

The data economy relies on value networks. In the data-driven economy, value streams are no longer bi-directional but involve several players exchanging different types of value. The party who is benefiting from a value-added service no longer needs to be the one who is paying for the service. Such value networks already exist in the Internet environment.

Most of the established players providing database solutions, such as Google, eBay, YouTube, Facebook and iTunes, are building up a growing user community by offering free services, which allows them to increase their income as each advertising company is paying a fee per click or user.

5 Data-Driven Innovation Framework and Success Stories

The economics of data has a strong impact on the development of data-driven business opportunities. For instance, data can be consumed an unlimited number of times without losing its value, and it can be reused as input for the production of different goods and services. However, its value still depends on complementary assets related to the capability to extract information out of the data (OECD 2015). Given the mentioned economic properties, disruptions through data are becoming more likely. In particular, due to network effects as well as the simplicity of how a variety of offerings with different value/price tags can be brought to the market, the success of data-driven innovation requires continuous alignment between the needs on the demand side and the opportunities on the supply side.

So how can data-driven business opportunities be screened? The data economy in general is a highly dynamic market. This is supported by the rapid growth of the European data markets, as well as recent technical breakthroughs that were made possible by the availability of large volumes of data, such as the Jeopardy demo by IBM Watson or Google Now or Siri. In addition, experts continue to highlight the wide range of commercial opportunities that can be realised by using the technologies available today.

Entrepreneurs bring new offerings to the market and should continuously scan the market's offerings to identify promising available technology components that can be reused to speed up the development time of their innovation. At the same time, although they are confronted with the highly dynamic market, they have to constantly investigate their own unique selling point and the competitiveness of their offering. To stay competitive in this fast-moving market, entrepreneurs need to continuously reassess what is part of their core offering and in which areas they are partnering with others.

The high-growth scenario[3] in the comprehensive European data market study (European Commission & Open Evidence 2017) is based on supply-demand dynamics that shift from technology push to demand pull. In other words, any means that provides guidance in match-making between market needs on the demand side and technical capabilities on the supply side helps to stimulate the development of data-driven innovation and in consequence the growth of the European data market.

To summarise, data-driven business opportunities should be described with a clear scope of offering per market segment (supply side) and reflect the ecosystem

---

[3]Which estimated 4% of GDP growth between 2016 and 2020.

Fig. 2 DDI Canvas with eight dimensions guiding the exploration of the relevant aspects of DDI

dynamics and benefits of network effects (demand side). In the next section, we present a high-level overview of the data-driven innovation framework which guides innovators to systematically explore and analyse the supply and demand sides of data-driven business opportunities by incorporating the particularities of data.

5.1 The Data-Driven Innovation Framework

The data-driven innovation (DDI) framework addresses the challenges of identifying and exploring data-driven innovation in an efficient manner. It guides entrepreneurs in scoping promising data-driven business opportunities by reflecting the dynamics of supply and demand through investigating the co-evolution and interactions between the scope of the offering (supply) and the context of the market (demand) in a systematic manner.

The DDI framework is based on a conceptual model in the form of an ontology with a set of categories and concepts describing all relevant aspects of data-driven business opportunities. Its categories are divided into supply side and demand side aspects. On the supply side the focus is on the development of new offerings. For a clearly defined value proposition, this includes the identification of and access to required data sources, as well as the analysis of underlying technologies. On the demand side the focus is on the dynamics of the addressed markets and associated ecosystems. The analysis includes the development of a revenue strategy, a way forward in how to harness network effects as well as an understanding of the type of business. As data-driven innovations are never done in isolation, the identification and analysis of potential development partners as well as partners in the ecosystem help to align/balance the supply and demand aspects in such a way that their competitive nature will stand out. Figure 2 illustrates the DDI Canvas that covers eight central dimensions to be explored when scoping data-driven innovation.

The DDI framework was developed and tested in the context of the Horizon 2020 BDVe project[4] and is backed by empirical data and scientific research encompassing

---

[4]Zillner. S. D 2.7 Annual Report on Opportunities (BDVe Deliverable), March 2020 Zillner. S. D 2.6 Annual Report on Opportunities (BDVe Deliverable), April 2019 and Zillner. S. et al.: D 2.5 Annual Report on Opportunities (BDVe Deliverable), March 2018.

a quantitative and representative study of more than 90 data-driven business opportunities. The results of the research study guided the fine-tuning and updating of the DDI framework and helped to identify success patterns of a successful data-driven innovation.

Currently the DDI framework is used to run workshops for projects of the BDV Public-Private Partnership, data-driven start-ups and SMEs, and with corporates. It consists of:

- **DDI Canvas** guiding users in exploring all relevant dimensions on the supply and demand sides of a data-driven innovation in a systematic manner.
- **DDI Navigator** and methods that support users in exploring each dimension at the required level of detail by investigating the aspects mentioned by guiding questions as well as by applying complementary methods.
- Specific **DDI Tools** will help the user to work through each of the eight DDI dimensions, producing a conclusive set of results that will guide a company-specific setup of new, data-based products and services.

More details can be found in Chapter "Big Data Value Creation by Example" of this book, at https://ddi-canvas.com/ or in Zillner and Marangoni (2020), Zillner (2019) or Zillner et al. (2018).

5.2 Examples of Success Stories

In the following section, we provide some examples of success stories of the aforementioned research study of data-driven start-ups to give the reader an impression of how clearly and precisely their supply and demand sides can be pitched.

Artomatix is a Dublin-based software company founded in 2014 that uses artificial intelligence to create realistic 3D art creations.

Artomatix's users are artists and developers of the video gaming industry that can benefit from a service that supports the realistic 3D art generation of textures and texturing. Previously, this tedious task was done manually but with the suite of tools provided by Artomatix, artists can now do the same task ten times faster.

The technology is based on computer graphics, Deep Learning and computer vision. It uses generative neuronal networks to "imagine" new details of a texture in a way a human would, i.e. it recognises objects in a video and can add texture and features automatically by relying on the "learned" knowledge that should be there.

The data used for training and developing the algorithm is video and image data. The software can be integrated with Photoshop and leading gaming engines like Unity and Unreal.

The company uses three different subscription models (Indie (revenue $< \$100$ K/year), Professional (revenue $< \$1$ M/year) and Enterprise (revenue $> \$1$ M/year)). Enterprises can license Artomatix's technology and build it into their existing process for an annual fee. The technology is offered as a data-driven service.

There are no network effects that need to be reflected. A short summary is provided in Fig. 3.

5.2.1 Selectionnist

Selectionnist is a France-based company founded in 2014 offering image recognition technology with the goal of connecting readers of print journals with the world's largest brands through an application or a chatbot. They aim to bridge the gap between offline content and online experience by offering an advanced match-making service to connect consumer and brands.

They address two different customer groups with different value propositions:

- Value proposition for consumer: they locate and potentially purchase a product they spot in a magazine just by snapping a picture of it.
- Value proposition for brands: brands can see in real time how readers interact with their editorial and advertising in print magazines.

Selectionnist's match-making algorithm is based on image recognition technology that continuously improves the images of brands' products in their databases (more brands) and the user request they receive. Thus, their offering is based on network effects on data level. The service is conceptualised as marketplace based on commission fee and with network effects on marketplace level. A short summary of the above explanations is provided in Fig. 4.

5.2.2 Arable

Arable is a US-based company founded in 2013 offering agriculture businesses a global solution for managing weather and crop health risks, delivering real-time, actionable insights from the field.

The target users are growers, advisors and businesses who aim to play a proactive role in the quality and longevity of their operations.

The agricultural business intelligence solution is based on in-field measurements allowing the production of real-time continuous visibility and predictive analytics in the areas of crop growth, harvesting time, yield and quality. The solution relies on field-level weather and crop monitoring devices (hardware that is part of the solution) that collect over 40 field-specific data metrics. To enable access to data from anywhere in real time, a cloud-based software platform based on a tiered SaaS offering (different levels of services) is combined with IoT hardware.

Arable sells licences for enterprise software to agribusinesses. As the prediction service improves with more data available, the solution of Arable is based on network effects on a data level. Figure 5 summarises the above-described findings.

ARTOMATIX

PROBLEM

Studio Artists spend a considerable amount of time doing repetitive tasks on 3D art creation instead of focusing on creativity

OFFERING

A service for new artistic workflows, enhancing video footage to new heights and creating stunning final textures in moments

ABOUT

- Headquarter: Dublin, Ireland
- Founded in 2014, acquired in 2019
- Funding in 2018: $2.6 ↑ 313%
- Funding in 2020: $10.8M
- Automating up to 80-90% of 3D artist's scan-based workflow

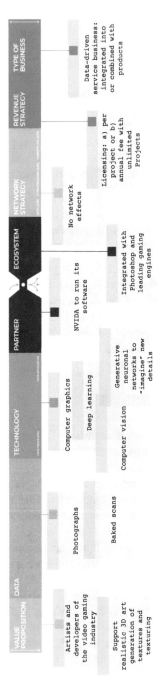

Fig. 3 DDI summary of Artomatix

SELECTIONNIST

PROBLEM

Readers of offline content cannot find the advertised products in an easy manner, being a time-consuming task

OFFERING

Image recognition technology with the aim to connect print readers with the world's largest brands through an application or a chatbot

ABOUT

- Headquarter: Paris, France
- Founded in 2014
- Funding in 2018: €2M
- Funding in 2020: €2M   = 0%
- Identify a product on a picture and match it with an online content

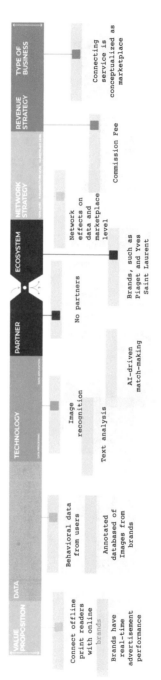

Fig. 4 DDI summary of Selectionnist

ARABLE

PROBLEM

Earth science is mostly in "model world" and little based on empirical data, affecting natural resources management and food waste throughout the supply chain

OFFERING

Global solution for agriculture businesses to managing weather risk and crop health, delivering real-time, actionable insights from your field

ABOUT

- Headquarter: Princeton, USA
- Founded in 2013
- Funding in 2018: $9.8'   = 0%
- Funding in 2020: $9.8h
- 1,000 devices have been deployed in 22 countries globally since 2017

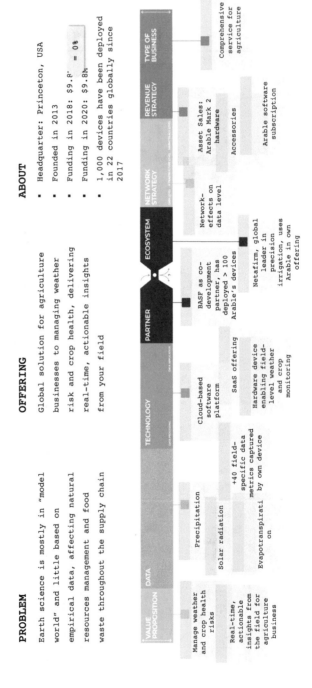

Fig. 5 5 DDI summary of Arable

6 Conclusion

Big data allows new value to be brought to existing and new businesses. To capture these new types of value, the scoping of data-driven business opportunities needs to incorporate multiple perspectives, ranging from user needs, data availability and technical capabilities to the sustainable establishments of partnerships and ecosystems.

The data-driven innovation framework offers a proven method for all members of the BDV ecosystem to provide guidance in exploring and scoping data-driven business opportunities. The comprehensive content can be used for industrial workshops and educational setups.

References

Ardichvili, A., Cardozo, R., & Ray, S. (2003). A theory of entrepreneurial opportunity identification and development. *Journal of Business Venturing, 18*(1), 105–123. https://doi.org/10.1016/S0883-9026(01)00068-4

BITKOM. (2013). *Management von Big-Data Projekten. Leitfaden des BITKOM AK Big Data.*

Brynjolfsson, E., & McAfee, A. (2012). Winning the race with ever-smarter machines. *MIT Sloan Management Review, 53*(2), 53–60.

European Commission, (IDC), I. D. C., & Open Evidence. (2017). *Final results of the European Data Market study measuring the size and trends of the EU data economy.*

OECD. (2015). Data-driven innovation: Big data for growth and well-being. *OECD digital economy papers*, p. 456. https://doi.org/10.1787/9789264229358-en

Peltoniemi, M., & Vuori, E. (2004). Business ecosystem as the new approach to complex adaptive business environments. *Proceedings of EBusiness research forum.*

Spinelli, S., & Adams, R. (2012). *New venture creation: Entrepreneurship for the 21st century.*

Zillner, S. (2019). *D2.6. Annual report in opportunities, BDVe deliverable 2.6.*

Zillner, S., & Marangoni, I. (2020). *D2.7 Annual report on opportunities; BDVe deliverable 2.7.*

Zillner, S., Curry, E., Metzger, A., Auer, S., & Seidl, R. (Eds.). (2017). *European big data value strategic research & innovation agenda.* Retrieved from big data value association website: www.bdva.eu

Zillner, S., Timan, T., & Kotterink, B. (2018). *D2.5 Annual report in opportunities.*

Zillner, S., Bisset, D., Milano, M., Curry, E., Hahn, T., Lafrenz, R., et al. (2020). *Strategic research, innovation and deployment agenda - AI, data and robotics partnership. Third Release (3rd).* Brussels: BDVA, euRobotics, ELLIS, EurAI and CLAIRE.

Innovation in Times of Big Data and AI: Introducing the Data-Driven Innovation (DDI) Framework

Sonja Zillner

Abstract To support the process of identifying and scoping data-driven innovation, we are introducing the *data-driven innovation (DDI) framework*, which provides guidance in the continuous analysis of factors influencing the demand and supply sides of a data-driven innovation. The DDI framework describes all relevant aspects of any generic data-driven innovation and is backed by empirical data and scientific research encompassing a state-of-the-art analysis, an ontology describing the central dimensions of data-driven innovation, as well as a quantitative and representative research study covering more than 90 data-driven innovations. This chapter builds upon a short analysis of the nature of data-driven innovation and provides insights into how to best screen it. It details the four phases of the empirical DDI research study and discusses central findings related to trends, frequencies and distributions along the main dimensions of the DDI framework that could be derived by percentage-frequency analysis.

Keywords Data-driven innovation · Business models · Data ecosystems · Value proposition · Collaboration · Platform economy

1 Introduction

To support the process of identifying and scoping data-driven innovation by reflecting the dynamics of supply and demand trends, we are introducing the *data-driven innovation (DDI) framework*, which provides guidance in the continuous analysis of factors influencing the demand and supply sides. The framework systematically addresses the challenges of identifying and exploring data-driven innovations. It guides start-ups, entrepreneurs and established companies alike in scoping

S. Zillner (✉)
Siemens AG, Munich, Germany
e-mail: sonja.zillner@siemens.com

© The Author(s) 2021
E. Curry et al. (eds.), *The Elements of Big Data Value*,
https://doi.org/10.1007/978-3-030-68176-0_12

promising data business opportunities by analysing the dynamics of both supply and demand.

The DDI framework is based on a conceptual model represented as ontology. The DDI ontology describes all relevant aspects of any generic data-driven business. On the supply side the focus is on the development of new offerings. For a clearly defined *value proposition*, this includes identifying and accessing required *data sources*, as well as the analysis of *underlying technologies*. On the demand side the focus is on understanding the dynamics of the addressed markets and associated ecosystems. This includes the development of a *revenue strategy*, a way forward to harness *network effects* as well as an understanding of the *type of business*. As data-driven innovations are never created in isolation, identifying potential *partners* and a *viable ecosystem* helps to align supply and demand in order to achieve a competitive advantage.

The DDI ontology and framework were developed and tested in the context of the Horizon 2020 BDVe project[1] and are backed by empirical data and scientific research encompassing a quantitative and representative research study covering more than 90 data-driven business opportunities. The objective of the empirical research study was to systematically analyse and compare successfully implemented data-driven business innovations.

By relying on the DDI ontology and framework, we now have a method in place that we can share with members of the big data value ecosystem to explore data-driven business opportunities. The DDI ontology and framework are complemented by a comprehensive set of methods and guiding questions that are used for industrial trainings and university lectures. The derived characteristics and patterns of successful data-driven innovation help entrepreneurs, innovators and managers to scope their data-driven business opportunities in such a way that industrial investment decisions will become more successful and sustainable.

In what follows, Sect. 2 aims to define the notion of data-driven innovation. Section 3 details the four phases of the empirical research study establishing the foundation for developing the DDI framework. Section 4 summarises the main findings of the empirical DDI research study and Section 5 concludes the chapter.

2   Data-Driven Innovation

Finding a way to identify and scope data-driven innovation requires an understanding of the business opportunities in general as well as of the characteristics of data-driven innovation and an appropriate way forward to scope them. The following section briefly describes the overall layout, characteristics and specific challenges for data-driven innovations.

---

[1]https://www.big-data-value.eu/

2.1   What Are Business Opportunities?

The term business opportunities is a broad concept that is used to describe the chance to address a particular market need through the creative combination of resources that allow the delivery of advanced value propositions (Ardichvili et al. 2003).

From this definition, we can derive that promising business opportunities are based on a smooth balancing of two perspectives, i.e. the mainly technical capabilities on the supply side with the market dynamics and user requests, motives and interests on the demand side.

This argument is supported by a study by Timmons and Spinelli (2007) showing that most successful entrepreneurs and investors continuously observe the demand side very carefully in order to understand what customers and marketplaces want and never lose track of it. The insights gained about the demand side is used to guide the scoping of offerings by combining innovative technology components with reusable and available assets in a way that fosters competitiveness.

We observe several economic properties that play a crucial role when developing of data-driven business opportunities. For instance, when re-using a data source as input for producing of data-driven offering, it will never lose its initial value. However, the value of the data is not given per se but depends on availability of complementary assets that allow to extract the relevant information from the raw data.

The mentioned economic properties of data are impacting the dynamics of the market. In particular, due to network effects and the increasing flexibility of how offerings are scoped and priced for the different customer segments, the success of data-driven innovation requires continuous alignment between the needs on the demand side and the capabilities on the supply side.

2.2   Characteristics of Data-Driven Innovation

Data-driven innovation refers to the use of data and analytics to improve and foster new products and processes, new organisational processes, and new markets and business models (OECD 2015). We observe several economic properties that play a crucial role when developing of data-driven business opportunities. For instance, when re-using a data source as input for producing of data-driven offering, it will never loose its initial value. However, the value of the data is not given perse but depends on availability of complementary assets that allow to extract the relevant information from the raw data.

The mentioned economic properties of data are impacting the dynamics of the market. In particular, due to network effects and the increasing flexibility of how offerings are scoped and priced for the different customer segments, the success of data-driven innovation requires continuous alignment between the needs on the demand side and the capabilities on the supply side.

2.3   How to Screen Data-Driven Innovation?

The data economy is perceived as highly dynamic market: This is supported by the rapid growth of the European data markets, recent technical breakthroughs and the continuous growth of data assets.

Same, same but different: It is expected that the development of data-driven offerings will speed up as the existing data technologies along the data value chain are getting reused, combined and aligned with each other. For instance, systems such as Watson that required development over several years with the involvement of a large team will in the future become available to ordinary software engineers.

This in consequence leads to situations where entrepreneurs aiming to bring new offerings to the market need to continuously scan market offerings in order to identify promising available technology components – such as specific algorithms, knowledge models or hardware assets – that can be reused to speed up the develop-ment time of their innovation. At the same time, they need to constantly investigate their own unique selling point and the competitive advantage of their offerings in a highly dynamic environment. In such settings, innovations are no longer implemented by one organisation alone but rather a population of organisations and entrepreneurs that copy from each other as much as possible to ensure that technological assets can be reused and combined.

Of course, it is still necessary to put in enough effort to ensure that they make a difference in the market with a unique offering. This can be compared to a swarm of birds flying in the same direction with each bird continuously observing where the others are flying to have enough distance to avoid collision, but at the same time to be close enough to benefit from the wind shadow (Baecker 2007). In this way entrepreneurs need to continuously reassess what is part of their core offering and in which areas they are partnering with others in order to stay competitive in a fast-moving market.

The matching of supply and demand is a key success criterion for data market growth: The high-growth scenario[2] in the comprehensive European data market study (IDC & OpenEvidece 2017) is based on supply-demand dynamics that shift from technology push to demand pull. In other words, any means that provides guidance in match-making between market needs on the demand side and technical capabilities on the supply side helps to stimulate the adoption of data-driven innovation and in consequence the growth of the European data market. This can become possible through a fully developed ecosystem that is generating positive feedback loops between data/technology companies and users.

Accordingly, data-driven business opportunities that are described with a clear scope of offering per market segment (*supply side*) and reflect the ecosystem

---

[2]Which estimated 4% of GDP growth between 2016 and 2020.

Fig. 1 The four phases of
the DDI research study

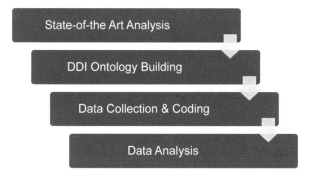

dynamics and benefits of network effects (*demand side*) are more likely to find a promising market fit. Given the dynamics of the growing data economy, the relation between the scope of offering on the supply side and the type of attributed value (e.g. price) on the demand side requires continuous reassessment. In consequence this leads to a *co-evolution between the supply side* (e.g. *the offering) and the demand side (e.g. adjacent ecosystems)* for each data-driven business opportunity.

To summarise, data-driven business opportunities should be described with a clear scope of offering per market segment (*supply side*) and reflect the ecosystem dynamics and benefits of network effects (*demand side*).

3 The "Making-of" the DDI Framework

This section describes the set-up of the DDI framework. The ontology and framework were developed in four phases (see Fig. 1).

By first *reviewing the literature* on existing proven methods and the theoretical concepts for scoping data-driven innovation/business opportunities, we could identify the relevant aspects of the data-driven innovation. The learnings from the literature review guided us in developing a *conceptual model* in the form of ontology describing the central aspects of supply and demand in data-driven ecosystems. Based on the conceptual model, data from a representative sample of data-driven start-ups could be collected and coded. Subsequently, the data was *analysed*, and best-practice insights and *patterns* identified.

3.1 State-of-the-Art Analysis

So as not to reinvent the wheel, we aimed to reuse and combine existing business modelling methodologies whenever possible – and to complement them with a *meta-analysis of demand- and supply-side trends* in order to guide the process of identifying data-driven offerings.

In our state-of-the-art analysis, we investigated to which extent existing frameworks, research results and methodologies can be used to describe the supply and demand sides of data-driven innovation. The DDI approach builds upon popular existing business modelling methodologies and related research, such as Osterwalder and Pigneur (2010), Nooren et al. (2014), Gassmann et al. (2014), Hartmann et al. (2014), Attenberger (2016) and Johnson et al. (2008).

We could reuse valuable content from the OECD (2015) to scope the actors in data ecosystems and learn about the characteristics and nature of data-driven innovation in general. From Adner (2006) we use findings about the handling of risks involved either when working with partners to develop innovations or when engaging with partners required to adopt the innovation. In our work we relied on findings about emerging disruptive business and market patterns (Hagel et al. 2015), as well as insights about the different strategic roles in the governance of ecosystems (Iansiti and Levien 2004). In addition, we used important concepts and findings from research about emerging platform businesses, such as Parker et al. (2016) and Choudary (2015).

The data and technologies along the *data value chain* are the central aspect of the supply side of data-driven business opportunities. To explore the data value chain, we relied on a simplified version of the DAMIAN methodology that we developed and prototyped in particular for the scoping of data-driven scenarios. This approach could be complemented with our findings in Cavanillas et al. (2016) and with methodologies for exploring the value proposition (Osterwalder et al. 2014) and co-innovation partners (Adner 2006).

3.2 DDI Ontology Building

Based on the above-mentioned literature review, the dimension of data-driven innovation could be identified. This leads to an initial version of a conceptual model as an ontology, covering relevant dimensions and concepts to describe data-driven innovations in a comprehensive manner. The objective of the DDI ontology is to cover all relevant aspects of data-driven innovations and establish the basis for analysing these aspects in an effective way. Recognising the findings of IDC and OpenEvidece (2017), the dimensions/concepts of the DDI ontology have been divided into two areas: the *supply side* and the *demand side*. Figure 2 gives an overview of all dimensions of the DDI ontology.

On the *supply side* the focus is on the development of new offerings. For a clearly defined value proposition, this includes the identification of and access to required *data sources* and the analysis of *underlying technologies*, as well as of all the

Fig. 2 Overview of all DDI dimensions on the supply and demand sides

partners that are required for the development and implementation of the data-driven innovation.

On the *demand side* the focus is on the dynamics of the addressed markets. The analysis includes the development of a *revenue strategy*, a way forward to harness *network effects* as well as an understanding of the *type of business*. As data-driven innovations are often built into established value chains, the partners in the *ecosystem* are analysed to understand under which conditions value chain partners are willing to adopt the innovation and thus will facilitate market access.

The initial version of the DDI ontology was continuously updated by incorporating lessons learned and insights gained by running DDI university lectures, seminars and workshops, as well as by performing a coding test run on a smaller set of 20 start-ups. For further details related to the different versions of the DDI ontology as well as the description of the final version of the DDI ontology, we refer to the following technical reports: Zillner et al. (2018), Zillner (2019) and Zillner and Marangoni (2020).

3.3   Data Collection and Coding

Based on three selection criteria, a representative sample set of data-driven innovation could be collected. In accordance with the dimensions described in the DDI ontology, the initial sample set of data-driven start-ups was enriched by findings from manual research (data coding).

3.3.1   Selection Criteria

To identify a representative data set, the following three selection criteria have been identified:

Focus on start-ups: Being well aware that data-driven innovations are developed in all types of organisation, i.e. in large, medium and small enterprises, we decided simply due to two practical reasons to focus in our study on data-driven start-ups only. First, as larger corporates and SMEs barely share information about their business or innovation designs and decisions, no public information was available. Second, as innovation activities in large corporates and SMEs are often influenced by existing infrastructures, legacy systems, prior systems and the existing customer base, organisational implications, such as changes in the sales channels, customer bases, migration issues, pricing models and processes, and customer expectations, need to be incorporated into the analysis. Those interdependencies with existing operations make it difficult to analyse data-driven innovation in isolation or to derive generic patterns.

Success criteria: To identify successful data-driven start-ups, we needed to define a *measurement for success*. We decided to choose start-ups with funding between

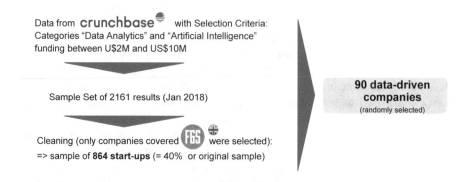

Fig. 3 Overview of the generation of the start-up data set

US$2 M and US$10 M[3] to cover the ones that had already convinced some ventures to invest in them, meaning that they would already have their product validated, but still are a "younger" start-up.

Technology focus: To identify data-driven start-ups, keywords/selection criteria such as *data analytics* and *artificial intelligence* seemed to be promising.

3.3.2 Sample Data Generation

To ensure high data quality, we decided to cross the data from two start-up databases. The initial database was Crunchbase,[4] an American-based platform for finding business information about private and public companies, and this served as the primary source for generating our sample data set. The second data source was F6S,[5] the largest platform for founders based in Europe.

The start-up data was extracted on 16 January 2018 from Crunchbase using the aforementioned filters:

- Categories "Data Analytics" and "Artificial Intelligence"[6]
- Funding between US$2 M and US$10 M

---

[3]The decision criteria for the values (between two and ten million dollars) were made in the light of venture capital theory. Although there is no consensus regarding the exact amount of money that determines each stage, we decided to follow the criteria used by Crunchbase: Angel is the first round, normally financed with less than US$10,000. The following stage is Seed, ranging from US $10,000 to US$2 M. Then there are the venture rounds that could have many series (A–Z), with A and B series normally valued between US$1 M and US$20 M.

[4]https://www.crunchbase.com/

[5]https://www.f6s.com/

[6]Crunchbase is using 46 categories to classify all of its companies.

Based on these filters, we could extract a sample set of 2161 data-driven companies.

From this larger sample set, we extracted a statistically valid sample set of 90 start-ups with entries in both databases. Figure 3 provides an overview of how the initial data set of start-ups was generated.

3.3.3   Coding of Data

The start-up data was coded in accordance with the categories of the data-driven innovation framework. For each start-up, relevant background information was manually searched and investigated to identify relevant statement(s) related to certain categories of the DDI framework.

To ensure reliability, the different categories of the DDI model were defined before the coding exercise started. To avoid coding errors, a test run of the coding exercise based on a manually selected sample of 20 start-ups was performed. After coding of this initial set of start-ups by two independent coders, all categories or concepts with a high percentage of disagreement in coding were discussed in detail and then *redefined or removed*.

The start-ups from the sample set were coded by three independent coders. For each start-up the three coders manually annotated a binary feature vector covering all DDI dimensions and concepts. In case a specific feature was present, it was annotated with "1"; in case it was not present, it was annotated with "0"; and in case no information could be found, it was indicated with "2".[7] This was done by searching the Internet for relevant statements indicating a specific feature of the DDI ontology.

For each start-up at least three websites (Crunchbase, F6S and company website) were consulted. Very often additional webpages, e.g. linked press releases, were analysed, and complementary Internet searches were conducted to ensure that all categories and concepts were addressed.

After having performed the manual annotations, the coders met online to compare coding results and to discuss and resolve disagreements. The result of the coding process was 90 binary feature vectors representing the presence or absence of each DDI category or concept for each start-up.

---

[7]Although the feature vector can be annotated with three values (0,1, 2), we still treat it as a binary feature vector, as the third value category "2" was only introduced for practical reasons, to indicate that for a specific feature the accomplished search did not reveal any related information. This helped us to monitor the progress of the coding exercise as well as to remove start-ups from the analysis.

3.4   Data Analysis

Based on the three previous phases, it was possible to generate a sample data set that had 90 variables (dimensions and categories of the DDI ontology) and 90 observations (start-ups) that were marked either by the presence of the variable (1) or by the absence of it (0). For example, one of the variables described whether a start-up was doing business in the B2B domain. For start-ups for which this was true, we marked a (1), and for start-ups that did not target B2B, we marked a (0). In the percentage-frequency analysis, we then counted how many start-ups were marked with (1) and divided this by the total number of observations for that variable. Using the same example, we could observe that 88 start-ups out of 90 were marked with (1), which means that 98% of companies target B2B customers.

The first method employed to assess which variables could shape data-driven business innovation was a percentage-frequency analysis. The goal of using this method was to understand how frequently a variable was observed in our data.

4   Findings of the Empirical DDI Research Study

To derive meaningful insights into trends, frequencies and distributions, a *classical statistical data analysis* was used. Based on a percentage-frequency analysis, many insightful findings along the main dimensions of the DDI framework could be identified. In the following subsections, we will summarise all findings derived from the percentage-frequency analysis. We will represent these findings by first discussing some generic findings before discussing the findings in relation to the dimension the DDI framework.

4.1   General Findings

It was important for us to find out whether the distinction between B2B and B2C has an influence on the design of data-driven innovation. In addition, we wanted to better understand the possible impact of the (non-)sector focus of data-driven innovations.

Target Customer: The majority of data-driven start-ups (78%) are addressing B2B markets. Only 2 out of 90 start-ups in our sample focused solely on end-customer markets. Start-ups addressing end-user needs prefer already established channels to deliver their offering to the users. They tend to rely on partnerships with established business partners to bring their offering to users. A second, quite frequent, strategy used by 19% of start-ups is positioning data-driven solutions as multi-sided market offering combining complementary offerings to align private and business needs.

Seventy-five per cent of our start-up sample have developed a clear *sector focus*. Companies with clear *sector focus* have a concrete customer segment in mind for whom a concrete value proposition is delivered. Those companies have a concrete customer segment(s) in mind for which a concrete value proposition is delivered.

> For example, *CloudMedx*[8] Inc. designs artificial intelligence-driven software for medical analytics. Clinical partners at all levels can derive meaningful and real-time insights from their data and intervene at critical junctures of patient care. Its underlying clinical AI computing platform uses healthcare-specific NLP and machine learning to generate real-time clinical insights at all points of care to improve patient outcomes. By relying on evidence-based algorithms and deep learning, a wide variety of structured and unstructured data being stored in clinical workflows can be understood and used for decision making.

In comparison, we also found start-ups that focus on technology with cross-domain impact. In general, their solution will be used by other intra- or entrepreneurs to build data-driven solutions for end users.

> For instance, the start-up *DGraph Labs*[9] is offering an open-source distributed graph database. The company is planning to release an enterprise version that is closed source, as well as a hosted version (as it is easier to run hosted services for customers than trying to help them debug every issue on their own). Customers are using the service to build their own sector-specific applications.

In summary, sector-specific data-driven offerings are much more frequent than technology-driven sector-agnostic solutions. This is due to the very different pre-processing challenges of data sources in the various sectors, as well as the higher possibilities of identifying target groups in concrete sector settings. Most sector-agnostic offerings are intermediate functionalities addressing developers to build customised solutions.

4.2 Value Proposition

To analyse the value proposition in the context of data-driven businesses, our main focus is on the different ways data is used to generate value. *Data value* refers to the insights that can be generated out of data and how this can be used in a particular user or business context. In accordance with its value and complexity, we distinguish four different types of analytics that are used for generating different types of insights, i.e. descriptive analytics explain what happened, diagnostic analytics highlight why something happens, predictive analytics forecast what will happen in the future, and prescriptive analytics identify optimal actions and strategies (Zillner 2019).

---

[8]http://www.cloudmedxhealth.com/

[9]https://dgraph.io/

Two out of every three start-ups rely on *data analytics* in general for generating insights. Among the start-ups using data analytics, 83% rely on descriptive analytics in their offering (i.e. every second start-up).

For instance, the start-up *Apptopia*[10] is using *descriptive analytics* to provide app analytics, data mining and business intelligence services. They collect, measure, analyse and provide user engagement statistics for mobile apps and visualise the aggregated data in classical dashboards. The unique selling point of their offering is the high number of data points they are able to integrate and visualise, i.e. they state that they rely on "more different data points than nearly any other app data provider in the world". The insights, which can be generated by descriptive data in this large data set, are of interest to the worldwide mobile app developer community as they allow them to compare their own app performance with competing or related apps. Whenever app developers are engaging with the *Apptopia* platform to benchmark their own apps, additional valuable data sets can be generated. By offering free-of-charge descriptive analytics-based dashboards, *Apptopia* are able to attract a large number of developers to use their platform, which again allows them to produce high-value data sets that can be sold to business customers.

Four out of ten start-ups in our sample set relied on *predictive analytics* to generate value for their users.

For instance, the start-up *Visiblee*[11] collects IP addresses and cookies of all website visitors and uses these to predict the identity of unknown visitors in real time. By relying on these real-time predictions, the company is able to increase the leads[12] threefold.

Compared to *descriptive and predictive analytics*, we can observe that diagnostics and prescriptive analytics are used less frequently. Only every fifth data-driven start-up is offering solution for automating manual tasks or activities, and match-making is observed in only 16% of cases.

To implement data-driven offerings, in general, several algorithms and approaches are combined. This is also true for the four different types of data analytics discussed earlier. In our sample, 4 out of 10 start-ups use more than 2 different types of data analytics, and 19% of start-ups rely even on 3 or more types of analytics to generate value.

For instance, *Eliq*[13] provides a comprehensive platform for the intelligent energy monitoring of utilities. The AI-powered app offers a wide range of insights:

- By relying on *descriptive analytics*, *Eliq* shows periodic energy consumption patterns that can be drilled down into different time frames, i.e. yearly, monthly, hourly, etc.).
- By relying on *diagnostics analytics*, *Eliq* helps users to identify potential "energy leaks" or potential sources of energy theft.
- By integrating external data sources, such as extreme weather change forecast, *Eliq* can inform users that their energy consumption is likely to change significantly (*predictive analytics*). Utilities benefit from such information as they can customise marketing communication accordingly.
- By relying on *prescriptive analytics*, the *Eliq* platform can not only inform users about increased energy consumption but also recommend strategies to overcome such high

---

[10]https://apptopia.com/

[11]https://www.visiblee.io/en/home/

[12]In a sales context leads refer to contacts with potential customers.

[13]https://eliq.io/

consumption scenarios, e.g. by upgrading or replacing devices with higher efficiencies. This allows utilities to establish a personalised and targeted user engagement.

Eliq is an example of a start-up that establishes a unique value proposition and competitive edge by offering a wide range of analytical services. We want to highlight that this is not a frequent pattern. The majority of start-ups (62%) is focusing on only one analytical offering.

4.3   Data

Data is *the* key resource for realising data-driven innovation. In general, we observe that the used data sources greatly influence the efforts in data pre-processing as well as the scope of the data-driven offering. In case a data-driven innovation is based on image data, we can conclude that an image segmentation algorithm needs to be in place. In accordance with how specific or domain specific the underlying image data set is, a new pre-processing image algorithm needs to be developed. Or in the case of personal data and of industrial or operational data, GDPR-compliant services and data privacy methods need to be in place, respectively.

For that reason, we recommend exploring the data assets early when scoping one's data-driven innovation. Data exploration will help to understand:

- Whether the envisioned value proposition can be realised. Very often, we face the situation that the data quality is not good enough to generate the needed insights.
- How much effort is needed to create data of high quality. Often the raw data is not yet the data quality needed. The good news is that there exist many approaches to increase the quality of data for this scoped purpose. However, the expected return always needs to be aligned with the efforts needed. Other projects in the Big Data Value Public-Private Partnership (BDV PPP) have reported similar experiences (Metzger et al. 2020).

In the following, we will give an overview of which data types and sources are used and how frequently in data-driven innovations.

A wide range of different types of data sources exist that are relevant for developing data-driven innovation. Although only 19% of start-ups were addressing B2C markets, personal data was still the most frequently (67%) used in the analysed data-driven offerings. This is a very impressive number given the fact that only a very low number of companies in our sample (19%) were addressing business-to-consumer markets. In consequence this also implies that a high percentage of start-ups addressing business customers in Europe[14] need to handle the constraints of the General Data Protection Regulation (GDPR).

---

[14]Our sample set is not restricted to European start-ups only, as we wanted to make sure that our analysis covers worldwide excellence. As we do not have precise numbers for European data companies, the sentence is formulated with some ambiguity.

For example, *Oncora Medical*[15] is using personal data to fight cancer. The US-based company collects data on cancer patients including information related to treatments and clinical outcomes through an intuitive software used by doctors. Their objective is to deliver predictions that can help design better radiation treatments for patients, as well as enabling precision medicine in radiation oncology. The data collected is personal data and is thus sensitive and has higher standards of protection.

Industrial data, i.e. any data assets that are produced or used in industrial areas, is a second type of data which has high data protection requirements. In comparison to personal data, industrial data is used only half as often. Organisations seem to be reluctant (in particular if they do not see the immediate value) to share their industrial and operational data with third parties, such as start-ups, because they are afraid to reveal relevant business secrets.

One successful example, *PlutoShift*,[16] offers a platform that is helping industrial customers to improve their operational efficiency by identifying inefficient patterns of energy usage by analysing customer data stored in the cloud and operational sensor data. With energy being a high-cost driver, *PlutoShift* can help industrial customers to reduce resource consumption and operating costs.

The second most popular types of data source are time-series and temporal data. Fifty-six per cent of start-ups in our sample rely on these types of data to generate value. The high frequency might be due to the popularity of using behavioural data that is tracked within each user interaction on the web and mobile devices and is thus very likely to cover time-series data. Another very frequently used data source is geo-spatial data (46%), and the usage of Internet of Things (IoT) data is seen in 30% of our sample.

4.4 Technology

The BDV Strategic Research and Innovation Agenda (SRIA) (Zillner et al. 2017) describes five technical priorities identified by the BDVA ecosystem and experts as strategic technical objectives. In our study, we were interested in which of these technical areas were most frequently covered when realising data-driven innovation.

Among the five technology areas listed in the BDV SRIA, *data analytics* is used most frequently. Eighty-two per cent of our start-up samples relied on some type of data analytics to implement data-driven value proposition. The usage of technologies in the *data management* area is seen in 41% of cases and is very much in line with offerings addressing the challenges of processing unstructured data sources. Solutions for *data protection* are the least frequently addressed research challenge with 13%. When looking at to which extent BDV SRIA technologies are used in

---

[15]https://oncoramedical.com/

[16]plutoshift.com; previously called Pluto AI

combination, we observed that more than half of the start-ups, precisely 59%, combine two or more technologies.

> Uplevel Security[17] is one example that combines *data management* with *data protection*. They redefine security automation by using graph theory for real-time alert correlation. Their product creates a dynamic security graph (data management) for an organisation based on incoming alerts, prior incident investigations and current threat intelligence (data protection). *Uplevel Security* then transforms the ingested data into subgraphs that continuously inform the main security graph. By automatically surfacing relationships, investigations no longer occur in isolation but begin with context.

Less frequently observed, 22% of the companies combine more than three technologies.

> One example of this is the medical company *CloudMedx*,[18] which started with the aim to make healthcare affordable, accessible and standardised for all patients and doctors. The company uses NLP and proprietary clinical contextual ontologies (data management) and deep learning (data analytics) to extract key clinical concepts from electronic health records, which serve as insights for physicians and care teams with the goal to improve clinical operations, documentation and patient care. In addition, *CloudMedx* is presenting the results to dedicated teams through a user-friendly platform that allows for interactive predictive and prescriptive analytics to assess current metrics and build a path forward with informed decisions.

4.5 Network Strategies

For digital and data-driven innovations, network effects are important phenomena to reflect. In our study, 57% of start-ups rely on network effects. A network effect occurs when a product or a service becomes more valuable to its users as more people use it (Shapiro and Varian 1999). Network effects are also known as demand-side economies of scale and predominately exist in areas where networks are of importance, such as online social networks or online dating sites. A social network or dating site is more appealing to its user when it is able to continuously attract and add more and more users. In consequence, harnessing network effects requires developing a broader network of users in order for the network or site to differentiate itself from its competitors. For that reason, the critical mass of users and timing are key success factors in a network economy.

Due to the high impact of the network effects, competitors starting from "ground zero" with no users in their network will face difficulties in entering the market success fully. In this context we are using the expression "network effect" to

---

[17]https://www.uplevelsecurity.com/
[18]https://www.cloudmedxhealth.com/

highlight the positive feedback (positive network externality[19]), i.e. the phenomena that already existing strengths or weaknesses are reinforced, might lead to extreme outcomes. In the most extreme case, positive feedback can lead to a winner-takes-all market (e.g. Google).

Network effects impact the underlying economics and operation of data-driven innovation. Instead of creating products that are early on the market and different from other offerings, the focus here is on scaling and scoping the demand perspective. Understanding network effects and their underlying market dynamics is crucial to successfully positioning data-driven products, services and businesses in the market. In doing so, data-driven innovation can harness network effects on *three different levels*.

First, data-driven businesses are relying on **network effects at data level**, if they are able to improve their offerings by the sheer amount of data they hold available. In our sample this was the case in 49% of start-ups.

For instance, the already mentioned company *Apptopia*[20] uses big data technology to collect, measure, analyse and provide user engagement statistics for mobile apps. The more app providers produce data being connected to the platform, the more valuable the service becomes. In order to gain more real-time data, they attract app developers to connect to their platform by providing free data analytics products. With this free-of-charge value proposition, developers benefit in registering their mobile apps on the platform while giving the platform the permission to analyse user engagement data of the mobile app. Professional and expensive subscription fee models for business customers, including Google, Pinterest, Facebook, NBCUniversal, Deloitte and others, benefiting from real-time engagement insights of mobile apps, complement the revenue strategy of this offering.

In this context, multi-sided business models are the usual way forward. Typically, a multi-sided business model brings together two or more distinct but interdependent groups of customers. Value is only created if all groups are attracted and addressed simultaneously. The intermediary, in our example the company *Apptopia*, generates value by facilitating interactions between the different customer groups, whereas the value increases when more users are attracted. The more app developers register on the platform, the more accurate the statistics become. With an increasing number of business customers, *Apptopia* then creates the required resources to invest in advanced functionalities for app developers.

Second, when businesses are providing a technical foundation for others to build upon, we can observe **network effects at infrastructure level.** In our sample these have been 12% of start-ups. Based on a layer of common components, third-party players are invited to develop and produce an increasing number of data-driven offerings.

This set-up is also known as product platforms (Hagel et al. 2015). A prominent example is the Android platform – it provides the technical foundation for others to

---

[19]For completeness we want also to mention the phenomena of negative network externalities which occur when more users make a product less valuable (e.g. traffic congestion). Negative network effects are also referred to as "congestion".

[20]https://apptopia.com/

build apps. This includes any type of tool and service that enables the plug-and-play building of data-driven offerings, e.g. (open) standards, de facto standards, APIs and standardised data models. The more functionalities are available that help others to build and position innovative offerings better, faster, etc., the more attractive the offering itself becomes. The infrastructure layer has little value per se unless other users and partners create value on top of it.

An example of this dynamic is the agricultural-robotics technology company *Skyx*.[21] This company is offering neither hardware nor agriculture end-customer applications, but a software that enables a modular swarm of autonomous drones for spraying. By providing a technology to plan and control the mission of drones in real time as well as to auto-pilot the entire fleet/swarm, it addresses the need for agri-spraying application developer applicators in building their solutions at a higher quality and at less cost by relying on a standardised approach. In addition, as the software is compatible with any commercially available hardware, the cost of connecting the wide range of drones can be significantly reduced. Thus, Skyx provides tools and connectors for agri-spraying application developers to build their own solutions. The more drone hardware can be connected, and the more spraying functionalities can be provided, the more attractive the overall offering for applicators.

Third, in cases where the number of marketplace participants is the key source of value, data-driven offerings can harness **network effects at marketplace level**. Offerings that are able to connect participants in their specific roles, such as buyer and seller, and consumer and producer, allow two participants to easily interact with each other.

The low number of network effects at marketplace level in our study (10%) indicates the difficulties and challenges in building them. The challenges are less at the technical level and more at the level of building critical size and balanced user communities. Several strategies to attract users from the different communities have been implemented by start-ups.

4.6 Revenue Strategy

We have been interested in the question of how data-driven businesses are making money. Is this different from traditional businesses? And can we identify some dominant revenue models?

Our first finding is that it was often difficult to find information about the type of revenue models used. Especially in cases when start-ups have been focusing on emerging technical advances, such as drones or autonomous driving, information about revenue models was – understandably – not available.

As emerging technology businesses are often seen as a risky investment or bet on the future in a market not yet established, the absence of revenue-related information is not surprising. This was the case for 10% of the companies analysed: We couldn't find or extract any information about the revenue model.

---

[21] https://www.skyx.solutions/

Our study confirmed the findings of Attenberger (2016) that revenue models have not changed through the usage of data technologies per se. The major difference to traditional businesses is that data-driven innovations rely on different types and combinations of revenue streams that are continuously changing over time in order to address the specific user needs of each customer segment. On the one hand, we observe *new forms of value propositions*, ranging from service offerings, to the bundling and unbundling of offerings, to intermediate offerings, to product differentiations through versioning, that allow the specific user needs to be addressed.

On the other hand, the majority of data-driven innovations have – in comparison to traditional businesses – a different cost structure. With data and data offerings being cheap to reproduce and deliver, the typical cost structure of data-driven innovations relies on fixed costs for the development of the offerings but low variable cost. This kind of cost structure leads to substantial economies of scale as with more offerings sold, the average costs of development decrease dramatically. In addition, as the reproduction and distribution costs are often marginal, the danger of price dumping and surplus of offerings in the competitive market is a frequent phenomenon. For instance, Aitken and Gauntlett (2013) counted more than 40,000 health apps in the app store being offered for free or for a very low price.

With this new cost structure for most data-driven innovations, organisations have a new flexibility to *adjust the equation between value proposition and price* in accordance with the user needs of various customer segments. In this context, companies elaborate the specific price level the targeted user group is willing to pay. The main objective for aligning the product version with the pricing version for each customer segment is to attract more users and interactions, as well as to grow the community.

The most frequently used revenue model in our study was the subscription model. We observed in this context a strong correlation between the spread and high adoption of software as a service (SaaS) approach, which brings a lot of flexibility when used for deploying data-driven innovations. The second most frequently used revenue model is the selling of services in which the person's time is paid for. These revenue models are very often used for open software offerings as well as when offerings are not standardised or off-the-shelf. Advertisement as a revenue model is rarely observed. In our sample, only 2% of start-ups are applying it. Although this might seem surprising, it merely reflects the high percentage of B2B models.

4.7 Type of Business

Data-driven innovations can disrupt existing value chains. However, at the same time, we observe a large number of "low hanging fruits", i.e. business opportunities in the scope of established processes (intern) or value chains (cross-organisational).

To classify data-driven business opportunities we will introduce four strategies with a significant impact on markets and associated value chains:

(a) Providing new value to customer with established market position
(b) Developing a new marketplace/ecosystem
(c) Leverage an existing ecosystem by scoping a niche offering
(d) Building technology assets that ensure a future competitive advantage

The following remarks describe the four strategies in detail and illustrate them with an example from our sample of start-ups.

In general, this classification is based on approaches available for the classification of traditional business opportunities. One important work in this context is Ardichvili et al. (2003), who classified business opportunities into two dimensions: *value creation capability* and *value sought.* Although both dimensions have at first glance a good mapping to the DDI supply and demand side, they did not reflect the changing nature of underlying business ecosystems. As already discussed at the beginning of this chapter, data-driven innovations are rarely developed alone but rely on the collaboration between many partners in the value chain.

When positioning data-driven offerings in the market, it is also necessary to reflect the associated business strategy and innovation ecosystem.

Data-driven services are often associated with the strategy of "**Finding a new business partner**". This strategy tries to focus on one single customer (segment) and his or her business processes. Based on a detailed understanding of his/her business processes (including the pain points, happiness points and unaddressed user needs), new values/services for specific user needs are built. As the service is heavily focusing on this one specific partner, the overall market and business ecosystem is only observed in an indirect manner. In our study, the *data-driven service business* was the most frequently observed approach (with 78%) to position offerings in the market.

> For instance, the company *Arable* provides an agricultural solution based on in-field measurements as a software-as-a-service (SaaS)-based service offering. To enable growth, advisors and businesses are invited to play a proactive role in ensuring high quality and longevity of their agricultural operations. As a consequence, the company can derive real-time, actionable monitoring and predictions related to weather risk and crop health by means of a tiered SaaS offering with different levels of services combined with IoT businesses. The tier I service includes reporting, integrating and visualisation, whereas the tier II services include predictions and advanced analytics.

Compared to data-driven services, the second type of business strategy – **developing a data-driven marketplace** – is significantly more complex as a new marketplace/ecosystem needs to be built up. Only 16% of companies in our sample relied on this approach. Market participants on the supply as well as on the demand side need to be attracted. In addition, it is necessary to ensure that a critical number of participants are providing their assets and at the same time a critical number of participants are requesting them.

The growth of the marketplace needs to be balanced on both sides – the supply and demand sides – in order to retain its attractiveness. It seems that organisations have been developing very different strategies to attract the different participant groups, e.g. by providing necessary IT services and analytics services, and offering services for free.

One example of this strategy is *Zizoo*,[22] a Vienna-based company that established a global boat rental platform. Zizoo is building a global digital booking platform and website connecting suppliers (charter companies) to travellers worldwide, similar to "Booking.com for Boats". When the building of this marketplace started, the founders of the company were entering a market (the boat rental market) which was 10 years behind any other travel sector. As the majority of boat charter companies had not yet been digitalised, they needed to put a lot of effort into attracting the supply side to join their emerging marketplace. For instance, they offered charter companies a powerful inventory management tool and business intelligence for free. As they are making boat holidays affordable and accessible to everyone (bookings start at €20 a day), they were also able to attract the demand side.

Another strategy is to identify an existing healthy ecosystem that is already in place which gives the opportunity to position one's own offering as a niche application. The so-called niche player leverages *an existing ecosystem* by scoping a niche offering in accordance with the defined constraints of the dominant or key player of the ecosystem. Typical examples of such strategies are the thousands of apps offered in the iOS or Android ecosystems for mobiles. In our sample we could observe this in 12% of cases.

> One good example of this strategy is *AIMS Innovation*.[23] This start-up develops AI and machine learning technologies to give the world's largest companies deep insights into and control of their most business-critical processes – such as safely distributing electricity, shipping thousands of daily orders to ecommerce customers or delivering the results of medical tests to doctors quickly and reliably. They are positioning their offering in the Microsoft ecosystem. According to their website, they offer the only artificial intelligence solution in IT operations covering all core Microsoft enterprise technologies.

The last type of business category is the **emerging technology business** that anticipates a future ecosystem or market. In our study this was seen in 9% of the sample. As the market is not yet settled and the technology is often in a very early stage, it is scoped as investment in the future. Thus, revenue strategies cannot be implemented. The main focus of emerging technology businesses is building capabilities/assets ensuring a future competitive advantage.

> For instance, the company *Carfit*[24] is working on creating the most comprehensive library of car vibrations. They collect and generate systematically data related to noise, vibration or harshness. An enhanced data analytics algorithm is in place to incorporate automotive domain expertise. The company is aiming at a car vibration tracking device that can help to lower car maintenance costs and increase the efficiency and transparency of the car's operations. But the self-diagnostic and predictive maintenance platform only brings real value to end users when vehicles are moving autonomously. Thus, the company is addressing a future market (as today drivers are in general good at detecting abnormal noises in their car). However, when cars are moving autonomously the need for remote monitoring will become critical.

---

[22]https://www.zizoo.com/

[23]https://www.aims.ai/

[24]https://car.fit/

5   Conclusion

The data-driven innovation (DDI) framework addresses the challenges of identifying and exploring data-driven innovation in an efficient manner. It guides entrepreneurs systematically in scoping promising data-driven business opportunities by reflecting the dynamics of supply and demand through investigating the co-evolution and interactions between the scope of the offering (supply) and the context of the market (demand). The DDI framework consists of eight dimensions that are divided into a supply side (value proposition, data, technology and partners) and a demand side (ecosystem, network strategy, revenue strategy and type of business).

The DDI framework was developed and tested in the context of the BDVe project and is backed by empirical data and scientific research encompassing a quantitative and representative study of more than 90 data-driven business opportunities.

The data-driven innovation framework offers a proven method for all members of the BDV ecosystem to provide guidance in exploring and scoping data-driven business opportunities. The comprehensive content can be used for industrial workshops and educational set-ups.

References

Adner, R. (2006). Match Your Innovation Strategy to Your Innovation Ecosystem Match Your Innovation Strategy to Your Innovation Ecosystem. *Harvard Business Review, 84*(4), 98–107. https://doi.org/10.1007/978-1-4614-3858-8_100487

Ardichvili, A., Cardozo, R., Ray, S., & S., R. (2003). A theory of entrepreneurial opportunity identification and development. *Journal of Business Venturing, 18*(1), 105–123. https://doi.org/10.1016/S0883-9026(01)00068-4

Aitken, M., & Gauntlett, C. (2013). *Patient apps for improved healthcare: From novelty to mainstream.* IMS Institute for Healthcare Informatics.

Attenberger, Y. (2016). *Business Models and Big Data.* Master Thesis at the Munich School of Management, LMU Munich.

Baecker, D. (2007). *Studien zur nächsten Gesellschaft* 1. Aufl., Frankfurt am Main: Suhrkamp.

Cavanillas, J. M., Curry, E., & Wahlster, W. (2016). *New horizons for a data-driven economy: A roadmap for usage and exploitation of big data in Europe.* https://doi.org/10.1007/978-3-319-21569-3

Choudary, S. P. (2015). *Platform Scale—How a new breed of startups is building large empires with minimal investment.* Singapore: Platform Thinking Labs.

Gassmann, O., Frankenberger, K., & Csik, M. (2014). Revolutionizing the business model—St. Gallen business model navigator. *Management of the Fuzzy Front End of Innovation, 18*(3), 89–97. https://doi.org/10.1007/978-3-319-01056-4_7

Hagel, J., Brown, J. S., Wooll, M., & De-Maar, A. (2015). *Patterns of disruption—Anticipating disruptive strategies in a world of unicorns, black swans and exponentials* (p. 28). Texas: Deloitte University Press.

Hartmann, P. M., Zaki, M., Feldmann, N., & Neely, A. (2014). Big data for big business? A taxonomy of data-driven business models used by start-up firms. *Cambridge Service Alliance Whitepaper*, 1–29. https://doi.org/10.1016/j.im.2014.08.008

Iansiti, M., & Levien, R. (2004). *The keystone advantage.* Boston: Harvard Business School Press.

IDC, & OpenEvidece. (2017). *The European data market study: Final report.* 275. Retrieved from http://datalandscape.eu/study-reports/european-data-market-study-final-report

Johnson, M. W., Christensen, C. M., & Kagermann, H. (2008). Reinventing your business model. *Harvard Business Review, 86*(12). https://doi.org/10.1111/j.0955-6419.2005.00347.x

Metzger, A., Franke, J., & Jansen, T. (2020). Ensemble deep learning for proactive terminal process management at duisport. In Brockke, Mendling, & Rosemann (Eds.), *Business process management cases.* (Vol. 2) Springer.

Nooren, P., Koers, W., Bangma, M., Berkers, F., & Boertjes, E. (2014). Media-internet-telecom value web introducing the damian method for systematic analysis of the interdependencies between media-internet-telecom value web. *TNO report.*

OECD. (2015). Data-driven innovation: Big data for growth and well-being. In *OECD digital economy papers* (p. 456). Paris: OECD. https://doi.org/10.1787/9789264229358-en

Osterwalder, A., & Pigneur, Y. (2010). *Business model generation: A handbook for visionaries, game changers, and challengers.* Toronto: Self Published.

Osterwalder, A., Pigneur, Y., Bernarda, G., & Smith, A. (2014). Value proposition design. In *Strategyzer series.* https://doi.org/10.1017/CBO9781107415324.004

Parker, G. G., Van Alstyne, M. W., & Choudary, S. P. (2016). Platform revolution: how networked markets are transforming the economy--and how to make them work for you. *W.W. Norton & Company*, 256. https://doi.org/0393249131

Shapiro, C., & Varian, H. R. (1999). *Information rules: A strategic guide to the network economy.* Harvard Business School Press. https://doi.org/10.1145/776985.776997

Timmons, J. A., & Spinelli, S. (2007). New venture creation: entrepreneurship for the 21st Century. In *Business*, vol. 2009. Retrieved from http://www.amazon.com/dp/0071276327

Zillner, S. (2019). *D2.6. Annual report in opportunities, BDVe deliverable 2.6.*

Zillner, S., & Marangoni, I. (2020). *D2.7 Annual report on opportunities; BDVe deliverable 2.7.*

Zillner, S. et al. (eds) (2017). *Big data value strategic research and innovation agenda.*

Zillner, S., Timan, T., & Kotterink, B. (2018). *D2.5 Annual report in opportunities.*

Recognition of Formal and Non-formal Training in Data Science

Ernestina Menasalvas, Nik Swoboda, Ana Moreno, Andreas Metzger, Aristide Rothweiler, Niki Pavlopoulou, and Edward Curry

Abstract The fields of Big Data, Data Analytics and Data Science, which are key areas of current and future industrial demand, are quickly growing and evolving. Within Europe, there is a significant skills gap which needs to be addressed. A key activity is to ensure we meet future needs for skills and align the supply of educational offerings with the demands from industry and society. In this chapter, we detail one step in this direction, a programme to recognise Data Science skills. The chapter introduces the data skills challenge and the importance of formal and non-formal education. It positions data skills within a framework for skills and education, and it reviews key projects which have advanced the data skills agenda. It then introduces recognition frameworks for formal and non-formal Data Science training, and it details a methodology to achieve consensus between interested stakeholders in both academia and industry, and the platforms needed to be deployed for the proposal. Finally, we present a case study of the application of recognition frameworks within an online educational portal for students.

Keywords Big Data · Data skills recognition · Skill badges · Skill labels · Education hub

E. Menasalvas · N. Swoboda · A. Moreno (✉)
Universidad Politécnica de Madrid, Madrid, Spain
e-mail: emenasalvas@fi.upm.es; nswoboda@fi.upm.es; ammoreno@fi.upm.es

A. Metzger · A. Rothweiler
paluno, University of Duisburg-Essen, Essen, Germany

N. Pavlopoulou · E. Curry
Insight SFI Research Centre for Data Analytics, NUI, Galway, Ireland

© The Author(s) 2021
E. Curry et al. (eds.), *The Elements of Big Data Value*,
https://doi.org/10.1007/978-3-030-68176-0_13

1   Introduction

Nowadays, fields like Big Data, Data Analytics and Data Science have drawn a considerable amount of attention from industry. In order to boost the data-driven economy in Europe, the data needs required by industry keep growing; therefore, the main challenge is bridging the gap between these industrial needs and the availability of skilled data scientists.

The popularity of data-oriented fields has an impact on the creation of a plethora of degrees in universities and online courses that offer a wide range of skill sets to aspiring data scientists. Therefore, the data skills needed by industry can be acquired through formal learning (e.g. undergraduate or graduate university degrees) or non-formal learning (e.g. e-learning or professional training).

Nevertheless, the availability of a plethora of resources does not suggest a direct link between industry and future data scientists, resulting in a range of challenges for the gap to be bridged, defined below:

- Given the constant technological and societal changes, the needs may also quickly change; therefore, it is vital to identify the current industrial needs or trends and adjust the educational offerings according to those altered needs.
- Given the plethora of available formal and non-formal programmes, there is a need to provide a platform and living repository that will give more targeted and filtered access to these resources to potential data scientists or professionals that want to enhance their skills.
- A programme needs to be defined that will be able to provide recognition of skills of data scientists acquired through both formal and non-formal education.
- A framework needs to be defined that will align the current industrial needs with the Data Science curricula and skills provided by formal and non-formal institutions.

This chapter explores the ways in which Europe could build a strong and vibrant big data economy by tackling the challenges above through the enhancement of the benefits that educational institutions and existing skills recognition initiatives have to offer. Specifically, some directions towards the desirable result involve the creation of the Big Data Value Education Hub (EduHub) and the Big Data Value (BDV) Data Science Badges and Labels.

The EduHub is a platform that provides access to Data Science and Data Engineering programmes offered by European universities as well as on-site/online professional training programmes. The aim of the platform is to facilitate knowledge exchange on educational programmes and meet current industrial needs.

BDV Data Science Badges and Labels are skills recognition programmes for skills acquired by formal and non-formal education, respectively. The initial stage of the badges contained the types and requirements for the system by leveraging existing work by the European Data Science Academy[1] (EDSA) and EDISON[2] projects, which were European Union (EU) projects related to Data Science skills. Later, the programmes were enhanced by gathering feedback from academia and industry and by proposing methodologies to bring together interested stakeholders (from both academia and industry) for the design and deployment of the badges and labels, as well as their evaluation and feedback.

This chapter also explores a practical view of how this platform and the skills recognition programme can work in isolation as well as together in order to bridge the industry with academia. This is presented via a pilot of the BDV Data Science Analytics Badge that is currently issued by two universities and the way the badges as well as the educational programmes which issue them can be accessed in the EduHub.

1.1 The Data Skills Challenge

In order to leverage the potential of BDV, a key challenge for Europe is to ensure the availability of highly and correctly skilled people who have an excellent grasp of the best practices and technologies for delivering BDV within applications and solutions (Zillner et al. 2017). In addition to meeting the technical, innovation and business challenges as laid out in this chapter, Europe needs to systematically address the need to educate people so that they are equipped with the right skills and are able to leverage BDV technologies, thereby enabling best practices. Education and training will play a pivotal role in creating and capitalising on BDV technologies and solutions.

There was a need to jointly define the appropriate profiles required to cover the full data value chain. One main focus should be on the individual needs linked to company size. Start-ups, SMEs and big industries have individual requirements in Data Science. We distinguish between three different profiles, (1) to cover the hardware- and software-infrastructure-related part, (2) the analytical part and (3) the business expertise.

The educational support for data strategists and data engineers is, however, far too limited to meet the industry's requirements, mainly due to the spectrum of skills and technologies involved. By transforming the current knowledge-driven approach into an experience-driven one, we can fulfil industry's needs for individuals capable of shaping the data-driven enterprise. Current curricula are furthermore highly siloed, leading to communication problems and suboptimal solutions and implementations. The next generation of data professionals needs this wider view in order to deliver the data-driven organisation of the future:

---

[1] http://edsa-project.eu/

[2] http://edison-project.eu/

- **Data-intensive engineers:** Successful data-intensive engineers control how to deal with data storage and management. They are experts on distributed computing and computing centres; hence they are mostly at the advanced system administrator levels. They have the know-how to operate large clusters of (virtual) machines, configure and optimise load balancing, and organise Hadoop clusters, and know about Hadoop Distributed File System and Resilient Distributed Datasets, etc.
- **Data scientists:** Successful data scientists will require solid knowledge in statistical foundations and advanced data analysis methods, combined with a thorough understanding of scalable data management, with the associated technical and implementation aspects. They will be the specialists that can deliver novel algorithms and approaches for the BDV stack in general, such as advanced learning algorithms and predictive analytics mechanisms. They are data-intensive analysts. They need to know statistics and data analysis; they need to be able to talk to data-intensive engineers, but should be relieved from system administrator problems; and they need to understand how to transform problems into appropriate algorithms which may need to be modified slightly. Data scientist benchmarks select and optimise these algorithms to reach a business objective. They also need to be able to evaluate the results obtained, following sound scientific procedures. A data scientist curriculum would ideally provide enough insight into the Data Engineering discipline to steer the selection of algorithms, not only from a business perspective but also from an operational and technical perspective. For this, Europe needs new educational programmes in Data Science as well as ideally a network between scientists (academia) and industry that will foster the exchange of ideas and challenges.
- **Data-intensive business experts:** These are the specialists that develop and exploit techniques, processes, tools and methods to develop applications that turn data into value. In addition to technical expertise, data-intensive business experts need to understand the domain and the business of the organisations. This means they need to bring in domain knowledge and are thus working at the intersection of technology, application domains and business. In a sense, they thereby constitute the link between technology experts and business analysts. Data-intensive business experts will foster the development of big data applications from an "art" into a disciplined engineering approach. They will thereby allow the structured and planned development and delivery of customer-specific big data solutions, starting from a clear understanding of the domain, as well as the customer's and user's needs and requirements.

In order to successfully meet the skills challenge, it is critical that industry works with both higher education institutes and education providers to identify the skill requirements that can be addressed with the establishment of:

- New educational programmes based on interdisciplinary curricula with a clear focus on high-impact application domains.
- Professional courses to educate and re-skill/up-skill the current workforce with the specialised skillsets needed to be data-intensive engineers, data scientists and

data-intensive business experts. These courses will stimulate lifelong learning in the domain of data and in adopting new data-related skills.

- Foundational modules in Data Science, Statistical Techniques, and Data Management within related disciplines such as law and the humanities.
- A network between scientists (academia) and industry that leverages innovation spaces to foster the exchange of ideas and challenges.
- Datasets and infrastructure resources, provided by industry, that enhance the industrial relevance of courses.

1.2 Formal and Non-formal Learning

To provide a more enhanced educational support to tackle the skills challenges defined above, both formal[3] and non-formal[4] learning can be considered as they contribute to the lifelong learning of data scientists – the continual training of data scientists throughout their careers. While formal systems are often focused on initial training, a lifelong learning system must include a variety of formal and non-formal learning together. This is necessary to meet the individual's need for continuous and varied renewal of knowledge and the industry's need for a constantly changing array of knowledge and competences.

Here, we will consider non-formal education to include any organised training activity outside of formal education (undergraduate or graduate university degrees). Non-formal training includes both e-learning and traditional professional training. These courses can be of widely different durations and include training provided by employers, traditional educational institutions and other third parties.

Therefore, in Data Science non-formal education plays a crucial role and complements formal training, by allowing practitioners to up-skill and re-skill to adapt to new Data Science requirements.

---

[3]"Education that is institutionalised, intentional and planned through public organisations and recognised private bodies and, in their totality, make up the formal education system of a country. Formal education programmes are thus recognised as such by the relevant national educational authorities (UNESCO)" (http://uis.unesco.org/en/glossary-term/formal-education).

[4]"Education that is institutionalised, intentional and planned by an education provider. The defining characteristic of non-formal education is that it is an addition, alternative and/or a complement to formal education within the process of the lifelong learning of individuals (UNESCO)" (https://unevoc.unesco.org/home/TVETipedia+Glossary/filt=all/id=185).

2   Key Projects on Data Skills

Previous EU projects have already worked on Data Science skills. The two main initiatives in this context have been the EDISON project and the EDSA project analysed below.

2.1   The EDISON Project

The EDISON project defined the EDISON Data Science Framework (EDSF). The definition of the whole framework was based on the results of extensive surveys. Its four components are as follows:

- The Data Science Competence Framework (CF-DS) provides the definition of Competences for Data Science according to the e-CF 3.0. These competences are represented in five competence groups:

  - Data Science Analytics
  - Data Science Engineering
  - Domain Knowledge and Expertise
  - Data Management
  - Research Methods

  For each of these groups, several component competences are given at three levels of proficiency (associate, professional, expert). For example, for the Data Science Analytics competence group, six component competences have been defined. Two of them are:

  - DSDA01: Effectively use a variety of Data Analytics techniques, such as machine learning (including supervised, unsupervised, semi-supervised learning), data mining and prescriptive and predictive analytics for complex data analysis through the whole data lifecycle.
  - DSDA02: Apply designated quantitative techniques, including statistics, time series analysis, optimisation and simulation to deploy appropriate models for analysis and prediction.

- The Data Science Body of Knowledge (DS-BoK) provides, for each competence group, the identification of knowledge areas and knowledge units.
- The Data Science Model Curriculum (MC-DS) provides, for each competence group and individual competence, the learning outcomes required to obtain the competence. These outcomes are given for each of the three levels of proficiency.
- The Data Science Professional Profiles (DSPP) provides a listing of 22 professional profiles in Data Science grouped in 6 categories: managers, professional (data handling/management), professional (database), technicians and associate professionals, and clerical support workers (general and keyboard workers). The

framework also identifies the relevance of each competence group for each professional profile.

2.2 The EDSA Project

One of the aims of the EDSA project was to propose a curriculum for Data Science. That curriculum was based upon what the EDSA consortium identified as core Data Science knowledge rather than the skills that might be needed for a particular job in Data Science. This curriculum was validated through various surveys.

The EDSA curriculum consists of 15 core Data Science topics. Each of these topics has learning objectives, descriptions as well as resources and materials, which were also produced as part of the EDSA project. The 15 topics that make up the core EDSA curriculum were divided into 4 stages: Foundations, Storage and Processing, Analysis, and Interpretation and Use. Table 1 shows an example of the documentation provided by EDSA for a topic, in this case for the Data-Intensive Computing Topic.

3 The Need for the Recognition of Data Skills

With the development of new technologies and the digital transformation of our economy, the labour market has also evolved. Nowadays, applicants for a job are no longer asked to submit a traditional paper résumé; this information is presented digitally, that is, recruiters and headhunters search the Internet (on an international level) for candidates who have the required skills, and some assessment of candidates can be done online. Moreover, the labour market is constantly evolving, and the required skills and qualifications change rapidly over time. Adequately adapting to these changes is essential for the success of employers, learning institutions and governmental agencies related to education. In this section, we will discuss mechanisms for recognising skills in the EU, with a focus on the internationalisation, digitalisation and flexibility of these credentials and their application to Data Science. We begin with a brief review of the main challenges we hope to address.

How Can We Standardise Credentials Throughout Europe?
Although political institutions in the EU have strived to coordinate and standardise diplomas and other forms of credentialing in higher education, the variety of educational systems in the EU and the lack of an adequate system to recognise learning and skills have contributed to great differences in the economic and social outcomes of the member states. The many different educational and training systems in Europe make it difficult for employers to assess the knowledge of potential employees. There is no automatic EU-wide recognition of academic diplomas; students can only obtain a "statement of comparability" of their university degree. The statement of comparability details how the student's diploma compares to the

Table 1 Material developed by EDSA for a data-intensive computing-related course[a]

Scalable machine learning and deep learning

The course studies the fundamentals of distributed machine learning algorithms and the fundamentals of deep learning. It covers the basics of machine learning and introduces techniques and systems that enable machine learning algorithms to be efficiently parallelised. The course complements courses in machine learning and distributed systems, with a focus on both deep learning and the intersection between distributed systems and machine learning. The course prepares the students for master's projects and Ph.D. studies in the area of Data Science and distributed computing.

The main objective of this course is to provide the students with a solid foundation for understanding large-scale machine learning algorithms, in particular deep learning and their application areas.

Intended learning outcomes

Upon successful completion of the course, the student will:

Be able to re-implement a classical machine learning algorithm as a scalable machine learning algorithm

Be able to design and train a layered neural network system

Syllabus and topic descriptions

Main topics:

Machine learning (ML) principles

Using scalable data analytics frameworks to parallelise machine learning algorithms

Distributed linear regression

Distributed logistic regression

Distributed principal component analysis

Linear algebra, probability theory and numerical computation

Convolutional networks

Sequence modelling: recurrent and recursive nets

Applications of deep learning

Detailed content

Introduction:

Brief history and application examples of deep learning and large-scale machine learning: at Google and in industry, ML background, brief overview of deep learning, understanding deep learning systems, linear algebra review, probability theory review

Distributed ML and linear regression:

Supervised and unsupervised learning, ML pipeline, classification pipeline, linear regression, distributed ML, computational complexity

Gradient descent and Spark ML:

Optimisation theory review, gradient descent for least squares regression, the gradient, large-scale ML pipelines, feature extraction, feature hashing, Apache Spark and Spark ML

Logistic regression and classification:

Probabilistic interpretation, multinomial logistic classification, classification example in Tensorflow, quick look in Tensorflow

Feedforward neural nets and backprop:

Numerical stability, neural networks, feedforward neural networks, feedforward phase, backpropagation

Regularisation and debugging:

A flow of deep learning, techniques for training deep learning nets, regularisation, why does deep learning work?

. . . ..

Existing courses:

Scalable machine learning and deep learning at the Royal Institute of Technology, KTH

Scalable machine learning, edX, https://courses.edx.org/courses/BerkeleyX/CS190.1x/1T2015/info

(continued)

Table 1 (continued)

Distributed machine learning with Apache Spark, edX https://www.edx.org/course/distributed-machine-learning-apache-uc-berkeleyx-cs120x
Deep learning systems, University of Washington, http://dlsys.cs.washington.edu/
Scalable machine learning, University of Berkeley, https://bcourses.berkeley.edu/courses/1413454/
Existing materials:
Ian Goodfellow and Yoshua Bengio and Aaron Courville. Deep learning, MIT Press
Spark ML pipelines, http://spark.apache.org/docs/latest/ml-pipeline.html
Spark ML overview, https://www.infoq.com/articles/apache-sparkml-data-pipelines

[a]https://edsa-project.eu/edsa-data/uploads/2015/02/EDSA-2017-P-D23-FINAL.pdf

diplomas of another EU country.[5] Something similar happens with the recognition of professional qualifications as the mobility of Europeans between member states of the EU often requires the full recognition of their professional qualifications (training and professional experience). This is accomplished through an established procedure in each European country.[6]

Directives 2005/36/EC and 2013/55/UE on the recognition of professional qualifications establish guidelines that allow professionals to work in another EU country different from the one where they obtained their professional qualification, on the basis of a declaration.

These directives provide three systems of recognition:

- Automatic recognition – for professions with harmonised minimum training conditions, i.e. nurses, midwives, doctors (general practitioners and specialists), dental practitioners, pharmacists, architects and veterinary surgeons
- General system – for other regulated professions such as teachers, translators and real estate agents
- Recognition on the basis of professional experience – for certain service providers such as carpenters, upholsterers, beauticians, etc.

Additionally, the European professional card (EPC) has been available since 18 January 2016 for five professional areas (general care nurses, physiotherapists, pharmacists, real estate agents and mountain guides). It is an electronic certificate issued via the first EU-wide fully online process for the recognition of qualifications. Unfortunately, these existing mechanisms do not easily accommodate many professions including that of Data Science.

---

[5]http://europa.eu/youreurope/citizens/education/university/recognition/index_en.htm

[6]http://europa.eu/youreurope/citizens/work/professional-qualifications/recognition-of-professional qualifications/index_en.htm

How Can Data Science Credentials Be Digital, Verifiable, Granular and Quickly Evolving?

Traditionally, skills and credentials were conveyed via a résumé on paper and other paper-based credentials. Nowadays, this information can be shared via the Internet in web pages, on social media and in many other forms. The digitalisation of credentials not only allows easier access but also offers new possibilities like:

- The online verification of the validity of the credentials
- Greater granularity in the definition of the credentials
- The expiration of credentials requiring their periodic renewal, which could take into account changes in the demands for skills
- Access to the evidence used in the awarding of credentials

Future schemes for the recognition of skills need to adapt to and accommodate these new demands.

How Can Non-formal Learning in Data Science Be Recognised?

The educational landscape is rapidly changing. The great emphasis which was previously placed on formal university training is slowly eroding. The role of both informal and non-formal learning is increasing, and skills recognition schemes need to contemplate these changes. The BDVe[7] proposed BDV Data Science Badges as a skills recognition tool for formal education and BDV Data Science Labels for non-formal education.

As mentioned, our work on data skills recognition aimed to address these challenges. To do so, the needs of the different stakeholders participating in the process, formal and non-formal education providers, as well as students and industry also play a very relevant role.

4 BDV Data Science Badges for Formal Education

4.1 Methodology

The recognition strategy proposed by the BDVe for formal education science is based on the use of Open Badges.

Open Badges are images that can be included in a curriculum, uploaded to platforms like LinkedIn and shared on social media. They contain metadata to allow:

- The online verification of their authenticity and ownership
- Reviewing information regarding requirements to receive the badge
- Access to details regarding the organisation who issued the badge
- Viewing when the badge was issued and when it expires
- Downloading evidence of the acquisition of skills

---

[7]https://www.big-data-value.eu/

Table 2 Key aspects of the BDV badge recognition schema

Who defines the requirements for the badge?	A collection of experts, including representatives from industry and academia, establish both the types and the requirements of the badges included in the programme. They also define the process for applying to issue badges. This group of experts will periodically meet to review the programme. Based upon progress reports, they can propose changes and improvements.
Who issues the badge?	Interested institutions/educators can apply to issue a badge. This application includes submitting evidence that shows that they provide their students with the skills required by the badge.
Who decides if an institution can issue badges?	A group of experts defined by representatives from industry/academia evaluate applications received to issue badges. Reviewers are assigned applications to assess and decide whether the applicant programme meets the established standards to issue badges. Applications can be rejected, accepted conditionally for 1 year or accepted for 4 years.
How does an institution issue a badge?	Students in a Data Science programme authorised to issue BDV Badges acquire and demonstrate their Data Science skills through their studies. Students in the programme can submit an application to their programme to receive a badge. The programme reviews badge applications, and if the applicant has met the requirements of the badge, then it issues the student that badge. Badges are individualised and contain metadata including the requirements to earn the badge and evidence of the student's achievements.
How are the badges displayed?	Students can display their badges online: in their CV, on social networks, etc. Interested employers can verify that a badge is valid and use its metadata to access relevant information regarding the earners of a badge.

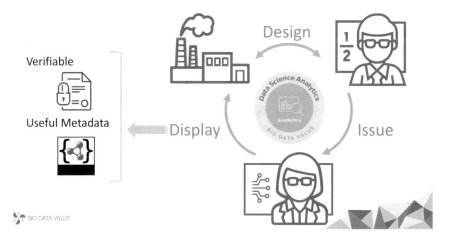

Fig. 1 BDV Badges – application and issuing process

The key aspects of the Open Badges recognition schema proposed by the BDVe are detailed in Table 2.

Figure 1 represents graphically the BDV Badge programme proposed. The badges will be designed by a committee of experts from both industry and academia. Institutions will be responsible for issuing the badges (once a review process has been successfully passed) to their students, and they will be able to display their badges online, so employers will have access to the content and thereby verify the Data Science knowledge of the students.

4.2   Badge Overview

Based on the EDISON framework, we initially proposed the creation of one group of badges for each competence group, with each group of badges having three levels of proficiency (basic, intermediate and expert). To make the proposal more accessible to a wider audience, we chose to use the term "required skills" in place of "learning outcomes".

Thus, the following is the initial collection of BDV Data Science Badges:

- Data Science Analytics Badge
- Data Engineering Badge
- Data Science Management Badge
- Business Process Management Badge
- Data Science Research Method and Project Management Badge

With the aim of verifying the comprehensibility and utility of this proposal, we conducted an evaluation process which involved both industry and academia. In order to get detailed feedback and make this assessment process effective, in the initial stage, we focused only on the Data Science Analytics Badge. We obtained feedback from 12 companies from industry. The aims were to obtain information about the relevance of the different required skills to their hiring practices and to ensure that the descriptions of the required skills were easy to understand. Fifteen universities were contacted to participate in several rounds of the evaluation. The aim was to get feedback about the review process (specifically the kinds of material to be requested of badge applicants) and about the requirements of the badge. Additionally, the members of the Big Data Value Association (BDVA) Skills and Education Task Force provided feedback on the initial version of the badges as well as on the comments gathered from industry and academia.

Based on the results of the assessment process, the three levels of proficiency (basic, intermediate and expert) were replaced by two levels (academic and professional) having the same required skills. The academic level requires knowledge and training which can be acquired in an academic context, while the professional level requires real professional practice.

Table 3 BDV Data Science Analytics Badge skills

Data Science Analytics Badge v1.0
Required skills
DSA.1. Identify existing requirements to choose and execute the most appropriate data discovery techniques to solve a problem depending on the nature of the data and the goals to be achieved
DSA.2. Select the most appropriate techniques to understand and prepare data prior to modelling to deliver insights
DSA.3. Assess, adapt and combine data sources to improve analytics
DSA.4. Use the most appropriate metrics to evaluate and validate results, proposing new metrics for new applications if required
DSA.5. Design and evaluate analysis tools to discover new relations in order to improve decision-making
DSA.6. Use visualisation techniques to improve the presentation of the results of a Data Science project in any of its phases

Fig. 2 Data Science Analytics Badges with academic and professional levels (v1.0)

The description of some of the requirements was also modified, providing the final version of the BDV Data Science Analytics Badge shown in Table 3. Images of both the academic and professional badges are shown in Fig. 2.

Figure 3 shows how the Data Science Analytics Badge of one student could be visualised.

4.3 Platform

As mentioned, the proposed recognition framework works with Open Badges. In this section, we address the badge-issuing platform selected. First, we will consider some details of v2.0 of the Open Badge Standard.

The most recent version of the technical specifications for Open Badges (v2.0) was published on 12 April 2018.[8] An Open Badge must contain three pieces of linked metadata in JSON-LD:

---

[8]https://www.imsglobal.org/sites/default/files/Badges/OBv2p0Final/index.html

BDV DATA SCIENCE ANALYTICS - ACADEMIC LEVEL

Issued by:

Created by: Big Data Value eCosystem Project

Issued on: 2020

Expires on: 2024 (days)

Earner:

Data Science Analytics is use of techniques to gather, organize and process a collection of raw data to discover new and useful information and support the making of conclusions and decisions based upon that data. Earners of this badge have demonstrated their ability to apply Data Science Analytics techniques in an academic setting.

Q   Check this badge ...

CRITERIA

Open criteria.

BDV Data Science Analytics Badge v1.0 - Academic Level

General Description

This badge is part of the BDV Data Science Skills Recognition Program organized in a collaboration between the BDVe Project and the BDVA Skills Task Force.

The BDV Data Science Analytics badge serves to recognize an academic level of understanding in Data Analytics. Earners of this badge have shown that they can use appropriate techniques to discover new relations in a collection of data and thereby give insight into that data set to support decision making in an academic setting. This badge includes skills related to Machine Learning, Data Mining, Statistical Methods and Algorithms.

This badge was issued by an educational program which meets the educational standards established by the BDV Data Science Skills Recognition Program.

Required Skills

To earn this badge, a student must acquire and demonstrate their competence in the following data science skills in an academic setting.

- DSA 1 Identify existing requirements to choose and execute the most appropriate data discovery techniques to solve a problem depending on the nature of the data and the goals to be achieved
- DSA 2 Select the most appropriate techniques to understand and prepare data prior to modeling to deliver insights
- DSA 3 Assess, adapt, and combine data sources to improve analytics
- DSA 4 Use the most appropriate metrics to evaluate and validate results, proposing new metrics for new applications if required
- DSA 5 Design and evaluate analytics tools to discover new relations in y order to improve decision making
- DSA 6 Use visualization techniques to improve the presentation of the results of a data science project in any of its phases.

Credits

This work was partially funded by the project ID 732630 under H2020 EU 2.1.1 . INDUSTRIAL LEADERSHIP - Leadership in enabling and industrial technologies - Information and Communication Technologies.

The description, requirements and types of badges used in the BDVe Data Science Skills Recognition Program were based upon the Edison EDSF and include changes suggested by the data science community.

(Additional Criteria can be added here.)

EVIDENCE

 sampleEvidenceFile.pdf
https://openbadgepassport.com/file...

Fig. 3 Data Science Analytics Badge of one student

Table 4 Requirements defined for platforms issuing BDV Data Science Badges

Functional requirements	FR1 – Badges are designed by a central body and then shared only with approved issuers (who cannot modify the shared badges)
	FR2 – Approved institutions can issue instances of the badges without the intervention of the central body
	FR3 – Approved institutions can only access the personal information of their own students
	FR4 – The hosting platform offers a badge application process for students at an approved institution. This application includes the upload of evidence
	FR5 – The hosting platform provides space to store student evidence files
	FR6 – The central body can view statistics regarding all issued BDV Badges (but not individual personal data)
	FR7 – The central body can modify or revoke existing badges as well as design new badges
	FR8 – The central body can revoke permission to issue a badge
Platform requirements	TR1 – Open Badge 2.0 certified
	https://www.imsglobal.org/cc/statuschart/openbadges
	https://www.imsglobal.org/sites/default/files/Badges/OBv2p0Final/cert/index.html
	TR2 – Meets requirements of GDPR
	TR3 – Company and/or infrastructure located in Europe
	TR4 – Perpetual badge hosting agreement (badges are not deleted if the issuing contract is discontinued)
	TR5 – Issued badges hashed emails and are signed by each approved institution

- Issuer Profile – This resource describes who issued the badge. Usually, one profile is created for each organisation, but it is possible to have multiple issuers (e.g. different departments within the same university).
- BadgeClass – This resource contains information regarding the badge itself and must include information such as the issuer, a description of the badge and the criteria used to issue the badge.
- Assertion – This represents one particular badge (a BadgeClass) issued to one particular person. People can be identified in a number of ways (telephone number, URL), but many badge platforms only accept email identifiers.

From this standard and other considerations specific to the BDV Badge programme, we developed two lists of requirements for the badge-issuing platform. These are summarised in Table 4.

Finally, all Open Badge v2.0-certified badge-issuing platforms were evaluated according to the previous requirements. The issuing platforms assessed were those listed at https://www.imsglobal.org/cc/statuschart/openbadges on 1 February 2019. From them, one that is based in the EU was chosen, which also fulfils the previous criteria.

Text	LOW⁶	MEDIUM	HIGH	
Colour code	Green	Amber	Red	
			>25% of RIs	>30% of RIs
Fat	≤ 3.0g/100g	> 3.0g to ≤ 17.5g/100g	> 17.5g/100g	> 21g/portion
Saturates	≤ 1.5g/100g	> 1.5g to ≤ 5.0g/100g	> 5.0g/100g	> 6.0g/portion
(Total) Sugars	≤ 5.0g/100g	> 5.0g to ≤ 22.5g /100g	> 22.5g/100g	> 27g/portion
Salt	≤ 0.3g/100g	> 0.3g to ≤ 1.5g/100g	>1.5g/100g	>1.8g/portion

Fig. 4 Example of the application of the UK guidelines for Front of Pack Labels (Source: (Department of Health 2016)). (Public sector information licensed under the Open Government Licence v3.0.)

5   BDV Data Science Labels for Non-formal Education

5.1   Methodology

In recent years the offerings of non-formal training in Data Science in the form of online courses, massive open online courses, in-company training, etc., from both official academic institutions and other non-academic institutions, have greatly increased.

Though the needs of stakeholders in the Data Science ecosystem when considering non-formal education are similar to those of formal education, there are a few issues worth highlighting:

- Students interested in Data Science training can quickly find a huge variety of options, and therefore face difficulties when trying to pick from this overwhelming supply. Which courses are more highly valued by the industry and what is the right course for their experience and expectations are just a couple of issues that arise.
- Employers that need to evaluate non-formal training also face the problem of how to compare the wide variety of different types of courses. For example, how rigorous are the different programmes in terms of duration, quality, evaluation of the students, identity verification during assessment activities, etc.?
- Educators offering these courses also face difficulties related, for example, to how to stand out from other courses, that is, how to clearly communicate their offer, attract students, ensure the quality of their training, etc.

In other contexts, standardised labelling systems are used to systematically provide information to help to characterise and compare different products in the same category. For example, Fig. 4 shows the UK guidelines for Front of Pack Labels, which could be used to, for example, compare different kinds of breakfast cereal.

With this idea of a standardised nutritional labelling system as an inspiration, a labelling system for characterising non-formal training in Data Science was proposed. The aim is to provide a labelling system to highlight educational value, which can be useful for the different stakeholders involved in the process (students, industry and course providers).

To develop this proposal, we have followed a process similar to that used for formal training, in the sense of obtaining a consensus from the stakeholders involved in the process about the content of the labels. For that aim, we have gathered feedback through different activities, such as an online seminar for BDVA members, internal feedback collected from BDVe members and feedback from course providers. This process has led us to define the content of the criteria to be included in the label, as we will explain in the next section.

5.2 Label Overview

The labelling system for non-formal training aims to promote and encourage the recognition of Data Science skills acquired through non-formal training. This new system is designed to achieve the following goals:

- Increase transparency – The labelling system should provide an easy-to-understand representation of the most relevant aspects of a Data Science course. The labels should assist students in the filtering of the vast offer to identify the courses best suited for their needs. The labels should also help employers to assess the relevancy of a course to a position. Lastly, the labels should encourage educational providers to readily provide the information which students and employers most need.
- Simplify the comparison – By standardising the presentation of the essential features of courses, the side-by-side comparison of different courses should be easier for everyone.
- Encourage practices which contribute to quality – By highlighting in the label the principal aspects of non-formal education which contribute to quality training, both students and employers can more easily assess the quality of a course. Also, training providers will be encouraged to adopt practices which increase the quality of their offerings.
- Ensure the alignment of training with industrial needs – The label should encourage educational providers to contrast their training with current industrial needs. This should help students and employers assess whether the training meets their needs. It should also promote changes in the content offered by educators to meet the needs of the industry.

From initial interviews with educational providers and employers, a list of criteria which could be used as the basis for the label has been identified.

Table 5 contains an example of these criteria for an imaginary online course containing the preliminary set of criteria which we are proposing. The appropriate graphical design will need to be produced. Then, the corresponding educational label can be provided along with the course information. Note that this labelling system

Table 5 Preliminary criteria of an online course for BDV Data Science Labels

Course title: introduction to data analytics		
C1 – In what language or languages is the course offered	English	
C2 – Total cost of the course	1000 euros	
C3 – Main audience of the course (programmers, data analytics, manager)	Programmers	
C4 – Expertise level expected for new students (beginner, intermediate, advanced)	Beginners	
C5 – What kind of training is provided; specify the hours of training in each of the following groups	In person: 0 h Online with the possibility of interacting with the trainer: 30 h Online with static (non-interactive) content: 120 h	
C6 – What kind of assessment is performed (no testing, in person, online testing/with or without identification verification)	Online testing without identification verification	
C7 – What is the content of the course – specify the number of hours of training and homework dedicated to acquiring each of the Data Science analytics skills	DSA.1. Identify existing requirements to choose and execute the most appropriate data discovery technique to solve a problem depending on the nature of the data and the goals to be achieved.	20
	DSA.2. Select the most appropriate techniques to understand and prepare data prior to modelling deliver insights.	50
	DSA.3. Asses, adapt, and combine data sources to improve analytics.	70
	DSA.4. Use the most appropriate metrics to evaluate and validate results, proposing new metrics for new applications if required.	0
	DSA.5. Design and evaluate analysis tools to discover new relations in order to improve decision-making.	0
	DSA.6. Use visualisation techniques to improve the presentation of the results of a data science project in any of its phases.	10

does not require any platform to be implemented, as it consists only of an image with the corresponding educational information.

6 Pilot and Use Case

To showcase how the skills recognition methodologies proposed above can be applied to bridge industry with academia, the BDVe conducted a pilot of the Data Science Analytics Badge with the results displayed on the EduHub, which is a platform that contains information about educational programmes as well as their offered BDV Badges.

6.1 BDV Badge Pilot

A pilot of the entire Data Science Analytics Badge application process was conducted in order to validate the process to be followed by the universities applying to issue the badge, as well as the review process. Institutions aiming to issue the badge must provide evidence to show that their students have acquired the corresponding skills. Table 6 shows for the first skill of the Data Science Analytics Badge the information to be provided, so reviewers can check the degree to which this skill is acquired by the students.

Each application form must be reviewed by two reviewers. A final decision is made if the recommendations of the two reviewers coincide. If the two reviewers are not able to reach a consensus, a third reviewer is asked to participate in the process. Each reviewer provides recommendations. The reviewer can recommend that the applicant programme be able to issue the badge for 4 years, that the badge-issuing period of the programme be limited, and that the programme will be required to resubmit another application to issue badges in the following year or that the institution is not able to issue the badge as major drawbacks have been found regarding the acquisition of the required skills.

Reviewers participating in this process must agree to the Code of Conduct for Badge Issuing Application Reviewers, available at https://www.big-data-value.eu/

Table 6 Extract of the application form with information about DSA.1

DSA.1. Identify existing requirements to choose and execute the most appropriate data discovery techniques to solve a problem depending on the nature of the data and the goals to be achieved					
Name of the course or activity	Optional/ mandatory	The average number of hours that students spend in the course/ activity acquiring the skill	Evidence used to evaluate the acquisition of the skill (exam, assignment,…)	Name of the evidence file	The minimum level required to demonstrate the acquisition of the skill
Course 1		X hours	Exam 1, questions 2-3		
			Assignment 4		
			…		
			Add as many rows as needed		
Course 2		Y hours			
Add as many rows as needed					
Minimum number of hours that the student dedicates to acquiring the skill					
Comments (optional)					

skills/skills-recognition-program/call-for-academic-level-data-science-analytics-badge-issuers/?et_fb=1&PageSpeed=off.

The pilot resulted in two institutions being able to issue the Data Science Analytics Badge: one application was accepted, and another application was accepted with comments regarding improvements that could be submitted within the following year.

The institutions and programmes that were granted the right to issue the badge were:

- M.Sc. in Big Data Analytics, Universitat Politècnica de València (Spain)
- Data and Web Science M.Sc. Programme, Aristotle University of Thessaloniki (Greece)

6.2   BDV Education Hub

The BDV Education Hub (EduHub) is designed to help users find the right programme of study or special training course among the many education and training opportunities in the big data area.

Accessible via http://bigdataprofessional.eu/, the EduHub is an online platform that offers a living repository for knowledge about European educational offerings related to big data. The EduHub covers programmes of all areas of the BDV Reference Model (see Chap. 3), including data processing, data management, data analytics, data visualisation and data protection.

The EduHub inventories European master's and Ph.D. programmes, as well as European training programmes (both online and on-site) in the field of Big Data and Data-Driven AI. At the time of writing, the EduHub included over 360 European educational offerings (217 European M.Sc. programmes, 12 European Ph.D. programmes as well as 133 professional trainings). The programmes are carefully selected to reflect their focus on BDV, thereby helping interested students and professionals to find the matching skilling and up-skilling offerings. While the master's programmes are targeted for undergraduate students and the Ph.D. programmes for graduate students, the professional training is targeted for professionals looking for reconversion towards Data Science, as well as employees/ employers looking for up-skilling opportunities.

The EduHub reflects the intention of the BDVA to promote the education of European citizens in this important key area (Zillner et al. 2017). The European Digital Skills and Jobs Coalition recognises these efforts and lists the EduHub as part of the European Digital Skills and Jobs Coalition's Pledge Viewer, a tool for creating, viewing and managing pledges reflecting an organisation's commitment to equip Europeans with the skills they need for life and work in the digital age.

The EduHub also serves as a platform to advertise and make visible the BDV Badges that are awarded to university programmes (see above). Figure 5 shows an example of how the badges are shown together with the key information about the university programme.

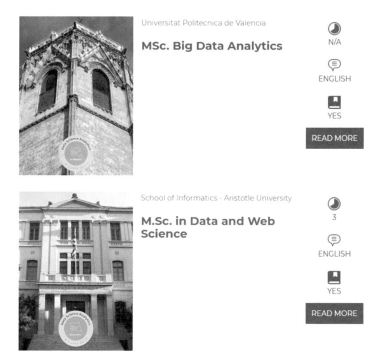

Fig. 5 Screenshot of the BDV EduHub showing awarded BDV Badges

7 Conclusion

Given the considerable amount of attention drawn lately to fields like Big Data, Data Analytics and Data Science, there is an ever-growing need for skilled data scientists by the industry. However, in order to create a vibrant data-driven economy in Europe, it is vital to find ways to bridge the gap between the industrial needs and skills offered by formal or non-formal education. This chapter explored how challenging this goal is as the current knowledge-driven approaches need to be transformed into experience-driven ones via re-definition of the roles and skills of data professionals. This could be achieved by the collaboration of industry and educational providers (formal or non-formal) to define the necessary skills requirements that need to be obtained by future data professionals. The chapter explored steps in that direction that involve the creation of an education platform and a skills recognition programme. Specifically, the EduHub was described, which is a platform that provides access to Data Science and Data Engineering programmes offered by European universities as well as on-site/online professional training programmes, and its aim is to facilitate knowledge exchange on educational programmes and meet current industrial needs. Additionally, the BDV Data Science Badge and Label recognition programmes were analysed for skills acquired by formal and non-formal training, respectively. The aim of the programmes is not only to provide a form of skills recognition but also to align the current industrial needs with the Data Science curricula and skills. Finally, a more practical view was given on how the EduHub and the skills recognition programmes can work in isolation as well as together by demonstrating a pilot on the Data Science Analytics Badge that is currently issued by two universities, and how the badges as well as the educational programmes to which they are issued can be accessed in the EduHub.

Acknowledgements Research leading to these results received funding from the European Union's Horizon 2020 research and innovation programme under grant agreement no. 732630 (BDVe). This publication has emanated from research supported in part by a research grant from Science Foundation Ireland (SFI) under Grant Number SFI/12/RC/2289_P2, co-funded by the European Regional Development Fund.

References

Department of Health. (2016). Guide to creating a Front of Pack (FoP) nutrition label for pre-packed products sold through retail outlets. In *Food standards agency*.
Zillner, S., Curry, E., Metzger, A., Auer, S., & Seidl, R. (Eds.). (2017). *European big data value strategic research & innovation agenda*. Retrieved from Big Data Value Association website: www.bdva.eu

The Road to Big Data Standardisation

Ray Walshe

Abstract This chapter covers the critical topic of standards within the area of big data. Starting with an overview of standardisation as a means for achieving interoperability, the chapter moves on to identify the European Standards Development Organizations that contribute to the European Commission's plan for the Digital Single Market. The author goes on to describe, through use cases, exemplar big data challenges, demonstrates the need for standardisation and finally identifies the critical big data use cases where standards can add value. The chapter provides an overview of the key standardisation activities within the EU and the current status of international standardisation efforts. Finally, the chapter closes with future trends for big data standardisation.

Keywords Standardisation · Strategy · Policy · European Commission · Reference architecture · Use cases · Big data · Future directions

1 Introduction

This chapter starts with an introduction to standardisation and the importance of adopting standardised services and products to effectively drive common services around the world. It identifies big data use cases for the purpose of building reference architecture. These use cases help to gather input and priority requirements more effectively to foster interoperability between legacy and new systems. Next, the chapter describes big data standardisation activities and their adoption at different levels. It discusses the trends in big data standardisation and details future plans that would leverage digital solutions to open up new opportunities and boost development. It explains that big data standards are likely to evolve with further research and

R. Walshe (✉)
ADAPT SFI Centre for Digital Content, Dublin City University, Dublin, Ireland
e-mail: Ray.Walshe@DCU.ie

© The Author(s) 2021
E. Curry et al. (eds.), *The Elements of Big Data Value*,
https://doi.org/10.1007/978-3-030-68176-0_14

the development of new technologies, tools and services. Finally, the chapter summarises the path to standardisation.

2   About Standardisation

In everyday life, at work, at play, at rest, we routinely use products, tools, techniques, processes and systems that are designed, tested, deployed, maintained and evolved using agreed global best practice. This agreed global best practice is the core of standardisation. It is what citizens look for when trying to determine product quality, safety, durability and interoperability. If one views standardisation as a critical input to products, services and tools, then quality and confidence are the tangible outputs.

Standards are everywhere and make it possible to carry out everyday activities as they impact our services such as communications, technology, media, healthcare, food, transport, construction and energy. Some standards have stood the test of time, being around for hundreds if not thousands of years (*Through History with Standards* 2020). The Sumerians in the Tigris/Euphrates valley devised a calendar, not very dissimilar to our modern calendar, 5000 years ago. They divided the year into 30-day months and the days into 12 h and each hour into 30 min.

Adopting standards helps ensure regularity, safety, reliability and environmental care. Standardised products and services are perceived as more dependable, raising user confidence, sales and new technology adoption. Standards are used by regulators and legislators for protecting consumer interests and to support government policies. They play a central role in the European Union's policy for a single market. Standards-compliant products and services enable devices to work together, and standardisation provides a solid foundation upon which to develop new technologies and to enhance existing practices. Standards open up market access, provide economies of scale, encourage innovation and increase awareness of technical developments and initiatives.

Standards provide the foundation for a greater variety of new products with new features and options. In a world without standards, products may be dangerous, of inferior quality, incompatible with others, lock in customers to one supplier and lead to manufacturers devising their own standards for every application or product.

The need for international standardisation in the provision of goods and services to consumers should be evident from the above and is also supported by many factual examples of success based on standards development.

The GSM™ mobile communication technology and its successors (3G, 4G) which were led by the European Telecommunications Standards Institute (ETSI) are good examples of standardisation. GSM was originally envisaged as a telecom solution for Europe, but the technologies were quickly adopted and have been deployed worldwide. Thanks to standardisation, international travellers can communicate and use common services anywhere in the world.

2.1   ICT Standardisation and the European Union

The EU supports an effective and coherent standardisation framework, which ensures that standards are developed in a way that supports EU policies and competitiveness in the global market.

Regulations on European standardisation set the legal framework in which the different actors in the standardisation system can operate. These actors are the European Commission, the European Standardization Organizations, industry, small and medium-sized industries (SMEs) and societal stakeholders.

The Commission is empowered to identify information and communications technology (ICT) technical specifications (European Commission 2020a) to be eligible for referencing in public procurement. Public authorities can therefore make use of the full range of specifications when buying IT hardware, software and services, allowing for greater competition and reducing the risk of lock-in to proprietary systems.

The Commission financially supports the work of the three European Standardization Organizations: ETSI, CEN and CENELEC.

2.1.1   ETSI: The European Telecommunications Standards Institute

ETSI, the European Telecommunications Standards Institute, produces globally applicable standards (Dahmen-Lhuissier 2020) for information and communications technologies (ICT), including fixed, mobile, radio, converged, broadcast and Internet technologies. These standards enable the technologies on which business and society rely. The ETSI standards for GSM™, DECT™, smart cards and electronic signatures have helped to revolutionise modern life all over the world.

ETSI is one of the three European Standardization Organizations officially recognised by the European Union and is a not-for-profit organisation with more than 800 member organisations worldwide, drawn from 66 countries and 5 continents. Members include the world's leading companies and innovative R&D organisations.

ETSI is at the forefront of emerging technologies, addressing the technical issues which will drive the economy of the future and improve life for the next generation.

2.1.2   CEN: The European Committee for Standardization

CEN, the European Committee for Standardization (CEN 2020), is an association that brings together the national standardisation bodies of 33 European countries. CEN is also one of three European Standardization Organizations (together with CENELEC and ETSI) that have been officially recognised by the European Union and by the European Free Trade Association (EFTA) as being responsible for developing and defining voluntary standards at European level.

CEN provides a platform for the development of European standards and other technical documents in relation to various kinds of products, materials, services and processes. It supports standardisation activities in relation to a wide range of fields and sectors including air and space, chemicals, construction, consumer products, defence and security, energy, the environment, food and feed, health and safety, healthcare, ICT, machinery, materials, pressure equipment, services, smart living, transport and packaging.

2.1.3 CENELEC: The European Committee for Electrotechnical Standardization

CENELEC is the European Committee for Electrotechnical Standardization (CENELEC 2020) and is responsible for standardisation in the electrotechnical engineering field. It prepares voluntary standards which help facilitate trade between countries, create new markets, cut compliance costs and support the development of a single European market. It creates market access at European level but also at international level, adopting international standards wherever possible, through its close collaboration with the International Electrotechnical Commission (IEC) (CENELEC n.d.), under the Dresden Agreement.

In the global economy, CENELEC fosters innovation and competitiveness, making technology available industry-wide through the production of voluntary standards. Its members, its experts, the industry federations and consumers help create European standards to encourage technological development, to ensure interoperability and to guarantee the safety and health of consumers and provide environmental protection. Designated as a European Standardization Organization by the European Commission, CENELEC is a non-profit technical organisation set up under Belgian law. It was created in 1973 as a result of the merger of two previous European organisations: CENELCOM and CENEL.

EU-funded research and innovation projects also make their results available to the standardisation work of several standards-setting organisations.

2.1.4 The European Multi Stakeholder Platform on ICT Standardisation

The European Multi Stakeholder Platform (MSP) (European Commission 2013a) on ICT standardisation was established in 2011. It advises the Commission on ICT standardisation policy implementation issues, including priority-setting in support of legislation and policies, and the identification of specifications developed by global ICT standards development organisations. The Multi Stakeholder Platform addresses:

- Potential future ICT standardisation needs
- Technical specifications for public procurements

- Cooperation between ICT standards-setting organisations
- A multi-annual overview of the needs for preliminary or complementary ICT standardisation activities in support of the EU policy activities (the Rolling Plan (European Commission 2013b))

The MSP is composed of representatives of national authorities from EU member states and EFTA countries, of the European and international ICT standardisation bodies, and of stakeholder organisations that represent industry, small and medium-sized enterprises and consumers. It meets four times per year and is co-chaired by the European Commission Directorate-General for Internal Market (European Commission 2016), Industry, Entrepreneurship and SMEs and CONNECT (*Communications Networks, Content and Technology*, 2015).

The Platform also Advises on the Elaboration and Implementation of the Rolling Plan on ICT Standardisation (European Commission 2020a)
The Rolling Plan (RP) provides a multi-annual overview of the needs for preliminary or complementary ICT standardisation activities in support of the EU policy activities. It is aimed at the broader ICT community stakeholders and outlines how practically support will be provided. It contains a distinct view of the landscape of standardisation activities in a given policy area.

The Rolling Plan puts standardisation in the policy context, identifies EU policy priorities where standardisation activities are needed, and covers ICT infrastructures and ICT standardisation horizontals. It references legal documents, available standards and technical specifications, as well as ongoing activities in ICT standardisation. The addenda to the Rolling Plan may be published alongside the Rolling Plan in order to keep current with new developments in the rapidly changing ICT sector.

Mission of the Multi Stakeholder Platform on ICT Standardisation (European Commission 2020d)
The Platform is an Advisory Expert Group on all matters related to European ICT standardisation and its effective implementation:

- Advise the Commission on its ICT standardisation work programme.
- Identify potential future ICT standardisation needs.
- Advise the Commission on possible standardisation mandates.
- Advise the Commission on technical specifications in the field of ICT with regard to its referencing in public procurement and policies.
- Advise the Commission on cooperation between standards developing organisations.

The 2016 Rolling Plan on ICT standardisation (European Commission 2020b) [13] covers all activities that can support standardisation and prioritises actions for ICT adoption and interoperability.

The Plan Offers Details on the International Contexts for each Policy
- Societal challenges: e-health, accessibility of ICT products and services, web accessibility, e-skills and e-learning, emergency communications and e-call

- Innovation for the Digital Single Market: e-procurement, e-invoicing, card/Internet and mobile payments, eXtensible Business Reporting Language (XBRL) and Online Dispute Resolution (ODR)
- Sustainable growth: smart grids and smart metering, smart cities, ICT environmental impact, European Electronic Toll Service (EETS) and Intelligent Transport System (ITS)
- Key enablers and security: cloud computing, (open) data, e-government, electronic identification and trust services including e-signatures, radio-frequency identification (RFID), Internet of things (IoT), network and information security (cybersecurity) and e-privacy

This latest Rolling Plan describes all the standardisation activities undertaken by Standard Setting Organizations (SSOs). This ensures an improved coherence between standardisation activities in the EU. This is the first time that the European Standardization Organizations and other stakeholders were involved in drafting the RP, and this improved process is a stronger guarantee that activities of standardisation-supporting EU policies in the ICT domain will be aligned.

3   Identifying Big Data Use Cases

In June 2013, the National Institute of Standards and Technology (NIST) Big Data Public Working Group (NBD-PWG) began forming a community of interested parties from all sectors, including industry, academia and government, to develop a consensus on big data definitions, taxonomies, secure reference architectures, security and privacy requirements, and ultimately a standards roadmap. Part of the work carried out by the working group identified big data use cases in NIST "Big Data Interoperability Framework: Volume 3, Use Cases and General Requirements", which would serve as exemplars to help develop a Big Data Reference Architecture (BDRA).

The NBD-PWG defined a use case as "a typical application stated at a high level for the purposes of extracting requirements or comparing usages across fields". They began by collecting use cases from publicly available information for various big data architecture examples. This process returned 51 use cases across nine broad areas (i.e. application domains). This list was not intended to be exhaustive, and other application domains will be considered. Each example of big data architecture constituted one use case. The nine application domains were Government Operation; Commercial; Defence; Healthcare and Life Sciences; Deep Learning and Social Media; Ecosystem for Research; Astronomy and Physics; Earth, Environmental and Polar Science; and lastly Energy.

3.1   Use Case Summaries

The initial focus of the NBD-PWG Use Case and Requirements Subgroup was to form a community of interest from industry, academia and government, with the goal of developing a consensus list of big data requirements across all stakeholders. This included gathering and understanding various use cases from diversified application domains.

The tasks assigned to the subgroup include the following:

- Gather input from all stakeholders regarding big data requirements, a goal that turned into the gathering of use cases.
- Analyse/prioritise a list of challenging general requirements derived from use cases that may delay or prevent the adoption of big data deployment.
- Develop a comprehensive list of big data requirements.

The report was produced by an open collaborative process involving weekly telephone conversations and information exchange using the NIST document system. The 51 use cases came from participants in the calls (subgroup members) and from others informed of the opportunity to contribute. The use cases are organised into nine broad sectors/areas (application domains) listed below with the number of use cases in parentheses and sample examples:

- Government Operation (4): National Archives and Records Administration, Census Bureau
- Commercial (8): Finance in cloud, cloud backup, Mendeley (citations), Netflix, web search, digital materials, cargo shipping (as in UPS)
- Defence (3): Sensors, image surveillance, situation assessment
- Healthcare and Life Sciences (10): Medical records, graph and probabilistic analysis, pathology, bioimaging, genomics, epidemiology, people activity models, biodiversity
- Deep Learning and Social Media (6): Self-driving cars, geolocate images, Twitter, crowd sourcing, network science, NIST benchmark datasets
- Ecosystem for Research (4): Metadata, collaboration, language translation, light source experiments
- Astronomy and Physics (5): Sky surveys (and comparisons to simulation), LHC at CERN, Belle Accelerator II in Japan
- Earth, Environmental and Polar Science (10): Radar scattering in atmosphere, earthquake, ocean, Earth observation, ice sheet radar scattering, Earth radar mapping, climate simulation datasets, atmospheric turbulence identification, subsurface biogeochemistry (microbes to watersheds), AmeriFlux and FLUXNET gas
- Energy (2): Smart Grid, Home energy management

4 Big Data Standards: The Beginning

Achieving big data goals set out by business and consumers will require the interworking of multiple systems and technologies, legacy and new. Technology integration calls for standards to facilitate interoperability among the components of the big data value chain (Adolph 2013). For instance, UIMA, OWL, PMML, RIF and XBRL are key software standards that support the interoperability of data analytics with a model for unstructured information, ontologies for information models, predictive models, business rules and a format for financial reporting. The standards community has launched several initiatives and working groups on big data. In 2012, the Cloud Security Alliance established a big data working group with the aim of identifying scalable techniques for data-centric security and privacy problems. The group's investigation is expected to clarify best practices for security and privacy in big data and also to guide industry and government in the adoption of those best practices. The US National Institute of Standards and Technology (NIST) kicked off its big data activities with a workshop in June 2012 and a year later launched a public working group. The NIST (NIST 2020) working group intends to support and secure an effective adoption of big data by developing consensus on definitions, taxonomies, secure reference architectures and a technology roadmap for big data analytic techniques and technology infrastructures.

4.1 NIST Big Data Public Working Group

The NIST developed a Big Data Interoperability Framework (Grady et al. 2014) which consists of seven volumes, each of which addresses a specific key topic, resulting from the work of the NBD-PWG. The seven volumes are as follows.

4.1.1 Volume 1, Definitions

The Definitions volume addresses fundamental concepts needed to understand the new paradigm for data applications, collectively known as big data, and the analytic processes collectively known as data science. Big data has had many definitions and occurs when the scale of the data leads to the need for a cluster of computing and storage resources to provide cost-effective data management. Data science combines various technologies, techniques and theories from various fields, mostly related to computer science and statistics, to obtain actionable knowledge from data.

4.1.2 Volume 2, Taxonomies

Taxonomies were prepared by the NIST Big Data Public Working Group (NBD-PWG) Definitions and Taxonomy Subgroup to facilitate communication and improve understanding across big data stakeholders by describing the functional components of the NIST Big Data Reference Architecture (NBDRA). The top-level roles of the taxonomy are System Orchestrator, Data Provider, Big Data Application Provider, Big Data Framework Provider, Data Consumer, Security and Privacy, and Management. The actors and activities for each of the top-level roles are outlined as well. The NBDRA taxonomy aims to describe new issues in big data systems but is not an exhaustive list. In some cases, the exploration of new big data topics includes current practices and technologies to provide needed context.

4.1.3 Volume 3, Use Cases and General Requirements

The Use Cases and General Requirements document was prepared by the NIST Big Data Public Working Group (NBD-PWG) Use Cases and Requirements Subgroup to gather use cases and extract requirements.

The use cases are, of course, only representative, and do not represent the entire spectrum of big data usage. All of the use cases were openly submitted, and no significant editing was performed. While there are differences in scope and interpretation, the benefits of free and open submission outweighed those of greater uniformity.

4.1.4 Volume 4, Security and Privacy

The Security and Privacy document was prepared by the NIST Big Data Public Working Group (NBD-PWG) Security and Privacy Subgroup to identify security and privacy issues that are specific to big data. Big data application domains include healthcare, drug discovery, insurance, finance, retail and many others from both the private and public sectors. Among the scenarios within these application domains are health exchanges, clinical trials, mergers and acquisitions, device telemetry, targeted marketing and international anti-piracy. Security technology domains include identity, authorisation, audit, network and device security, and federation across trust boundaries.

4.1.5 Volume 5, Architectures White Paper Survey

The Architectures White Paper Survey was prepared by the NIST Big Data Public Working Group (NBD-PWG Reference Architecture Subgroup to facilitate understanding of the operational intricacies in big data, and to serve as a tool for developing system-specific architectures using a common reference framework. The Subgroup surveyed published big data platforms by leading companies or individuals supporting the big data framework and analysed the material. This effort revealed a remarkable consistency of big data architecture. The most common themes occurring across the architectures surveyed are outlined below.

- Big Data Management: Structured, semi-structured and unstructured data; velocity, variety, volume and variability; SQL and NoSQL; distributed file system
- Big Data Analytics: Descriptive, predictive and spatial; real time; interactive; batch analytics; reporting; dashboard
- Big Data Infrastructure: In-memory data grids; operational database; analytic database; relational database; flat files; content management system; horizontal scalable architecture

4.1.6   Volume 6, Reference Architecture

The NIST Big Data Public Working Group (NBD-PWG) Reference Architecture Subgroup prepared this NIST Big Data Interoperability Framework: Reference Architecture, to provide a vendor-neutral, technology- and infrastructure-agnostic conceptual model and examine related issues. The conceptual model, referred to as the NIST Big Data Reference Architecture (NBDRA), was crafted by examining publicly available big data architectures representing various approaches and products. Inputs from the other NBD-PWG subgroups were also incorporated into the creation of the NBDRA. It is applicable to a variety of business environments, including tightly integrated enterprise systems, as well as loosely coupled vertical industries that rely on cooperation among independent stakeholders. The NBDRA captures the two known big data economic value chains: information, where value is created by data collection, integration, analysis and applying the results to data-driven services; and the information technology (IT), where value is created by providing networking, infrastructure, platforms and tools in support of vertical data-based applications.

4.1.7   Volume 7, Standards Roadmap

The Standards Roadmap summarises the deliverables of the other NBD-PWG subgroups (presented in detail in the other volumes of this series) and presents the work of the NBD-PWG Technology Roadmap Subgroup. In the first phase of development, the NBD-PWG Technology Roadmap Subgroup investigated existing standards that relate to big data and recognised general categories of gaps in those standards.

4.2   ISO/IEC JTC1's Data Management and Interchange Standards Committee (SC32)

ISO/IEC JTC1's data management and interchange standards committee (SC32) has a study on next-generation analytics and big data (ANSI [UNITED STATES] 2020). The W3C has created several community groups on different aspects of big data.

At the June 2012 SC32 Plenary in Berlin, the SC32 Chair, Jim Melton, appointed an ad hoc committee from all four SC32 working groups: WG1 E-business, WG2 Metadata, WG3 Database Languages and WG4 Multimedia.

The original request from JTC1 referenced a report by the US industry analyst Gartner Group where both "next-generation analytics" and "big data" are identified as strategic technologies.

4.2.1   Next-Generation Analytics

Analytics is growing along three key dimensions:

- From traditional offline analytics to in-line embedded analytics. This has been the focus for many efforts in the past and will continue to be an important focus for analytics.
- From historical data to explain what happened to analysing historical and real-time data from multiple systems to simulate and predict the future.
- Over the next 3 years, analytics will mature along a third dimension, from structured and simple data analysed by individuals to the analysis of complex information of many types (text, video, etc.) from many systems supporting a collaborative decision process that brings multiple people together to analyse, brainstorm and make decisions.

Analytics is also beginning to shift to the cloud and exploit cloud resources for high performance and grid computing.

In 2011 and 2012, analytics increasingly focused on decisions and collaboration. The next step was to provide simulation, prediction, optimisation and other analytics, not simply information, to empower even more decision flexibility at the time and place of every business process action.

4.2.2   Big Data

The size, complexity of formats and speed of delivery exceed the capabilities of traditional data management technologies; the use of new or exotic technologies is required simply to manage the volume alone. Many new technologies are emerging, with the potential to be disruptive (e.g. in-memory Data Base Management System [DBMS]). Analytics has become a major driving application for data warehousing, with the use of MapReduce outside and inside the DBMS, and the use of self-service data marts. One major implication of big data is that in the future users will not be able to put all useful information into a single data warehouse. Logical data warehouses bringing together information from multiple sources as needed will replace the single data warehouse model.

5 Big Data Standards Work

5.1 IEEE Big Data

Governance and metadata management poses unique challenges with regard to big data paradigm shift. The governance lifecycle needs to be sustainable from creation, maintenance, depreciation, archiving and deletion due to the volume, velocity and variety of big data changes, and can be accumulated whether the data is at rest, in motion or in transactions.

To facilitate and support the Internet of things, smart cities and other emerging technical and market trends, it is critical to have a standard reference architecture for Big Data Governance and Metadata Management (BDGMM) that is scalable and can enable the findability, accessibility, interoperability and reusability between heterogeneous datasets from various sources.

The goal of BDGMM is to enable data integration/mashup among heterogeneous datasets from diversified domain repositories and make data discoverable, accessible and usable through a machine-readable and actionable standard data infrastructure. The IEEE BDGMM was created jointly by the IEEE Big Data Initiative and the IEEE Standards Association.

5.2 ITU-T Big Data

Big data-driven networking (bDDN) and deep packet inspection (DPI): Deep packet inspection is essential for network operators to know the distribution of service/application traffic in the network.

- What enhancements to existing recommendations are needed to enable services/application identification/awareness/visibility and to enable traffic and resource optimisation based on deep packet inspection in future networks (including software-defined networking, network functions virtualisation, Internet of things, information-centric networking/content-centric networking and other candidate future network architecture and technology (e.g. IMT-2020))?

5.3   ISO/IEC JTC1 WG 9 Big Data Working Group

Standard ecosystems are required to perform analytics processing regardless of the dataset's needs in relation to the Vs (volume, velocity, variety, etc.) characteristics, underlying computing platforms and how big data analytics tools and techniques are deployed. Unified data platform architecture will support big data strategy across information management, analysis and search technology.

A standard ecosystem provides vendor, technology and infrastructure-agnostic platforms that will enable data scientists and researchers to share and reuse interoperable analytics tools and techniques. WG 9 works with academics, industry, government and various other stakeholders to understand the needs and foster such a standard big data ecosystem.

WG 9 has a three-pronged technical approach to achieve this standard ecosystem:

- Identify standard Big Data Reference Architecture (RA): this approach has already been captured in ISO/IEC 20547 to identify overall RA components and their interface descriptions.
- Identify standard Big Data Reference Architecture Interfaces: this would be a new project to investigate how data flows between RA components and define standard interfaces for such interactions. The goal is to use these validated standard interfaces to build big data applications.
- Identify standard Big Data Management Tools: this would be another new project to investigate how a collection of analytics tools and computing resources can be efficiently and effectively managed to enable standard big data enterprise computing. The goal is to provide system management tools to manage, monitor and fine-tune big data applications.

WG 9 produced the ISO/IEC 20546 (IS) Big Data Overview and Vocabulary committee draft (CD) in March 2016 with balloting results from 9 countries approved as presented, 5 countries approved with comments, 2 countries disapproved with comments and 15 countries choosing abstention. WG 9 spent two teleconferences (15 August and 30 August) reviewing, discussing and resolving all comments, and generated the Disposition of Comments and revised text for further contribution.

WG 9 produced the ISO/IEC 20547-2 Big Data Use Cases and Derived Requirements Provisional Draft Technical Report (51 use cases, 300+ pages) in July 2016 with a 2-month balloting period. All comments are expected to be reviewed, discussed and resolved at the 6th WG 9 November–December 2016 meeting.

For the 4th WG 9 meeting (7 March 2016, Ireland), WG 9 hosted a full-day programme with 16 speakers, 1 panel discussion and over 50 participants. For the 5th WG 9 meeting (11 July 2016, China), a half-day programme with 8 speakers and over 80 participants was conducted. Through outreach effort, and in addition to recruiting more big data experts, new opportunities and expansion of the big data standard foundation technologies such as Big Data Reference Architecture Standard Interface and Big Data Reference Architecture Standard Management were explored.

5.4   JTC1 SC42: *Artificial Intelligence*

5.4.1   Membership

31 Participating Members Australia SA; Austria ASI; Belgium NBN; Canada SCC; China SAC; Congo, the Democratic Republic of the OCC; Denmark DS; Finland SFS; France AFNOR; Germany DIN; India BIS; Ireland NSAI; Israel SII; Italy UNI; Japan JISC; Kenya KEBS; Korea, Republic of KATS; Luxembourg ILNAS; Malta MCCAA; the Netherlands NEN; Norway SN; Russian Federation GOST R; Saudi Arabia SASO; Singapore
SC; Spain UNE; Sweden SIS; Switzerland SNV; Uganda UNBS; United Arab Emirates ESMA; United Kingdom BSI; United States ANSI.

14 Observing Members Argentina IRAM, Benin ANM, Cyprus CYS, Hong Kong ITCHKSAR, Hungary MSZT, Lithuania LST, Mexico DGN, New Zealand NZSO, Philippines BPS, Poland PKN, Portugal IPQ, Romania ASRO, South Africa SABS, Ukraine DSTU.

5.4.2   Working Groups and Study Groups JTC1 SC42

The ISO/IEC standardisation committee JTC1/SC42 is structured as follows.

Working Group 1	On foundational standards that cope with AI concepts and AI terminology necessary for the full AI lifecycle.
Working Group 2	On big data that aims at vocabulary, framework and reference architecture for big data.
Working Group 3	Deals with requirements for trustworthy and bias-free AI systems that include assessment of the robustness of neural networks.
Working Group 4	Is oriented towards applications and use cases to demonstrate feasibility on AI standards.
Study Group 1	Investigates computational approaches comprising machine learning (ML) algorithms, reasoning approaches, NLP, etc.
Study Group 2	Investigates into aspects of trustworthiness and pitfalls, where the former aspects deal with system properties such as transparency, verifiability, explainability and controllability and the latter aspects deal with robustness, safety, security and privacy system properties.
New work item proposal	A new standardisation project NWIP 24300 is planned and is related to the AI process management for big data analysis (BDA) (ANSI [UNITED STATES] n.d.).

5.4.3   List of Published Standards in JTC1 SC42

Title	Lead editor	Co-editors
ISO/IEC 20546:2019 Information technology — Big data — Overview and vocabulary (Published)	Nancy Grady (USA)	David Boyd (USA)
ISO/IEC TR 20547-1, Information technology – Big Data Reference Architecture -- Part 1: Framework and Application Process	David Boyd (USA)	Su Wook Ha (KR), Ray Walshe (IE)
ISO/IEC TR 20547-2:2018, Information technology – Big Data Reference Architecture -- Part 2: Use Cases and Derived Requirements (Published)	Ray Walshe (IE)	Su Wook Ha (KR)
ISO/IEC 20547-3:2020, Information technology -- Big Data Reference Architecture -- Part3: Reference Architecture (Published)	Ray Walshe (IE)	David Boyd (USA), Liang Guang (CN), Toshihiro Suzuki (JP)
ISO/IEC TR 20547-5:2018(en), Information technology – Big Data Reference Architecture -- Part 5: Standards Roadmap (Published)	David Boyd (USA)	Toshihiro Suzuki (JP), Ray Walshe (IE)
ISO/IEC TR 24028:2020 Information technology — Artificial intelligence — Overview of trustworthiness in artificial intelligence	Jutta Williams (USA)	

5.4.4   List of Standards in Progress JTC1 SC42

ISO/IEC WD 5059	Software engineering — Systems and software Quality Requirements and Evaluation (SQuaRE)	Quality model for AI-based systems
ISO/IEC CD TR 20547-1	Information technology — Big data reference architecture — Part 1:	Framework and application process (submitted for publication)
ISO/IEC CD 22989	Artificial intelligence	Concepts and terminology
ISO/IEC CD 23053	Information Technology	Framework for Artificial Intelligence (AI) Systems Using Machine Learning (ML)
ISO/IEC CD 23894	Information Technology — Artificial Intelligence	Risk Management
ISO/IEC AWI TR 24027	Information technology — Artificial intelligence (AI)	Bias in AI systems and AI-aided decision making
ISO/IEC CD TR 24029-1	Artificial Intelligence (AI)	Assessment of the robustness of neural networks — Part 1: Overview
ISO/IEC CD TR 24030	Information technology — Artificial Intelligence (AI)	Use cases

(continued)

ISO/IEC AWI TR 24368	Information technology — Artificial intelligence	Overview of ethical and societal concerns
ISO/IEC AWI TR 24372	Information technology — Artificial intelligence (AI)	Overview of computational approaches for AI systems
ISO/IEC AWI 24668	Information technology — Artificial intelligence	Process management framework for big data analytics
ISO/IEC AWI 38507	Information technology — Governance of IT	Governance implications of the use of artificial intelligence by organisations

6 Trends and Future Directions of Big Data Standards

6.1 Public Sector Information, Open Data and Big Data

A key issue for leveraging data value and data value chains in this era of continuously increasing volumes of big data and open data (European Commission 2015) is the need for interoperability. Standardisation at different levels such as metadata, data formats and licensing is essential to enable broad data integration, data exchange and interoperability with the overall goal to foster data-driven innovation. This refers to both structured and unstructured data, as well as data from different domains as diverse as geospatial data, statistical data, weather data, Public Sector Information (PSI) and research data.

On 25 April 2018, the European Commission adopted the "data package" measures to improve the availability and reusability of data (European Commission 2020c), including government data and publicly funded research results, and to foster data sharing in business-to-business (B2B) and business-to-government (B2G) settings. Data availability is crucial to enable companies to leverage the potential of data-driven innovation or develop solutions using artificial intelligence.

The key elements of the Directive on open data and the reuse of public sector information (recast of Directive 2003/98/EC (EUR-Lex 2020a) amended by Directive 2013/37/EU (EUR-Lex 2020b)) are:

- Enhancing access to and reuse of real-time data
- Lowering charges for the reuse of public sector information
- Allowing for the reuse of new types of data, including data resulting from publicly funded research
- Minimising the risk of excessive first-mover advantage in regard to certain data
- "High-value datasets" belonging to six thematic categories (geospatial, Earth observation and environment, meteorological, statistics, companies and company ownership, mobility) to be made available mandatorily free of charge

6.2 European Commission-Funded Standards Projects

Ongoing European projects ELITE-S and StandICT.eu support the training and creation of the next generation of standardisation experts needed for the Digital Single Market.

ELITE-S is a Horizon 2020 Marie Skłodowska-Curie COFUND Action based at the ADAPT Centre at Dublin City University and its Irish academic partners. It is a postdoctoral fellowship programme for intersectoral training, career development and mobility offering 16 prestigious 2-year fellowships in technology and standards development to address five EU priority areas: 5G, Internet of things, cloud computing, cybersecurity and data technologies. Experienced researchers from any country enhance their qualifications and diversify their competencies by conducting a research project at a host institution in Ireland in any of the current research and technology application areas of the programme.

StandICT.eu, "Supporting European Experts Presence in International Standardisation Activities in ICT", addresses the need for ICT standardisation and defines a pragmatic approach and streamlined process to reinforce EU expert presence in the international ICT standardisation scene. Through a Standards Watch, it analyses and monitors the international ICT standards landscape and liaise with Standards Development Organizations (SDOs) and Standard Setting Organizations (SSOs), key organisations such as the EU Multi Stakeholder Platform for ICT standardisation, as well as industry-led groups, to pinpoint gaps and priorities matching EU Digital Single Market objectives. It provides support for European specialists:

- To contribute to ongoing standards development activities and attend SDO and SSO meetings
- To support the prioritisation of standardisation activities and build a community of standardisation experts
- To support knowledge exchange and collaboration and reinforce European presence in international ICT standardisation

6.3 The Big Data Value Association (BDVA)

The Big Data Value Association (BDVA) is a private, industry-led non-profit association with the mission of boosting European big data value research, development and innovation and fostering a positive perception of big data value. The aim is to maximise the economic and societal benefit to Europe, its businesses and its citizens, enabling Europe to take the lead in the global data-driven digital economy (Zillner et al. 2017).

BDVA membership is composed of large industries, SMEs and research organisations to support the development and deployment of the EU Big Data Value Public-Private Partnership with the European Commission representing the private side. The BDVA organises its work in Task Forces, where its members engage and influence, and it aims to be the European big data reference point.

The BDVA is open to new members to further enrich the data value ecosystem and play an active role. These include data users, data providers, data technology providers and researchers. Membership of the Association gives the following benefits:

- Part of the European big data industry initiative which will have a high impact on the deployment of big data technologies and thus business competitiveness and economic growth
- Influencing big data challenges and needs in the following years by contributing to the Strategic Research and Innovation Agenda (SRIA)
- Direct access to discussions with EU Commission and member states, thus gaining access to and influencing strategic direction
- Networking and partnering with industrial and research partners in the European data value chain, to set up collaborative research and innovation activities

6.4 European Commission Standardisation Ongoing Activities

The success of Europe's digital transformation (European Commission 2020f) will depend on tools, techniques, services and platforms to ensure trustworthy technologies and to give businesses the confidence and means to digitise. The Data Strategy (European Commission 2020e) and the White Paper on Artificial Intelligence (European Commission 2020g) published by the European Commission endeavour to put people first in developing technology, while continuing to defend and promote European values and rights in the design, development and deployment of technology in the real economy.

The European strategy for data aims to ensure Europe's global competitiveness and data sovereignty by creating a Digital Single Market for data. Common European data spaces will ensure that more data becomes available for use in the economy and society, while keeping companies and individuals who generate the data in control.

Data is an essential resource for economic growth, competitiveness, innovation, job creation and societal progress in general. Standardisation and its impact on the economy has already been well documented (Jakobs 2017) (Blind et al. 2012). Citizens will benefit from these data-driven applications through improved health care, safer and cleaner transport systems, new products and services, reduced costs of public services, and improved sustainability and energy efficiency.

Data availability will drive innovation and necessitate practical, fair and clear rules on data access and use, which comply with European values and rules such as personal data protection.

To ensure the EU's leadership in the global data economy, this European strategy for data intends to:

- Adopt legislative measures on data governance, access and reuse
- Open up high-value publicly held datasets across the EU for free
- Invest €2 billion in a European high-impact project to develop data processing infrastructures, data sharing tools, architectures and governance mechanisms for thriving data sharing and to federate energy-efficient and trustworthy cloud infrastructures and related services
- Enable access to secure, fair and competitive cloud services
- Empower users to stay in control of their data and invest in capacity building for small and medium-sized enterprises and digital skills
- Foster the roll-out of common European data spaces in crucial sectors such as industrial manufacturing, green deal, mobility and health

As part of data strategy, the European Commission has published a report on business-to-government (B2G) data sharing. The report, which comes from a high-level Expert Group (European Commission 2018), contains a set of policy, legal and funding recommendations that will contribute to making B2G data sharing in the public interest a scalable, responsible and sustainable practice in the EU.

6.5   Open Consultation AI White Paper and Data Strategy

The European Commission has adopted a new digital strategy for a European society powered by digital solutions that puts people first, opens up new opportunities for businesses and boosts the development of trustworthy technology. The Commission also presented a White Paper on Artificial Intelligence setting out its proposals to promote the development of AI in Europe whilst ensuring respect of fundamental rights.

Commission President Ursula von der Leyen stated: "Today we are presenting our ambition to shape Europe's digital future. It covers everything from cybersecurity to critical infrastructures, digital education to skills, democracy to media. I want that digital Europe reflects the best of Europe – open, fair, diverse, democratic and confident".

The Commission published on 15th December 2020 the proposal for a Regulation on a Single Market For Digital Services (Digital Services Act) and on 3rd December 2020 its European Democracy Action Plan to empower citizens and build more resilient democracies across the EU. The Regulation on electronic identification and trust services for electronic transactions in the internal market (eIDAS Regulation) allows use of national electronic identification schemes (eIDs) to access public services available online in other EU countries. The EU aims to enhance cyber defence cooperation and cyber defence capabilities, building on the work of the European Defence Agency. Europe will also continue to build alliances with global partners, leveraging its regulatory power, capacity building, diplomacy and finance to promote the European digitalisation model.

The White Paper on Artificial Intelligence was open for public consultation until 19 May 2020. The Commission is also gathering feedback on its data strategy. Using the feedback received, the Commission will take further action to support the development of trustworthy AI and the data economy.

7 Future (Big) Data Standardisation Actions

Standards are living documents. They coevolve with technology and, as such, go through similar phases. ICTs, tools and services go through innovation cycles with ideation, research and development, standardisation and disruption. Standards documents go through ideation, consensus building, publication and obsolescence where in many cases obsolescence is a step change where a new technology will replace existing standards. (Big) Data-related technological changes are on the horizon for the short to medium term as we come to terms with the expected 463 GB/day of digital data by 2025. Future standards work in JTC1 includes the following.

7.1 *ISO/IEC JTC1: Data Usage Advisory Group—AG9*

- Frameworks for Data Sharing Agreements: To address the intersection of the value chain and data sharing.
- Decision to Share Issue: Where transformation of digital services requires data to be shared, exchanged or exploited to deliver benefits and value, and needs to determine on what basis the decision to use data should be authorised.
- Data Quality: Data quality is an important element of data usage. Further work is needed to determine if JTC1 data usage needs are met.
- Appropriate Use of Analytics Outputs: Whilst restrictions to data use are often cited as concerns related to privacy, many of the concerns relate to unintended consequences of the use of data.
- Terminology and Use Cases: Data use is relevant to many JTC1 standards. Standardised terminology and harmonised use cases are needed for wider data usage and to unlock the value of data sharing, exchange and exploitation.
- Metadata: AG9 recognised the importance of metadata definition and use, especially to facilitate the utility to underpin data usage, kinds of metadata, models of metadata and metamodels of repositories.

7.2 ISO/IEC JTC1 SC42 AI WG2 Data

SC42 WG2 Data is investigating the following data topics related to data, data analytics and machine learning:

- Data Quality (DQ)
- Data Quality: Overview, Terminology and Examples
- Data Quality: Measurement
- Data Quality: Management Requirements and Guidelines
- Data Quality: Process Framework
- Data Quality: Assurance – potential new part
- Data Quality: Governance – potential new part
- Big Data: Data Analytics – leverage 20547-3 Big Data Reference Architecture
- Big Data: Data Governance, Usage, Curation, Contextualisation
- Data Mining: Management

8 Summary

This chapter has outlined the case for standardisation, the path to big data standardisation and exemplar activities ongoing in big data standards ecosystems. Projects completed and under way nationally, within European and global initiatives, have been mentioned and sample big data use case scenarios are listed, and some of the initiatives in the evolution of big data standards are described.

The digital ecosystems are global and do not stop at state or regional boundaries. Standardisation is the glue that holds the digital ecosystems together, the gravity of the digital universe. Standardisation in data is central to cloud, big data, IoT, AI and smart city technologies. ISO/IEC JTC1 committees are developing such standards on AI and data, data usage and data interoperability. Standardisation is the foundation stone of certification, regulation and legislation, and in this global digital age, in order to achieve digital sovereignty, we need to synergise the relationships between digital standardisation, digital innovation and digital research.

Acknowledgements This chapter is supported in part by the ADAPT SFI Centre for Digital Content Technology, which is funded under the SFI Research Centres Programme (Grant 13/RC/ 2106) and is co-funded under the European Regional Development Fund.

References

Adolph, M. (2013). *Big data: Big today, normal tomorrow*. ITU-T Technology Watch Report 2013.
ANSI [UNITED STATES]. (2020). *ISO/IEC JTC 1/SC 32 – Data management and interchange*.
ANSI [UNITED STATES]. (n.d.). *ISO/IEC JTC1/SC42 WG2 N1504 NWIP 24300 IT AI process management framework for big data analysis*.

Blind, K., Jungmittag, A., & Mangelsdorf, A. (2012). *The economic benefits of standardisation. An update of the study carried out by DIN in 2000.*

CEN. (2020). *CEN: Who we are.*

CENELEC. (2020). *Welcome to CENELEC – European committee for electrotechnical standardization.*

CENELEC. (n.d.). *CENELEC – about CENELEC – Who we are – Global partners.*

Communications Networks, Content and Technology. (2015).

Dahmen-Lhuissier, S. (2020). ETSI.

EUR-Lex. (2020a). EUR-Lex - 32003L0098 - EN - EUR-Lex.

EUR-Lex. (2020b). EUR-Lex - 32013L0037 - EN - EUR-Lex.

European Commission. (2013a). *European multi stakeholder platform on ICT standardisation – shaping Europe's digital future.*

European Commission. (2013b). *Rolling plan for ICT standardisation – shaping Europe's digital future.*

European Commission. (2015). *The 2016 rolling plan for ICT standardisation is released – internal market, industry, entrepreneurship and SMEs.*

European Commission. (2016). *About us – internal market, industry, entrepreneurship and SMEs.*

European Commission. (2018). *High-level expert group on artificial intelligence – shaping Europe's digital future.*

European Commission. (2020a). *Advanced technologies – Internal market, industry, entrepreneurship and SMEs.*

European Commission. (2020b). *DocsRoom.*

European Commission. (2020c). *Public sector information, open data and big data.*

European Commission. (2020d). *Register of Commission expert groups and other similar entities.*

European Commission. (2020e). *The European data strategy.*

European Commission. (2020f). *The European digital strategy – Shaping Europe's digital future.*

European Commission. (2020g). *White Paper on Artificial Intelligence: a European approach to excellence and trust.*

Grady, N. W., Underwood, M., Roy, A., & Chang, W. L. (2014). Big data: Challenges, practices and technologies: NIST Big Data Public Working Group workshop at IEEE Big Data 2014. 2014 IEEE International Conference on Big Data (Big Data), 11–15. https://doi.org/10.1109/BigData.2014.7004470

Jakobs, K. (2017). Two dimensions of success in ICT standardization – A review. *ICT Express, 3*, 85–89. https://doi.org/10.1016/j.icte.2017.05.008

NIST. (2020). *Big data information.*

Through History with Standards. (2020).

Zillner, S., Curry, E., Metzger, A., Auer, S., & Seidl, R. (Eds.). (2017). *European big data value strategic research & innovation agenda.* Retrieved from Big Data Value Association website: www.bdva.eu.

The Role of Data Regulation in Shaping AI: An Overview of Challenges and Recommendations for SMEs

Tjerk Timan, Charlotte van Oirsouw, and Marissa Hoekstra

Abstract In recent debates around the regulation of artificial intelligence, its foundations, being data, are often overlooked. In order for AI to have any success but also for it to become transparent, explainable and auditable where needed, we need to make sure the data regulation and data governance around it is of the highest quality standards in relation to the application domain. One of the challenges is that AI regulation might – and needs to – rely heavily on data regulation, yet data regulation is highly complex. This is both a strategic problem for Europe and a practical problematic: people, institutions, governments and companies might increasingly need and want data for AI, and both will affect each other technically, socially but also regulatory. At the moment, there is an enormous disconnect between regulating AI, because this happens mainly through ethical frameworks, and concrete data regulation. The role of data regulation seems to be largely ignored in the AI ethics debate, Article 22 GDPR being perhaps the only exception. In this chapter, we will provide an overview of current data regulations that serve as inroads to fill this gap.

Keywords Big data · Artificial intelligence · Data regulation · Data policy · GDPR

1 Introduction

It has been over 2 years since the introduction of the GDPR, the regulation aimed at harmonising how we treat personal data in Europe and sending out a message that leads the way. Indeed, many countries and states outside of Europe have since followed suit in proposing stronger protection on data trails we leave behind in digital and online environments. However, in addition to the GDPR, the European

T. Timan (✉) · M. Hoekstra
Strategy, Analysis & Policy Department, TNO, The Hague, The Netherlands
e-mail: tjerk.timan@tno.nl

C. van Oirsouw
Tilburg University, Tilburg, The Netherlands

E. Curry et al. (eds.), *The Elements of Big Data Value*,
https://doi.org/10.1007/978-3-030-68176-0_15

Commission (EC) has proposed and instated many other regulations and initiatives that concern data. The free flow of data agenda is meant to lead the way in making non-personal data usable across the member states and industries, whereas the Public Sector Information Directive aims to open up public sector data to improve digital services or develop new ones. Steps have also been made in digital security by harmonising cybersecurity through the NIS Directive, while on the other side law enforcement in both the sharing of data (through the e-Evidence Directive) and the specific ways in which it is allowed to treat personal data (Police Directive) has been developed. On top of this already complex set of data regulations, the new Commission has stated an ambitious agenda in which further digitisation of Europe is one of the key pillars, placing even more emphasis on getting data regulation right, especially in light of transitioning towards artificial intelligence.

Yet, however impactful and ahead-of-the-curve the regulatory landscape is, for day-to-day companies and organisations, often already part of a sector-specific set of regulations connected to data, it is not hard to see why for many states it has become difficult to know what law to comply with and how.[1] While there is no particular framework that specifically applies to (big) data, there are many frameworks that regulate certain aspects of it. In this chapter, we aim to give an overview of the current regulatory framework and recent actions undertaken by the legislator in that respect. We also address the current challenges the framework faces on the basis of insights gathered throughout the project[2] and using academic articles and interviews we held with both legal scholars and data practitioners, and multiple sessions and panels in both academic and professional conferences as a basis for this chapter.[3] One of the main challenges is to better understand the interaction between, and intersections of, data regulations and to look at how the different regulations around data interact and intersect. Many proposals have seen the light of day over the last couple of years, and, as stated, all these data-related regulations create a complex landscape that, especially for smaller companies and start-ups, is difficult to navigate. Complexity in itself should not be a concern; however, the world of data is complicated, as is regulating different facets of data. Uncertainty about data regulation and not knowing how to comply or what to comply with does leave its mark on the data-innovation landscape; guidance and clarification are key points of attention in bridging the gap between legal documents and data science practice. In this chapter, we also provide reflections and insight on recent policy debates, thereby contributing to a better understanding of the regulatory landscape and its several sub-domains. After discussing several current policy areas, we will end by providing

---

[1] See, for instance, the SMOOTH platform H2020 project, dedicated to helping SMEs in navigating the GDPR: https://smoothplatform.eu/about-smooth-project/.

[2] See for a recent view on the strategy by the novel Commission: http://www.bdva.eu/PositionDataStrategy.

[3] For an overview of activities, see https://www.big-data-value.eu/wp-content/uploads/2020/03/BDVe-D2.4-Annualpositionpaper-policyactionplan-2019-final.pdf, page 18.

concrete insights for SMEs on how data policy can help shape future digital innovations.

2 Framework Conditions for Big Data[4]

In previous work,[5] we have laid out a basis for looking at big data developments as an ecosystem. In doing so, we followed an approach presented by Lawrence Lessig in his influential and comprehensive publication *Code and Other Laws of Cyberspace* (Lessig, L., 2009). Lessig suggests online and offline enabling environment (or ecosystem) as the resultant of four interdependent, regulatory forces: law, markets, architecture and norms. He uses it to compare how regulation works in the real world versus the online world, in discussing the *regulability* of digital worlds, or cyberspace as it was called in 1999.[6]

In our work for the BDVe regarding data policy, we have worked along these axes in order to gather input and reflections on the development of the big data value ecosystem as the sum total of developments along these four dimensions. We have seen developments on all fronts, and via several activities throughout our interaction with the big data community. Some of the main challenges with respect to regulating data that we know from the academic debate also resonated in practice, such as the role and value of data markets and the sectoral challenges around data sharing. For example, ONYX,[7] a UK-based start-up operating in big data in the wind turbine industry, discussed their experience of vendor lock-in in the wind turbine industry and their involvement in a sector-led call for regulatory intervention from the EU. In another interview for the BDVe policy blog, Michal Gal provided an analysis of data markets and accessibility in relation to competitive advantages towards AI, for example.[8] On the level of architecture, some of the challenges concerning data sharing and 'building in' regulation can be found in the area of privacy-preserving technologies and their role in shaping the data landscape in Europe. In terms of norms and values, we want to reflect in this chapter on numerous talks and panels that delved into the topic of data ethics and data democracy. We will mainly focus on the regulatory landscape around data. In addition to norms (and values), markets and architecture, all remaining challenges in developing a competitive and value-driven Digital Single Market, there have been many legal developments in Europe that are

---

[4]Parts of this chapter appear in the public deliverable developed for the BDVe: https://www.big-data-value.eu/bdve-d2-4-annualpositionpaper-policyactionplan-2019-final/.

[5]See BDVe Deliverable D2.1, https://www.big-data-value.eu/bdve_-d2-1-report-on-high-level-consultation_final/.

[6]See BDVe Deliverable D2.1, p 18 and further: https://www.big-data-value.eu/bdve_-d2-1-report-on-high-level-consultation_final/.

[7]https://www.big-data-value.eu/the-big-data-challenge-insights-by-onyx-insights-into-the-wind-turbine-industry/

[8]https://www.big-data-value.eu/michals-view-on-big-data/

affecting and shaping the big data ecosystem. One of the main challenges we are facing right now is to see how, if at all, such a legal regime is up to the challenges of regulating AI and how this regulatory landscape can help start-ups in Europe develop novel services (Zillner et al. 2020).

3 The EU Landscape of Data Regulation

3.1 Data Governance Foundations

3.1.1 Data Governance and the Protection of Personal Data

Data is taking a central role in many day-to-day processes. In connecting data, ensuring interoperability is often the hardest part as the merging and connecting of databases takes a lot of curation time, as was stated by Mercè Crosas in an interview with the BDVe.[9] Therefore, it is important that data practices are arranged solidly by doing good data governance to avoid interoperability problems. In addition, data is an indispensable raw material for developing AI, and this requires a sound data infrastructure (High-Level Expert Group on Artificial Intelligence, 201) and better models on data governance. In a recent panel held during the BDV PPP Summit in June 2019 in Riga,[10] a researcher from the DigiTransScope project – a project in which an empirical deep-drive is made into current data governance models[11] – gave a definition of the concept of data governance, as follows: 'the kind of decisions made over data, who is able to make such decisions and therefore to influence the way data is accessed, controlled, used and benefited from'.[12] This definition covers a broad spectrum of stakeholders with varying interests in a big data landscape. More research is needed to find insights on the decision-making power of the different stakeholders involved so that a good balance is found between fostering economic growth and putting data to the service of public good. Concepts such as data commons (Sharon and Lucivero 2019) and data trusts have been emerging recently. Any kind of guidance should take all of these elements into account. It is important that all stakeholders are involved in the process of developing guidance, as otherwise the emergence and development of a true data economy are hampered.

In a data landscape, many different interests and stakeholders are involved. The challenging part about regulating data is the continuous conceptual flux, by which we mean that the changing meaning and social and cultural value of data is not easily captured in time or place. Yet, one can set conditions and boundaries that can aim to steer this conceptual flux and value of data for a longer foreseeable timeframe. One

---

[9]https://www.big-data-value.eu/the-big-data-challenge-recommendations-by-merce-crosas/

[10]See https://www.big-data-value.eu/ppp-summit-2019/.

[11]See https://ec.europa.eu/jrc/communities/en/community/digitranscope.

[12]https://ec.europa.eu/jrc/communities/en/community/digitranscope

of the most notable regulations passed recently is the General Data Protection Regulation (hereafter referred to as GDPR). With this regulation, and accompanying implementation acts in several member states, the protection of personal data is now firmly anchored within the EU. However, the distinction between personal and non-personal data has proven to be challenging to make in practice, even more so when dealing with combined datasets that are used in big data analytics. It has also recently been argued that the broad notion of personal data is not sustainable; with rapid technological developments (such as smart environments and datafication), almost all information is likely to relate to a person in purpose or in effect. This will render the GDPR a law that tries to cover an overly broad scope and it will therefore potentially lose power and relevance (Purtova 2018). In this vein, there is a need to continue developing notions and concepts around personal data and the types of data use.

For most big data analytics, privacy harm is not necessarily aimed at the individual but occurs as a result of the analytics itself because it happens on a large scale. EU regulation currently lacks in providing legal remedies for the unforeseen implications of big data analytics, as the current regime protects input data and leaves inferred data[13] out of its scope. This creates a loophole in the GDPR with respect to inferred data. As stated by the e-SIDES project recently,[14] a number of these loopholes can be addressed by court cases. The question remains as to whether and to what extent the GDPR is the suitable frame to curb such harms.

Despite many efforts to guide data workers through the meaning and bases of the GDPR and related data regulations such as the e-Privacy Regulation, such frameworks are often regarded by companies and governments as a hindrance to the uptake of innovation.[15] For instance, one of the projects within the BDV PPP found that privacy concerns prevent the deployment, operation and wider use of consumer data. This is because skills and knowledge on how to implement the requirements of data regulations are often still lacking within companies. The rapidly changing legal landscape and the consequences of potential non-compliance are therefore barriers to them in adopting big data processes. Companies have trouble making the distinction between personal and non-personal data and who owns which data. This was also reflected in a recent policy brief by TransformingTransport, which looked into many data-driven companies in the transport sector.[16] Additionally, these same companies experience trouble defining the purpose of processing beforehand, as within a big data context the purpose of processing reveals itself after processing. Mapping of data flows onto purposes of the data-driven service in

---

[13]Inferred data is data that stems from data analysis. The data on which this analysis is based was gathered and re-used for different purposes. Through re-use of data, the likelihood of identifiability increases.

[14]See e-SIDES, Deliverable D4.1 (2018).

[15]Big Data Value PPP: Policy4Data Policy Brief (2019), page 8. Available at https://www.big-data-value.eu/wp-content/uploads/2019/10/BDVE_Policy_Brief_read.pdf

[16]Transforming Transport, D3.13 – Policy Recommendations.

development presents difficulties, especially when having to understand which regulation 'fits' on different parts in the data lifecycle. On the other hand, sector-specific policies or best practices for sensitive personal data are perceived as assets by professionals because these give them more legal certainty, where they face big risks if they do not comply. In this sense, privacy and data protection can also be seen as an asset by companies. We feel that there is a need for governance models and best practices to show that the currently perceived dichotomy between privacy and utility is a false one (van Lieshout and Emmert 2018). Additionally, it is also important to raise awareness among companies in which scenarios concerning big data and AI are useful, and in which scenarios they are not.[17] One of the main challenges for law- and policymakers is to balance rights and establish boundaries while at the same time maximising utility (Timan and Mann 2019).

3.1.2 Coding Compliance: The Role of Privacy-Preserving Technologies in Large-Scale Analytics

One of the more formal/technical and currently also legally principled ways forward is to build in data protection from the start, via so-called privacy-by-design approaches (see, among many others, Cavoukian 2009 and Hoepman 2018). In addition to organisational measures, such as proper risk assessments and data access and storage policies, technical measures can make sure the 'human error' element in the risk assessment is covered.[18] Sometimes referred to as privacy-preserving technologies (PPTs), such technologies can help to bridge the gaps between the objectives of big data and privacy. Currently, many effective privacy-preserving technologies exist, although they are not being implemented and deployed to their full extent. PPTs are barely integrated into big data solutions, and the gap of deployment in practice is wide. The reasons for this are of a societal, legal, economic and technical nature. The uptake of privacy-preserving technologies is, however, necessary to ensure that valuable data is available for its intended purpose. In this way data is protected and can be exploited at the same time, dissolving the dichotomy of utility and privacy. To ensure this is achieved, PPTs need to be integrated throughout the entire data architecture and value chain, both vertically and horizontally. A cultural shift is needed to ensure the uptake of PPTs, as the current societal demand to protect privacy is relatively low. Raising awareness and education will be key in doing so. It is important that PPTs are not provided as an add-on but rather are incorporated into the product. There is wide agreement that the strongest parties have

---

[17]BigDataStack Project. Available at: https://bigdatastack.eu/

[18]Although obviously relying on technology only to solve data protection is not the way forward either, as in itself such technologies come with novel risks.

the biggest responsibilities concerning protecting privacy and the uptake of PPTs, as was also confirmed by the e-SIDES project (2018).[19]

Another point of discussion has been the anonymisation and pseudonymisation of personal data. It has also been argued that companies will be able to retain their competitive advantage due to the loophole of pseudonymised data, which allows for unfettered exploitation as long as the requirements of the GDPR are met.[20] Anonymised data needs to be fully non-identifiable and therefore risks becoming poor in the information they contain. Also, anonymisation and pseudonymisation techniques may serve as mechanisms to release data controllers/processors from certain data protection obligations related to breach-related obligations. Recent work done by the LeMO project found that anonymisation and pseudonymisation may be used as a means to comply with certain data protection rules, for instance with the accountability principle, measures that ensure the security of processing, purpose limitation and storage limitation. Pseudonymisation and anonymisation techniques can serve as a means to comply with the GDPR,[21] but at the same time, too far-reaching anonymisation of data can limit the predictability of big data analytics (Kerr 2012). However, as long as the individual remains identifiable, the GDPR remains applicable. It has been argued that, because of this, companies will be able to retain their competitive advantage by being able to unlimitedly exploit data as long as it is pseudonymised or anonymised.

3.1.3   Non-personal Data (FFoD)

In 2019, Regulation 2018/1807 on the free flow of non-personal data (FFoD) came into force, which applies to non-personal data and allows for its storage and processing throughout the EU territory without unjustified restrictions. Its objective is to ensure the free flow of data across borders, data availability for regulatory control and encouragement of the development of codes of conduct for cloud services. The FFoD is expected to eliminate the restrictions on cross-border data flows and their impacts on business, reduce costs for companies, increase competition (LeMO 2018),[22] increase the pace of innovation and improve scalability, thereby achieving economies of scale. This is all supposed to create more innovation, thereby benefiting the uptake of big data, in which the flow of non-personal data

---

[19]See the CJEU Google v. CNIL case (C-507/17). The CJEU decided that the right to be forgotten (RtBF, Article 17 GDPR) does not imply that operators of search engines (in this case Google) have an obligation to carry out global de-referencing if this RtBF is invoked because this would come into conflict with non-EU jurisdictions. It was also emphasised once more in this case that the right to data protection is not an absolute right.

[20]https://www.compliancejunction.com/pseudonymisation-gdpr/

[21]Specifically with the obligations of data protection by design and default, security of processing, purpose and storage limitation and data breach-related obligations.

[22]Especially in the cloud services market, start-ups increasingly rely on competitive cloud services for their own product or service.

will remain of continuing importance in addition to having solid data infrastructures. For instance, the GAIA-X Project addresses how open data plays a role in creating a data infrastructure for Europe.[23] Other more developed initiatives include European Industrial Data Spaces[24] or the MOBI network for opening up and sharing data around blockchains.[25]

The FFoD is the complementary piece of legislation to the GDPR as it applies to non-personal data. However, this distinction between the two regimes based on these concepts of personal and non-personal data is highly debated. The distinction is not easy to make in practice as datasets are likely to be mixed and consist of both personal and non-personal data. This is especially the case for big data datasets, as it is often not possible to determine which part of the set contains personal or non-personal data. This will result in it being impossible to apply each regulation to the relevant part of the dataset (LeMO 2018). In addition, as mentioned in the previous sections, these concepts are broad and subject to the dynamic nature of contextual adaptation. Whether data has economic value is not dependent on its legal classification. Hence, when facing opaque datasets, there is the risk of strategic firms on the basis of this legal classification, and they are likely to exploit the regulatory rivalry between the FFoD and the GDPR. The limitation of the FFoD to non-personal data is likely to be counterproductive to innovation, as personal data has high innovation potential as well (Graef et al. 2018). There is also further guidance needed where it concerns parallel/subsequent application of the GDPR and the FFoD, or where the two regimes undermine each other (Graef et al. 2018). Regardless of whether data is personal or non-personal, it is of major importance that it is secured. Hence, the following section addresses the EU regime on the security of data (Fig.1).

3.1.4   Security of Data

The Cybersecurity Act (Regulation (EU) 2019/881) was adopted to set up a certification framework to ensure a common cybersecurity approach throughout the EU. The aim of this regulation is to improve the security standards of digital products and services throughout the European internal market. These schemes are currently voluntary and aimed at protecting data against accidental or unauthorised storage, processing, access, disclosure, destruction, loss or alteration. The EC will decide by 2034 whether the schemes will become mandatory.

The NIS Directive (Directive (EU) 2016/1148) puts forward security measures for networks and information systems to achieve a common level of cybersecurity

---

[23]Project GAIA-X, 29/10/2019. See https://www.bmwi.de/Redaktion/EN/Publikationen/Digitale-Welt/das-projekt-gaia-x-executive-summary.html.

[24]https://ec.europa.eu/digital-single-market/en/news/common-european-data-spaces-smart-manufacturing

[25]Mobility Open Blockchain Initiative (MOBI); see www.dlt.mobi/.

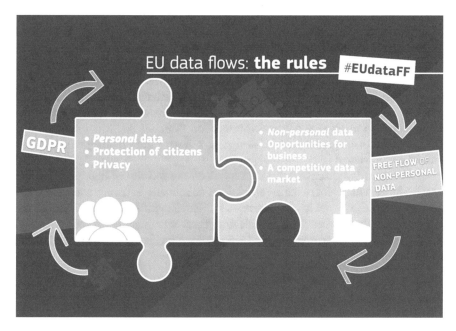

Fig. 1 The link between the GDPR and the FFoD (See https://ec.europa.eu/digital-single-market/sites/digital-agenda/files/newsroom/eudataff_992x682px_45896.jpg) (by European Commission licensed under CC BY 4.0)

throughout the European Union to improve the functioning of the internal market. The security requirements that the NIS Directive puts forward are of both a technical and organisational nature for operators of essential services and digital service providers. If a network or information system contains personal data, then the GDPR is most likely to prevail in case of conflict between the two regimes. It has been argued that the regimes of the GDPR and the NIS Directive have to be regarded as complementary (Markopoulou et al. 2019). Cyberattacks are becoming more complex at a very high pace (Kettani and Wainwright 2019[26]). The nature of the state of play is constantly evolving, which makes it more difficult to defend against attacks. Also, it has been predicted that data analytics will be used for mitigating threats but also for developing threats (Kettani and Wainwright 2019). The companies that can offer enough cybersecurity are non-European, and the number of solutions is very limited (ECSO 2017). Due to the characteristics of the digital world, geographical boundaries are disappearing, and a report by the WRR (the Dutch Scientific Council[27]) called for attention to cybersecurity at an EU level.

Some of the characteristics of cybersecurity make tackling this challenge especially difficult; fast-paced evolvement, lack of boundaries, the fact that

---

[26]https://www.ecs-org.eu/documents/uploads/european-cyber-security-certification-a-meta-scheme-approach.pdf

[27]https://www.wrr.nl/

infrastructures are owned by private parties and the dependence of society on these architectures are recurring issues (ECSO 2017[28]). Currently, cyber-strategies of SMEs mainly focus on the detection of cyber risks, but these strategies should shift towards threat prevention (Bushby 2019). Just like data and robotics, AI faces all of the possible cyberthreats, and every day threats are only further evolving. Cybersecurity will also play a key role in ensuring technical robustness, resiliency and dependability. AI can be used for sophisticated automated attacks and at the same time also to provide automated protection from attacks. It is important that cybersecurity is integrated into the design of a system from the beginning so that attacks are prevented.

This section has discussed the EU regime on the security of both personal and non-personal data. Cybersecurity attacks are continually evolving and pose challenges for those involved in a data ecosystem. Keeping different types of data secure is one aspect, but successfully establishing rights upon data is another. The next section addresses the interaction between data and intellectual property rights and data ownership.

3.1.5 Intellectual Property

Due to the fact that many different players are involved in the big data lifecycle, many will try to claim rights in (part of) the datasets to protect their investment. This can be done by means of intellectual property rights. If the exercise of such a right is not done for the right reasons, this can stifle the uptake of big data and innovation. This also holds true for the cases in which an intellectual property right does not exist yet is enforced by an actor that is economically strong.

3.1.6 Public Sector Information and the Database Directive

In January 2019, an agreement was reached on the revised Public Sector Information Directive (PSI Directive). Once implemented, it will be called the Open Data and Public Sector Information Directive. The revised rules still need to be formally adopted at the time of publication of this deliverable. Public bodies hold huge amounts of data that are currently unexploited. The access and re-use of raw data that public bodies collect are valuable for the uptake of digital innovation services and better policymaking. The aim of the PSI Directive is to get rid of the barriers that currently prevent this by reducing the market entry barriers, increasing the availability of data, minimising the risk of excessive first-mover advantages and increasing the opportunities for businesses.[29] This will contribute to the growth of the EU

---

[28]https://www.ecs-org.eu/documents/uploads/european-cyber-security-certification-a-meta-scheme-approach.pdf

[29]EC Communication 'Towards a common European data space', SWD (2018) 125 final.

economy and the uptake of AI. The PSI Directive imposes a right to re-use data, obliges public bodies to charge the marginal cost for the data (with a limited number of exceptions), stimulates the uptake of APIs, extends the scope to data held by public undertakings, poses rules on exclusive agreements and refers to a machine-readable format when making the data available. Although open data licences are stimulated by the PSI, they can still vary widely between member states. Another challenging aspect is the commercial interests of public bodies in order to prevent distortions of competition in the relevant market. Some of the challenges that the use of public sector information faces are related to standardisation and interoperability, ensuring sufficient data quality and timely data publication, and a need for more real-time access to dynamic data. In addition, the licences to use the data can still vary, as member states are not obliged to use the standard formats. Another challenge that the PSI Directive faces is its interaction with the GDPR, either because it prevents disclosure of large parts of PSI datasets or because it creates compliance issues. The GDPR is not applicable to anonymous data. In practice, however, it is very hard for data to be truly rendered anonymous, and it cannot be excluded that data from a public dataset, combined with data from third-party sources, (indirectly) allows for identification of individuals. The interaction between the GDPR and the PSI Directive is also difficult with respect to public datasets that hold personal data, especially because of the principle of purpose limitation and the principles of data minimisation (LeMO 2018). Another challenge is the relationship of the PSI Directive with the Database Directive (DbD), as public sector bodies can prevent or restrict the re-use of the content of a database by invoking its sui generis database right. How the terms 'prevent' and 'restrict' are to be interpreted is not clear yet. Exercise of these rights bears the risk of hindering innovation. Where it concerns data portability requirements, the interaction between the DbD, PSI Directive and the GDPR is not clear either (Graef et al. 2018).

In 2018, the Database Directive (hereafter: DbD) was evaluated for the second time. The DbD protects databases by means of copyright or by means of the substantial investment that was made to create it, the sui generis right. The outcome of the evaluation was that the DbD is still relevant due to its harmonising effect. The sui generis right does not apply to machine-generated data, IT devices, big data and AI. At the time of the evaluation, a reformation of the DbD to keep pace with these developments was considered too early and disproportionate. Throughout its evaluation, one of the challenges was measuring its actual regulatory effects.

3.1.7 Copyright Reform

As part of the Digital Single Market Strategy, the EU is revising the rules on copyright to make sure that they are fit for the digital age. In 2019, the Council of Europe gave its green light to the new Copyright Directive (European Parliament, 2019). The aim is to ensure a good balance between copyright and the relevant public body objectives, such as education, research innovation and the needs of persons with disabilities. It also includes two new exceptions for Text and Data

Mining (TDM), which allows for TDM for the purpose of scientific research[30] and the opt-out clause of Article 4 New Copyright Directive. This exception will be of special importance to the uptake of AI. In a big data context, it is difficult to obtain authorisation from the copyright holder of individual data. When a work is protected by copyright, the authorisation of the rights holder is necessary in order to use the work. In a big data context, this would mean that for every individual piece of data, the authorisation needs to be obtained from the rights holder. Also, not all data in a big data context is likely to meet the originality threshold for copyright protection, though this does not exclude the data from enjoying protection under copyright. This creates uncertainties on which data is protected and which data is not, and whether a work enjoys copyright protection can only be confirmed afterwards by a court as copyright does not provide a registration system. The copyright regime is not fully harmonised throughout the EU, and a separate assessment is required on whether copyright protection is provided. This bears the potential of having a chilling effect on the uptake of EU-wide big data protection. Regarding AI-generated works of patents, it is still unclear whether, and if so to whom, the rights will be allocated. The multi-stakeholder aspect plays a role here as well, and the allocation of rights is difficult.

The manner in which intellectual property rights on data will be exercised will have a significant impact on the uptake of big data and innovation in general. This will all be shaped by the interaction between the PSI Directive, the GDPR and the new Copyright Directive. These are all instruments to establish security on data in the form of a right, as this is currently lacking.

3.1.8   Data Ownership

There is no particular framework to regulate the ownership of data. Currently, the only means to establish ownership in data or protection of data is through the provisions of the GDPR, the DbD and the Trade Secrets Protection Directive, or by contracts through contract law. Whether there should be an ownership right in data has been widely debated in recent years, as this current framework does not sufficiently or adequately respond to the needs of all the actors involved in the data value cycle. At the same time, there is consensus that a data ownership right is not desirable, as granting data ownership rights is considered to create an over-protective regime with increased data fragmentation and high transaction costs[31]. The difficulty of assigning ownership to data lies in the nature of data, because it is neither tangible nor intangible, it is limitless and non-rivalrous, and its meaning and value are not static. Data has a lifecycle of its own with many stakeholders involved. This also implies that no stakeholder will hold exclusive ownership rights over the data. The lack of a clear regulatory regime creates high levels of legal uncertainty. Ownership

---

[30]Article 3 Directive (EU) 2019/790 e.

[31]https://ec.europa.eu/jrc/sites/jrcsh/files/jrc104756.pdf

is currently mainly captured by contractual arrangements. This situation is far from ideal, as it creates lock-in effects and power asymmetries between parties, and is non-enforceable against third parties. However, the fact that there is no legal form of ownership does not prevent a de facto form of ownership from arising either. The rise of data bargaining markets illustrates this. The de facto ownership of data does not produce an allocation that maximises social welfare. This results in market failures, strategic behaviour by firms and high transaction costs. There is a need for policies and regulations that treat 'data as a commodity'. This requires new architectures, technologies and concepts that allow sellers and buyers of data to link and give appropriate value, context, quality and usage to data in a sense that ensures ownership and privacy where necessary.[32] In the next section, we will elaborate how this plays out in the data economy.

3.1.9 Data Economy

The digital economy is characterised by extreme returns based on scale and network effects, network externalities and the role of data in developing new and innovative services. As a result, the digital economy has strong economies of scope with large incumbent players who are difficult to dislodge. In order to realise the European Digital Single Market, we need the conditions that allow for the realisation thereof. Moreover, AI and the IoT are dependent on data; the uptake of both will be dependent on the data framework.[33]

3.1.10 Competition

There have been many developments in the field of competition law that are of importance for the regulation of big data. The legal principles of competition law stem from a time when the digital economy did not even exist yet. It has been widely debated whether the current concepts of competition law policy are sufficient tools to regulate emerging technologies or whether new tools are needed. Currently, there is still a lot of legal uncertainty concerning the practical implementation of competition law related to the data economy due to its lack of precedents. The concepts of, among others, the consumer welfare standard, the market definition and the manner in which market power is measured need to be adapted or refined in order to keep up with the digital economy (European Commission, Report - Competition policy for the Digital Era,[34]). The question of whether big tech must be broken up was often

---

[32] BVD PPP Summit Riga 2019, Antonis Litke, Policy4Data and DataMarketplaces ICCS/NTUA.

[33] See also the recent DataBench recommendations: https://www.databench.eu/the-project/.

[34] Available at https://ec.europa.eu/competition/publications/reports/kd0419345enn.pdf

asked in competition policy debates. Facebook is currently under investigation by the US Federal Trade Commission for potentially harming competition, and Federal Trade Commission Chairman Joe Simons has stated in an interview with Bloomberg that he is prepared to undo past mergers if this is deemed necessary to restore competition. However, there are no precedents on breaking up big tech firms, and knowledge on how to do this if considered desirable is currently lacking.[35] The aim of some of the projects that are a part of the BDVA (Zillner et al. 2017) is to make sure that we as an EU landscape become stronger through data sharing, not by aiming to create another company that becomes too powerful to fail (e.g. GAFAM). The overall aim of DataBench[36] is to investigate the current big data benchmarking tools and projects currently in operation and to identify the main gaps and provide metrics to compare the outcomes that result from those tools. The most relevant objective mentioned by many of the BDVA-related projects is to build a consensus and reach out to key industrial communities. In doing so, the project can ensure that the activity of benchmarking of big data activities is related to the actual needs and problems within different industries. Due to rules imposed by the GDPR, the new copyright rules on content monitoring and potential rules on terrorist content monitoring,[37] and realising the complexity of tasks and costs that all such regulations introduce, for the moment only large international technology companies are equipped to take up these tasks efficiently. As of this moment, there is no established consensus on how to make regulation balanced, meaning accessible and enforceable.

Over the last couple of years, several competition authorities have been active with competition law in enforcement regarding big tech. For instance, the EC has started a formal investigation into Amazon as to whether they are using sales data (which becomes available as a result of using the platform) to compete unfairly.[38] In addition, several national competition authorities have taken action to tackle market failures causing privacy issues by using instruments of competition law.[39] For example, on 7 February 2019, the German Bundeskartellamt accused Facebook of abusing its dominant position (Art. 102 TFEU) by using exploitative terms and conditions for their services. The exploitative abuse consisted of using personal data which was obtained in breach of the principles of EU data protection law. The Bundeskartellamt used the standards of EU data protection law as a qualitative parameter to examine whether Facebook had abused its dominant position. The European Data Protection Board (EDPB) also stated that where a significant merger

---

[35]https://www.economist.com/open-future/2019/06/06/regulating-big-tech-makes-them-stronger-so-they-need-competition-instead

[36]https://www.databench.eu/

[37]The European Parliament voted in favour of a proposal to tackle misuse of Internet hosting services for terrorist purposes in April 2019: https://www.europarl.europa.eu/news/en/press-room/20190410IPR37571/terrorist-content-online-should-be-removed-within-one-hour-says-ep.

[38]https://ec.europa.eu/commission/presscorner/detail/en/IP_19_4291

[39]For instance, the Bundeskartellamt used European data protection provisions as a standard for examining exploitative abuse: (https://www.bundeskartellamt.de/SharedDocs/Meldung/EN/Pressemitteilungen/2019/07_02_2019_Facebook.html).

is assessed in the technology sector, longer-term implications of the protection of economic interests, data protection and consumer rights have to be taken into account. The interaction between competition law and the GDPR is unclear, and it seems like we are experiencing a merger of the regimes, to a certain extent.

It has been considered that if substantive principles of data protection and consumer law are integrated into competition law analysis, the ability of competition authorities to tackle new forms of commercial conduct will be strengthened. If a more consistent approach in the application and enforcement of the regimes is pursued, novel rules will only be necessary where actual legal gaps occur (Graef et al. 2018). It is also been argued that, even though there are shared similarities between the regimes of competition law, consumer protection law and data protection law because they all aim to protect the welfare of individuals, competition law is not the most suitable instrument to tackle these market failures (Ohlhausen and Okuliar 2015; Manne and Sperry 2015) because each regime pursues different objectives (Wiedemann and Botta 2019). Currently, the struggle of National Competition Authorities in tackling the market failures in the digital economy creates uncertainties about how the different regimes (of competition and data protection) interact, and this creates legal uncertainty for firms.

Even though competition authorities have been prominent players in the regulation of data, the lack of precedent creates much uncertainty for companies. The next section will discuss how data sharing and access, interoperability and standards play a role in this.

3.1.11   Data Sharing and Accessibility

Data is a key resource for economic growth and societal progress, but its full potential cannot be reaped when it remains analysed in silos (EC COM/2017/09). More industries are becoming digitised and will be more reliant on data as an input factor. There is a need for a structure within the data market that allows for more collaboration between parties with respect to data. Data access, interoperability and portability are of major importance to foster this desired collaboration. In this respect, data integrity and standardisation are reoccurring issues. Accessibility and re-use of data are becoming more common in several industries, and sector-specific interpretations of the concept could have spill-over effects across the data economy. There is a need for governance and regulation to support collaborative practices. Currently, data flows are captured by data-sharing agreements.

The complexity of data flows, due to the number of involved actors and the different sources and algorithms used, makes these issues complicated for the parties involved. The terms in data-sharing agreements are often rather restrictive in the sense that only limited access is provided. This is not ideal, as restriction in one part of the value chain can have an effect on other parts of the data cycle. Access to data is mainly restricted because of commercial considerations. An interviewee suggested that the main reason that full data access is restricted is that it allows the holder of the entire dataset to control its position on the relevant market, not because of the potential value that lies in the dataset.[40] Parties are often not aware of the importance of having full access to the data that their assets produce, resulting in the acceptance of unfavourable contractual clauses. The interviewee also suggested, however, that the real value creation does not lie in the data itself, but in the manner in which it is processed, for instance by combining and matchmaking datasets. In addition, there is a lack of certainty regarding liability issues in data-sharing agreements. Data-sharing obligations are currently being adopted in certain sectors and industries, for instance in the transport sector (LeMO 2018[41]), though due to the absence of a comprehensive legal framework, these still face numerous limitations. In some cases, the imposition of a data-sharing obligation might not be necessary as data plays a different role in different market sectors. It is worthwhile to monitor how the conditions imposed by the PSI Directive on re-use and access for public sector bodies play out in practice to see whether this could also provide a solution in the private sector (LeMO 2018).

The right to data portability of Article 20 GDPR (RtDP) is a mechanism that can facilitate the sharing and re-use of data, but regarding its scope and meaning, many areas are still unresolved. For instance, a data transfer may be required by the data subject where this is considered 'technically feasible', though what circumstances are considered to be 'technically feasible' by the legislator are not clear. In addition, there is no clarity on whether the RtDP also applies to real-time streams, as it was mainly envisaged in a static setting. There is also a strong need to consider the relationship between the right to data portability and IP rights, as it is not clear to what extent companies are able to invoke their IP rights on datasets that hold data about data subjects.[42] The interpretation of these concepts will make a big difference with respect to competition, as the right to data portability is the main means for data subjects to assay the counter-offers of the competitors for the services they use without the risk of losing their data. However, if competition law has to enforce the implementation and enforcement of interoperability standards that ensure portability, it will be overburdened in the long run.

The sharing and re-use of data require that effective standards are set across the relevant industry. Currently, the standardisation process is left to the market, but the efficient standards are still lacking, and this slows down data flows. Setting efficient standards will smoothen the process of data sharing and therefore also encourage it. Each market has its own dynamics, so the significance of data and data access will also be market dependent. In the standardisation process, it needs to be taken into account that a standard in one market might not work in another. Guidance on the creation of standards is needed to provide more legal certainty, because if this process is left to the market alone, this can result in market failures or standards that raise rivals' costs. The role of experts in the standardisation process is crucial, as

---

[40]This point has been made in an interview with ONYX InSight. See https://www.big-data-value.eu/the-big-data-challenge-insights-by-onyx-insights-into-the-wind-turbine-industry/.

[41]https://lemo-h2020.eu/

[42]See https://www.big-data-value.eu/spill-overs-in-data-governance/.

a deep understanding of the technology will lead to better standards. In addition, due to the multidisciplinary nature of many emerging technologies, the regulator should not address the issue through silos of law but have a holistic approach and work in regulatory teams consisting of regulatory experts that have knowledge of the fields relevant in setting the standard.[43]

Data access, interoperability, sharing and standards are important enabling factors for the data economy. The manner in which the data economy will be shaped will have an impact on commerce, consumers and their online privacy. The next section discusses these three points.

3.1.12 Consumers, e-Commerce and e-Privacy

In January 2018, the Payment Services Directive (PSD2) became applicable. This Directive was expected to make electronic payments cheaper, easier and safer. On 11 April 2018, the EC adopted the 'New Deal for Consumers' package. This proposal provides for more transparency in online marketplaces and extends the protection of consumers in respect of digital services, as they do not pay with money but with their personal data. The new geo-blocking regulation that entered into force will prohibit the automatic redirecting and blocking of access, the imposition of different general conditions to goods and services, and payment transactions based on consumer nationality. Furthermore, the EU has been working on the revision of the Civil Procedure Code regulation on consumer protection (Regulation (EC) 2017/2394), which entered into force on 17 January 2020. The new rules for VAT for the online sale of goods and services will enter into force in 2021. The Digital Services Act is a piece of legislation which is planned to tear up the 20-year-old e-Commerce Directive; it also targets Internet Service Providers (ISPs) and cloud services. It is likely to contain rules on transparency for political advertising and force big tech platforms to subject their algorithms to regulatory scrutiny (Khan and Murgia 2019). In the Communication on online platforms (Communication 2016 288[44]), the EC formulated principles for online platforms. These are about creating a level playing field, responsible behaviour that protects core values, transparency and fairness for maintaining user trust, and safeguarding innovation and open and non-discriminatory markets within a data-driven economy. Following this Communication, on 12 June 2020, the Regulation on platform-to-business relations (Regulation (EU) 2019/1150) was adjusted and is now applicable. The objective is to ensure a fair, predictable, sustainable and trusted online business environment within the internal market. Due to the scale and effects of platforms, this measure is taken at EU level instead of member state level. It applies to online intermediation services, business users and corporate website users, and it applies as soon as the business user or the corporate website user has an establishment within the EU. It sets

---

[43]https://www.big-data-value.eu/michals-view-on-big-data/
[44]https://ec.europa.eu/transparency/regdoc/rep/1/2016/EN/1-2016-288-EN-F1-1.PDF

requirements for the terms and conditions, imposes transparency requirements and offers redress opportunities.

The European Data Protection Supervisor has stressed the urgency for new e-privacy laws (Zanfir-Fortuna 2018), and since the publication of the previous deliverable in 2017, the e-Privacy Directive has been under review. Several governments and institutions have expressed their opinion on its current new draft. For example, the German government has stated that they do not support the current draft version as it does not achieve the objective of guaranteeing a higher level of protection than the GDPR,[45] and the Dutch Data Protection Authority has stated that cookie walls do not comply with EU data protection laws.[46] Furthermore, in October 2019, the Court of Justice of the European Union (CJEU) gave its decision in the Planet49 case (C-673/17, ECLI:EU:C:2019:801) and stated that the consent which a website user must give for the storage of and access to cookies is not valid when this consent is given by means of a pre-ticked checkbox. In addition, information that the service provider gives to the user must include the duration of the operation of cookies and whether or not third parties may have access to these cookies. This judgement will have a significant impact on the field of e-privacy and on big data in general as well, as a lot of the data that 'forms part of big data' was gathered and processed on the basis of pre-clicked consent-box cookies. Thus, this judgement will change how data should be processed from now on.[47] In extension thereof, the case Orange Romania (C-61/19) is currently pending at the CJEU for a preliminary ruling on what conditions must be fulfilled in order for consent to be freely given.

4   Conclusions

In this chapter, some of the main challenges and developments were addressed concerning the regulatory developments in (big) data. Where across the board the main development in Europe would be the GDPR, we have tried to show that many other regulatory reforms have taken place over the last years – regulations that, similar to the GDPR, affect the data ecosystem. In areas such as competition, IP, data retention, geographical data 'sovereignty' and accessibility, the shaping of data markets, cybersecurity and tensions between public and private data, among others, we have aimed to summarise the plurality of regulatory reform and, where possible,

---

[45]https://www.technologylawdispatch.com/2019/08/privacy-data-protection/update-on-eprivacy-regulation-current-draft-does-not-guarantee-high-level-of-protection-and-cannot-be-supported-german-government-states/

[46]https://autoriteitpersoonsgegevens.nl/nl/nieuws/ap-veel-websites-vragen-op-onjuiste-wijze-toestemming-voor-plaatsen-tracking-cookies (in Dutch).

[47]See https://pdpecho.com/2019/10/03/planet49-cjeu-judgment-brings-some-cookie-consent-certainty-to-planet-online-tracking/.

how they intersect or interplay. Moreover, aside from the novel proposals and developments from the regulator, we have also seen the first effects of the GDPR coming into force in the form of first fines handed out to companies and local governments.[48] and we have seen other major court decisions that will have a profound effect on the data landscape (e.g. the Planet49[49] decision on cookie regulation).

To summarise our findings, the challenging aspect of regulating data is its changing nature, meaning and value. There is a need for more research on how to shape data governance models and how to implement them. The GDPR is often regarded by companies as a hindrance to innovation, but privacy and data protection can also be regarded as an asset. The implementation of privacy-preserving technologies (PPTs) can help to bridge this gap, but a gap exists in terms of their implementation in practice. Anonymisation and pseudonymisation are often used as a means to comply with the GDPR. In practice, datasets are likely to consist of both personal and non-personal data. This creates difficulties in the application of both the GDPR and the FFoD to big data. The regulatory rivalry of the GDPR and FFoD is likely to be exploited. Clarity on parallel or subsequent application of the GDPR and the FFoD is needed. Regarding the security of data, several strategies have been implemented at EU level to tackle cybersecurity issues. The nature of cybersecurity challenges makes it difficult to tackle them. Looking ahead, cybersecurity will play a key role in the development of AI and as such is a key condition for AI to shape. Another key condition for big data and AI is the use of public sector data. Use of public sector information will be challenging due to the obstacles related to data governance, for instance ensuring interoperability. Where public sector information holds personal data, the PSI will face difficulties in the interaction with the GDPR. Public sector bodies can prevent the re-use of the content of a database by invoking the sui generis database right of the Database Directive. The interaction between the PSI Directive, the GDPR and the Database Directive is not clear yet where it regards data portability requirements. In a big data context, it remains uncertain which pieces of data enjoy copyright protection under the current regime, and, connected to this, the allocation of rights for AI-generated works remains unclear.

4.1 Recommendations for SMEs and Start-Ups

The previous section gave an overview of the current regulatory landscape. It addressed the foundations of data governance, intellectual property and the data economy, thereby also revealing the uncertainties and unclarities that these frameworks face in the light of big data. In this section, we will present some concrete

---

[48]See, for instance, enforcementtracker.com where all fines under the GDRP are being tracked.

[49]See C-673/17, ECLI:EU:C:2019:801.

insights and recommendations for SMEs and start-ups in how data policy can help shape future digital innovations.[50]

4.1.1 Potential of Privacy-Preserving Technologies

PPTs can help SMEs to bridge the gaps between the objectives of big data and privacy.[51] The GDPR is often regarded by companies as a hindrance to innovation, but privacy and data protection can also be regarded as an asset. PPTs have great potential for SMEs, because SMEs can use them to ensure that valuable data is available for its intended purpose and that their data is protected at the same time, dissolving the dichotomy of utility and privacy. However, it is important that PPTs are not provided as an add-on but are incorporated into the product.

4.1.2 Distinction Between Personal and Non-personal Data

Anonymisation and pseudonymisation of data are often used as a means to comply with the GDPR. However, SMEs should be aware that in practice, datasets are likely to consist of both personal and non-personal data. This creates difficulties in the application of both the GDPR and the FFoD to big data. As a result, the regulatory rivalry of the GDPR and FFoD is likely to be exploited.

4.1.3 Data Security

At the moment, SMEs mainly focus their cyber-strategies on the detection of cyber risks. However, it is of major importance that cyber-strategies of companies also focus on cyber defence. For example, if cybersecurity is integrated into the design of a system from the beginning, attacks can be prevented. SMEs should therefore shift their focus from the detection of cyber risks to threat prevention in order to keep their data fully secure.

4.1.4 Intellectual Property and Ownership of Data

Due to the nature of data, it is difficult to assign ownership. Data is neither tangible nor intangible, it is limitless and non-rivalrous, and its meaning and value are not static. Currently there is no particular framework to regulate the ownership of data.

---

[50]See also https://www.big-data-value.eu/the-big-data-challenge-3-takeaways-for-smes-and-startups-on-data-sharing-2/.

[51]See, for example, the SODA project, which enables multiparty computation (MPC) techniques for privacy-preserving data processing (https://www.soda-project.eu/).

The only means to establish ownership of data or protection of data is through the provisions of the GDPR, the DbD and the Trade Secrets Protection Directive, or through contracts by means of general contract law.

4.1.5 Use of Consumer Data: Importance of Transparency and Informed Consent

Consumer data plays an important role in the big data landscape. When companies collect consumer data, it is important that they are transparent towards consumers about what type of data they are collecting, and that consumers give informed consent. The previously mentioned Planet49[52] decision on cookie regulation is a case in point. The way forward for EU data companies aiming to use consumer data is to step from behind the curtain and be open about data practices and underlying algorithms. Taking citizens and consumers with them on a data journey, and truly developing inclusive digital services that take the necessary organisational and technical safeguards seriously from the start (and not after the fact), might seem to many business developers like the long and winding (and far more expensive) road. However, from the insights we have gathered from policymakers, data scientists and data workers, we strongly recommend looking at data policy not as a compliance-checklist exercise but as a strong attempt to create a human rights-based competitive and fair Digital Single Market.

Acknowledgements The research leading to these results received funding from the European Union's Horizon 2020 research and innovation programme under grant agreement no. 732630 (BDVe).

References

Botta, M., & Wiedemann, K. (2019). The interaction of EU competition, consumer, and data protection law in the digital economy: The regulatory dilemma in the Facebook odyssey. *The Antitrust Bulletin, 64*(3), 428–446.

Bushby, A. (2019). How deception can change cyber security defences. *Computer Fraud & Security, 2019*(1), 12–14.

Cavoukian, A. (2009). Privacy by design: The 7 foundational principles. *Information and privacy commissioner of Ontario, Canada, 5.*

European Parliament. (2019). *Directive (EU) 2019/790 of the European Parliament and of the Council of 17 April 2019 on copyright and related rights in the Digital Single Market and amending Directives 96/9/EC and 2001/29/EC.*

Graef, I., Husovec, M., & Purtova, N. (2018). Data portability and data control: lessons for an emerging concept in EU law. German Law Journal, 19(6), 1359–1398.

Hoepman, J. H. (2018). Privacy design strategies (the little blue book).

Kerr, O. S. (2012). The mosaic theory of the fourth amendment. *Michigan Law Review, 111*(3), 45.

---

[52]See C-673/17, ECLI:EU:C:2019:801.

Kettani, H., & Wainwright, P. (2019, March). On the top threats to cyber systems. In 2019 IEEE 2nd International Conference on Information and Computer Technologies (ICICT) (pp. 175–179). IEEE.

Lessig, L. (2009). *Code: And other laws of cyberspace*. ReadHowYouWant.com.

Manne, G. A., & Sperry, R. B. (2015). *The problems and perils of bootstrapping privacy and data into an antitrust framework, 2 CPI ANTITRUST CHRON. 1 (2015); Giuseppe colangelo & mariateresa maggiolino, data protection in attention markets: Protecting privacy through competition?, 8 J. EUR. C.*

Markopoulou, D., Papakonstantinou, V., & de Hert, P. (2019). The new EU cybersecurity framework: The NIS Directive, ENISA's role and the General Data Protection Regula-tion. Computer Law & Security Review, 35(6), 105336.

Purtova, N. (2018). The law of everything. Broad concept of personal data and future of EU data protection law. Law, Innovation and Technology, 10(1), 40–81.

Sharon, T., & Lucivero, F. (2019). Introduction to the special theme: The expansion of the health data ecosystem – Rethinking data ethics and governance. *Big Data & Society, 6*, 205395171985296. https://doi.org/10.1177/2053951719852969

Timan, T., & Mann, Z. Á. (2019). *Data protection in the era of artificial intelligence. Trends, existing solutions and recommendations for privacy-preserving technologies*. BDVA.

van Lieshout, M., & Emmert, S. (2018). RESPECT4U -- Privacy as innovation opportunity. In M. Medina, A. Mitrakas, K. Rannenberg, E. Schweighofer, & N. Tsouroulas (Eds.), *Privacy technologies and policy* (pp. 43–60). Cham: Springer International Publishing.

Wiedemann, K., & Botta, M. (2019). The interaction of EU competition, consumer and data protection law in the digital economy: The regulatory dilemma in the facebook odyssey. *The Antitrust Bulletin, 64*(3), 428–446.

Zillner, S., Curry, E., Metzger, A., Auer, S., & Seidl, R. (Eds.). (2017). *European Big Data Value Strategic Research & Innovation Agenda*. Retrieved from Big Data Value Association website: www.bdva.eu

Zillner, S., Bisset, D.,Milano,M., Curry, E., Hahn, T., Lafrenz, R., et al. (2020). Strategic research, innovation and deployment agenda - AI, data and robotics partnership. Third Release (3rd). Brussels: BDVA, euRobotics, ELLIS, EurAI and CLAIRE.

Part IV
Emerging Elements of Big Data Value

Data Economy 2.0: From Big Data Value to AI Value and a European Data Space

Sonja Zillner, Jon Ander Gomez, Ana García Robles, Thomas Hahn, Laure Le Bars, Milan Petkovic, and Edward Curry

Abstract Artificial intelligence (AI) has a tremendous potential to benefit European citizens, economy, environment and society and already demonstrated its potential to generate value in various applications and domains. From a data economy point of view, AI means algorithm-based and data-driven systems that enable machines with digital capabilities such as perception, reasoning, learning and even autonomous decision making to support people in real scenarios. Data ecosystems are an important driver for AI opportunities as they benefit from the significant growth of data volume and the rates at which it is generated. This chapter explores the opportunities and challenges of big data and AI in exploiting data ecosystems and creating AI value. The chapter describes the European AI framework as a foundation for deploying AI successfully and the critical need for a common European data space to power this vision.

Keywords Data ecosystem · Data spaces · Future directions · Big data value · Artificial intelligence

S. Zillner (✉)
Siemens AG, Munich, Germany
e-mail: sonja.zillner@siemens.com

J. A. Gomez
Universitat Politècnica de València, València, Spain

A. García Robles
Big Data Value Association, Bruxelles, Belgium

T. Hahn
Siemens AG, Erlangen, Germany

L. Le Bars
SAP, Paris, France

M. Petkovic
Philips and Eindhoven University of Technology, Eindhoven, The Netherlands

E. Curry
Insight SFI Research Centre for Data Analytics, NUI, Galway, Ireland

© The Author(s) 2021
E. Curry et al. (eds.), *The Elements of Big Data Value*,
https://doi.org/10.1007/978-3-030-68176-0_16

1 Introduction

Artificial intelligence (AI) has a tremendous potential to benefit European citizens, economy and society and already demonstrated its potential to generate value in various applications and domains. From a data economy point of view, AI means algorithm-based and data-driven systems that enable machines with digital capabilities such as perception, reasoning, learning and even autonomous decision making to support people in real scenarios. AI is based on a portfolio of technologies ranging from technologies for the perception and interpretation of information extracted from vast amounts of information data; software that draws conclusions and learns, adapts or adjusts parameters accordingly; and methods supporting human-based decision making or automated actions.

A critical driver for the emerging AI business opportunities is the significant growth of data volume and the rates at which it is generated. In 2014 the International Data Corporation (IDC) forecasted that in 2020 more than 16 zettabytes of useful data (16 trillion GB) will be made available, reflecting a growth of 236% per year from 2013 to 2020 (Turner et al. 2014). We know today that this forecast was far too low. According to a new update of the IDC Global Data Sphere[1] report, more than 59 zettabytes will be created, captured, copied and consumed. This growth is forecast to continue through 2024 with a 5-year compound annual growth rate (CAGR) of 26%. In consequence, this leads to an exponential growth, i.e. the amount of data being created over the next 3 years will be greater than the amount of data created over the past 30 years. The IDC report revealed that productivity/embedded data will be the fastest growing type of data with a CAGR of 40.3% from 2019 to 2024.

This chapter expands on a recent position paper (Zillner et al. 2018) from the Big Data Value Association community aligning it with recent developments on the European strategies for AI and data. It explores the potential of big data and AI in exploiting data ecosystems and creating new opportunities in AI application domains. It also addresses the ethical challenges associated with AI. It reflects on the need to develop trustworthy AI to mitigate conflicts and to avoid the adverse impact of deploying AI solutions. The European AI framework is described as a foundation for deploying AI successfully. The framework captures the processes and standards to deliver value that is acceptable to the users and citizens based on trust. Finally, the chapter describes the critical role of data and the need for common European data space to strengthen competitiveness across Europe.

---

[1]Worldwide Global DataSphere Forecast, 2020–2024: The COVID-19 Data Bump and the Future of Data Growth (Doc; #US44797920), IDC Report, https://www.idc.com/getdoc.jsp?containerId=IDC_P38353

2   The AI Value Opportunity

The current data explosion, combined with recent advances in computing power and connectivity, allows for an increasing amount of big data to be analysed anytime, anywhere. These technical advances enable addressing industrial relevant challenges and foster developing intelligent industrial application in a shorter time and with higher performance. AI will increase value creation from big data and its use to rapidly emerging B2B, B2G, G2C, G2B and B2C scenarios in many AI application domains. Machines and industrial processes which are supported by AI are augmenting human capacities in decision making and providing digital assistance in highly complex and critical processes.

Established industrial players are starting to implement AI in a wide range of industrial applications, such as complex image recognition, primarily for interpreting computed tomography (CT) and magnetic resonance imaging (MRI); autonomously learning, self-optimising industrial systems such as those used in gas turbines and wind farms; accurate forecasts of copper prices and expected power grid capacity utilisation; physical, autonomous systems for use in collaborative, adaptive, flexible manufacturing as part of Industry 4.0; and many more. At their heart, many of these AI systems are powered by using data-driven AI approaches such as deep learning. Exploiting data ecosystems is essential for AI (Curry and Sheth 2018).

In addition to the above, the EU Big Data Value Public-Private Partnership (BDV PPP) has established 32 projects with their respective experimentation playgrounds for the adoption of big data and AI solutions. In particular, the BDV PPP lighthouse projects play a fundamental role in piloting and showcasing value creation by big data with new data-driven AI applications in relevant sectors of great economic and societal value for Europe (Zillner et al. 2017). These projects demonstrate the essential role of data for AI, a few examples of which are as follows.

DataBio Data-Driven Bioeconomy takes on a major global challenge of how to ensure that raw materials for food, energy and biomaterials are sufficient in the era of climate change and population growth. Through big data and AI, DataBio is significantly enhancing raw material production in agriculture, forestry and fishery in a sustainable way. With its 26 pilots, DataBio strives to demonstrate annual increases in productivity ranging from 0.4% in forestry to 3.7% in agriculture and fishery (through savings in vessel costs). This makes up for a productivity gain of 20% over 5 years in agriculture and fishery. Big data pipelines and AI techniques are used in multiple pilots using the DataBio platform deployed in multiple clouds. The platform gathers Earth observation data from satellites and drones as well as IoT sources from in situ sensors in fields and vehicles. It manages and analyses the generated big data and presents it to the end users. These include farmers, foresters, fishers and many other stakeholders, supporting their operational decision making in a user-friendly way by providing them guidance in critical daily questions, such as what and where to grow, crop or fish; how to fight diseases; or when and how to harvest, cut or fish.

TransformingTransport Demonstrates in a realistic, measurable and replicable way the transformation that data-driven AI solutions can bring to the mobility and logistics market in Europe. Mobility and logistics are two of the most used industries in the world – contributing to approximately 15% of GDP and employment of over 11 million people in the EU-28 zone, i.e. 5% of the total workforce. The freight transport activities are projected to increase, since 2005, to 40% in 2030 and 80% in 2050. This will transform the current mobility and logistics processes to significantly higher efficiency and more profound impact. Structured into 13 different pilots, which cover areas of significant importance for the mobility and logistics sectors in Europe, TransformingTransport validates the technical and economic viability of big data-driven solutions for reshaping transport processes and services across Europe. To this end, TransformingTransport exploits access to industrial datasets from over 160 data sources, totalling over 164 terabytes of data. Initial evidence from TransformingTransport shows that big data-driven solutions using AI may deliver 13% improvement of operational efficiency[2]. The data-driven solutions in this project entail both traditional AI technology for descriptive analytics (such as support vector machines) and deep learning methods employed for predictive analytics (such as recurrent neural networks). With today's promising results using AI technology (e.g. 40% increase of prediction accuracy), we expect such AI solutions of advanced analytics as enablement to automated decision support for operational systems. These will establish the next level of efficiency and operational improvements in the mobility and transport sectors in Europe.

BigMedilytics In 2014, the EU-28 total healthcare expenditure was 1.39 trillion €. Spending is expected to increase to 30% by 2060, primarily due to a rapidly ageing population who typically suffer from chronic diseases. These figures indicate that current trends within the EU's healthcare sector are very unsustainable. The BigMedilytics Healthcare Lighthouse project demonstrates how the application of AI technologies on big data can help disrupt the healthcare sector so that quality, cost and access to care can all be improved. Market reports predict a CAGR of 40–50% for AI in healthcare, with a market size reaching to 22 billion by 2022 €. The project applies data-driven AI technologies over 12 pilots which focus on three main themes: (1) population health, (2) oncology and (3) industrialisation of healthcare. These themes effectively cover major disease groups, which cause 78% in mortality. AI-based methods together with privacy-preserving techniques are deployed to analyse large integrated datasets of more than 11 million patients, which cover a great range of key players in the healthcare sector (i.e. healthcare providers, healthtech companies, pharma and payers). The aim is to derive insights which can ultimately improve the efficiency of care providers while ensuring a high quality of care and protecting patients' privacy.

---

[2]According to the ALICE ETP, a 10% efficiency improvement will lead to EU cost savings of 100 B€.

Boost 4.0 Roland Berger[3] reveals that big data could see the manufacturing industry add a gross value worth 1.25 T€ or suffer a loss of 605 B€ in lost value if it fails to incorporate new data, connectivity, automation and digital customer interface enablers in their digital manufacturing processes. European Data Market (EDM) Monitoring 2018 reports manufacturing as data market value leader with 14B€. However, the manufacturing industry is losing up to 99% of the data value since evidence cannot be presented at the speed decisions are made. Boost 4.0 reflects on this challenge, leveraging a European industrial data space for connected Smart Factory 4.0 that requires collecting, analysing, transporting and storing vast amounts of data. The Factory 4.0 will use such industrial data spaces to drive efficiencies through the advanced use of data-driven AI capabilities. First, connecting workforce, assets and things to the Internet will enable the leveraging of predictive maintenance to reduce equipment downtime by 50% and increase production by 20%. Second, integration with non-production departments enables new business insights with savings of around 160 B€ only for the top 100 European manufacturers thanks to improved zero-defect manufacturing and the ability to adjust production in real time. Lastly, improved data visibility among companies enables collaborative business models.

DeepHealth Healthcare is one of the most important sectors for the EU economy, as previously highlighted by the BigMedilytics project. In order to contribute to the adoption and use of AI and data technologies in the health sector within the EU, the DeepHealth project has two main goals: one at the technological level and the other at the economical level. The objective at the technological level is the development of two software libraries that aim to be at the core of European data-driven AI-based solutions/applications/systems regardless of the sector. In the case of the DeepHealth project, the use of both libraries is focused on healthcare as the 14 use cases are based on medical datasets. These two libraries are the European Deep Learning Library and the European Image Processing Library. Both libraries will make intensive use of hybrid HPC + big data architectures to process data by parallelising algorithms to learn from data and to process digital images. The integration of both libraries into software platforms will considerably reduce the time for training deep learning-based models and contribute to the other objective concerning economy, which is to increase the productivity of IT experts (ML practitioners and data scientists) working in the health sector. IT experts giving support to doctors and other medical personnel are usually faced with the problem of image manipulation (i.e. transformations, segmentation, labelling and extraction of regions of interest) where they need to use a set of different libraries and toolkits from different developers to define a pipeline of operations on images. Installing and configuring different libraries and toolkits is repetitive hard work. The DeepHealth project focuses on facilitating the daily work of IT experts by integrating all the necessary functionalities into a toolkit, including the two libraries and a front-end for using them. The toolkit, one of the

---

[3]https://www.rolandberger.com/publications/publication_pdf/roland_berger_digital_transformation_of_industry_20150315.pdf

outcomes of this project, will facilitate the definition of pipelines of operations on images and testing distinct Deep Neural Network (DNN) topologies.

3 AI Challenges

The challenges for the adoption of AI range from new business models that need to be developed, trust in AI that needs to be established, ecosystems that are required to ensure that all partners are on board as well as access to the state-of-the-art AI technology. The following subsection will detail all these aspects.

3.1 Business Models

With the recent technical advances in digitalisation and AI, the real and the virtual worlds are continuously merging, which, again, leads to entire value-added chains being digitalised and integrated. For instance, in the manufacturing domain, all the way from the product design through to on-site customer services is digitalised. The increase in industrial data combined with AI technologies triggers a wide range of new technical applications with new forms of value propositions that shift the logic of how business is done. To capture these new types of value, data-driven AI-based solutions for the industry will require new business models. The design of data-driven AI-based business models needs to incorporate various perspectives ranging from customer and user needs and their willingness to pay for new AI-based solutions to data access and the optimal use of technologies while taking into account the currently established relationships with customers and partners. Successful AI-based business models are often based on strategic partnerships with two or more players establishing the basis for sustainable win-win situations through transparent ways of sharing resources, investments, risks, data and value.

3.2 Trust in AI

With AI disruptive potential, there are significant ethical implications on the use of AI and autonomous machines and their applications for decision support. Future AI research needs to be guided by new and established ethical norms. Although the current AI methods have already achieved encouraging results and technical break-throughs, results in individual cases show some concerning signs of unpredictable behaviour. Recent studies showed that the state-of-the-art deep neural networks are vulnerable to adversarial examples or are unable to cope with new unknown situations. To overcome those shortcomings, for any critical applications (where "critical" needs to be defined with clarity), one should be able to explain how AI applications came to a specific result ("explainable AI"). Explainability will ensure the commitment of industrial users to measurable ethical values and principles when using AI. One should foster responsible technological development (e.g. avoid bias) and enhance transparency in such exercise. Explainable AI should provide transparency about input data as well as the "rationale" behind the algorithm usage leading to the specific output. The algorithm itself need not necessarily be revealed in this case.

The purpose of AI, data analytics, machine and deep learning algorithms is not only to boost the effectiveness and quality of the services which are delivered to the client but also to ensure that no negative impact is brought as a result of deploying AI solutions in critical applications. For instance, ensuring that AI-powered systems treat different social groups fairly is a matter of growing concern for societies. FAT-ML, i.e. Fairness, Accountability and Transparency in Machine Learning, is an emerging important multidisciplinary field of research (Barocas and Selbst 2016; Carmichael et al. 2016). Related areas including big data for social good, humanistic AI and the broader field of AI ethics have only recently started exploring complex multi-faceted problems, e.g. fostering the creation of social and human-centred values by adding new parameters and enhanced objective functions and restrictions.

Trusted AI involves the simultaneous achievement of objectives that are often in conflict. One critical challenge stems from the ever-increasing collection and analysis of personal data and the crucial requirement for protecting the privacy of all involved data subjects as well as protecting commercially sensitive data of associated organisations and enterprises. There are some approaches attempting to address this issue, including security-oriented (e.g. machine learning on encrypted data with secure computation technologies), privacy-enhancing (e.g. detect privacy risks and alert users) and distributed processing (e.g. federated machine learning) ones. As all privacy approaches add cost and complexity to AI systems, the optimal trade-offs without adding considerable complexity are important research challenges to be addressed. A critical problem is presented by the difficulty to allocate and distribute liabilities and responsibilities across assemblages of continuously evolving autonomous systems with different goals and requirements. While existing risk-based, performance-driven, progressive and proportionate regulatory approaches have promised a more flexible, adaptive regulatory environment, stakeholders are increasingly struggling to deal with the complexities of multi-level, multi-stakeholder and multi-jurisdictional environments within which AI is being developed. Multidisciplinary efforts at both international and regional levels are therefore required to ensure the establishment of an enabling environment where trust and safety of AI are dealt with from a global governance perspective. Existing tools from other domains, such as regulatory sandboxing, testing environments for autonomous vehicles and so forth, could serve as incubators for establishing new policy; legal, ethical and regulatory norms; and measures of trusted AI in Europe.

3.3 Ecosystem

For developing sustainable data-driven AI businesses, it will be central to consider a value-network perspective, i.e. looking at the entire ecosystem of companies involved in value networks. The ecosystems will be increasingly shaped by platform providers who offer their platform based on open standards to their customers. European economic success and sustainability in AI will be driven by ecosystems which need to have a critical size. Speed is a necessity for the development of these ecosystems.

Data sharing and trading are essential ecosystem enablers in the data economy, although secure and personal data present particular challenges for the free flow of data (OECD 2014; Curry 2016). The EU has made considerable efforts in the direction of defining and building data-sharing platforms. However, there is still a significant way to go to guarantee AI practitioners' access to large volumes of data necessary for them to compete. Further actions must be carried out to develop data for AI platforms, such as awareness campaigns to foster the idea of sharing their data in companies and research centres, and incentives for parties to join data exchange/sharing initiatives. To overcome barriers to data sharing for AI, frameworks for data governance are needed to be established that will enable all parties to retain digital sovereignty over their data assets. Obviously, data sharing must be done, from the legal point of view, by preserving privacy by anonymising all the attributes referring to people, and respecting commercial interests (IPR, competition, ownership) by providing solutions to deal with technical and legal challenges such as data governance and trust-enhancing protocols for data sharing/exchange, decentralised storage and federated machine learning. And from the technical perspective, data sharing is done by (1) designing information systems (i.e. databases) in order to ensure the future use of the datasets with minimal efforts in terms of cleaning data or defining ontologies, by (2) transforming and mapping data sources taking into account the variety and heterogeneity of data in order to gain interoperability and (3) by ensuring the veracity of shared data according to quality standards.

Open AI platforms will play a central role in the data economy at three different levels: (1) definition of protocols and procedures for uploading datasets into data-sharing platforms, (2) definition of standard APIs for different libraries (AI/ML, image processing, etc.) and (3) the design and development of a web-based user interface to allow data scientists to upload data, to define pipelines of transformations to apply to data before training and testing AI models, and to choose among a wide range of AI techniques to run on the same data to carry out comparative studies. Successful European Open AI platforms require the contribution of many agents, such as universities, research centres, large companies and SMEs.

By relying on data-sharing platforms, data innovation spaces, Open AI platforms and digital innovation hubs (DIH), industrial collaborations between large and small players can be supported at different levels: technical, business model and ecosystem while, at the same time, ensuring data and technology access for SMEs and start-ups.

To complement technical and legal infrastructures for the free and controlled flow of industrial data, the building and nurturing of industrial ecosystems fostering data-driven industrial cooperation across value chains and therefore networks will have a critical impact.

Enabling data-driven AI-based business models across value chains and beyond organisational boundaries will significantly maximise the impact of the data economy to power European AI industries. Mechanisms that overcome the lack of data interoperability and foster data sharing and exchange need to be defined and implemented. Notwithstanding, the creation of and compliance with binding international standards is of central importance to the sustainability of solutions, and thus it is a competitive strength. Preferably these standards should be global – because only global standards ultimately lead to success in a world that is more and more networked and where multinational companies make significant contributions to national GDPs.

3.4 Technology

Success in industrial AI application relies on the combination of a wide range of technologies, such as:

Advanced Data Analytics: Many data analytics techniques require adaptation for running more efficiently when working with large datasets. These improvements rely on the development of new algorithms and new ways of transforming data. Additionally, with self-adjusting AI systems, machines will become self-operating by making decisions according to specific contexts to dynamically optimise performance, beyond the level of efficiency the same AI systems can reach when adjusted by humans.

Hybrid AI: To derive value from domain knowledge, methods from both symbolic AI and statistical AI need to be combined to give the maximum potential and usability of AI-based applications. This combination of knowledge graphs and statistical AI techniques supports AI solutions concerning (1) data quality issues, (2) better integration and use of training data, (3) explainable AI (no black-box solutions) and, finally, (4) the mutual fertilisation of semantic technologies and AI techniques towards self-optimising machines.

Distributed AI/Edge Analytics: The increasing number of intelligent devices at the edge is one of the critical elements for AI. We are now at the point where the collective computing power at the edge (outside data centres) is surpassing the centralised computing capacity. As computing capabilities in the cloud and at the edge are increasingly intertwined, we will see the emergence of distributed AI, i.e. new research approaches that will bring the AI at the core of most future data analytics-based applications.

Hardware Optimised to AI: Specialised hardware devices and architectures have an increasingly strong impact both on the AI learning process on applications with large datasets and on the predicting/inference task, in particular when fast decisions and actuation matter. The designs of powerful and affordable systems on both sides of the AI data flow are an important research topic. Nevertheless, AI algorithms need to be optimised to the specific hardware capabilities.

Multilingual AI: Humans use language to express, store, learn and exchange information. AI-based multilingual technologies can extract knowledge out of tremendous amounts of written and spoken language data. Processing of multilingual data empowers a new generation of AI-based applications such as question answering systems, high-quality neural machine translation, speech processing in real time and contextually and emotionally aware virtual assistants for human-computer interaction.

4 Towards an AI, Data and Robotics Ecosystem

The Big Data Value Association (BDVA) and the European Robotics Association (euRobotics) have developed a joint Strategic Research, Innovation and Deployment Agenda (SRIDA) for an AI, Data and Robotics Partnership in Europe (S Zillner et al. 2019). This is in response to the Commission Communication on AI published in December 2018. Deploying AI successfully in Europe requires an integrated landscape for its adoption and the development of AI based on Europe's unique characteristics. In September 2020 the BDVA, CLAIRE, ELLIS, EurAI and euRobotics are pleased to announce the official release of the joint Strategic Research Innovation and Deployment Agenda (SRIDA) for the AI, Data and Robotics Partnership which unifies the strategic focus of each of the three disciplines engaged in creating the Partnership.

Together these associations have proposed a vision for an AI, Data and Robotics Partnership: "The Vision of the Partnership is to boost European industrial competitiveness, societal wellbeing and environmental aspects to lead the world in developing and deploying value-driven trustworthy AI, Data and Robotics based on fundamental European rights, principles and values".

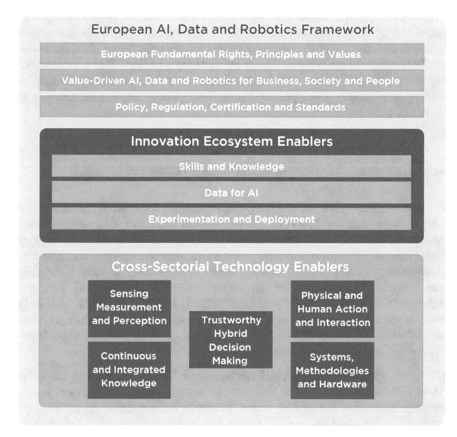

Fig. 1 European AI, Data and Robotics Framework and Enablers (Zillner et al. 2020) (by European Commission licensed under CC BY 4.0)

To deliver on the vision of the AI, Data and Robotic Partnership, it is important to engage with a broad range of stakeholders. Each collaborative stakeholder brings a vital element to the functioning of the Partnership and injects critical capability into the ecosystem created around AI, Data and Robotics by the Partnership. The mobilisation of the European AI, Data and Robotics Ecosystem is one of the core goals of the Partnership. The Partnership needs to form part of a wider ecosystem of collaborations that cover all aspects of the technology application landscape in Europe. Many of these collaborations will rely on AI, Data and Robotics as critical enablers to their endeavours. Both horizontal (technology) and vertical (application) collaborations will intersect within an AI, Data and Robotics Ecosystem.

Figure 1 sets out the context for the operation of the AI, Data and Robotics. It clusters the primary areas of importance for AI, Data and Robotics research, innovation and deployment into three overarching areas of interest. *European AI, Data and Robotics Framework* represents the legal and societal fabric that underpins the impact of AI on stakeholders and users of the products and services that businesses will provide. The AI, Data and Robotics *Innovation Ecosystem Enablers* represent the essential ingredients for effective innovation and deployment to take place. Finally, the *Cross-Sectorial AI, Data and Robotics Technology Enablers* represent the core technical competencies that are essential for the development of AI, Data and Robotics systems. The remainder of this section offers a summary of the European AI, Data and Robotics Framework, which is the core of the SRIDA (Zillner et al. 2020) developed by the BDVA, euRobotics, ELLIS, EurAI and CLAIRE.

4.1 European AI, Data and Robotics Framework

AI, Data and Robotics work within a broad framework that sets out boundaries and limitations on their use. In specific sectors, such as healthcare, they operate within the ethical, legal and societal contexts and within regulatory regimes that can vary across Europe. Products and services based on AI, Data and Robotics are shaped by certification processes and standards and impact on users to deliver value compatible with European rights, principles and values. Critical to deploying AI, Data and Robotics is its acceptance by users and citizens, and this acceptance can only come when they can assign trust. This section explores this European AI, Data and Robotics Framework (Zillner et al. 2020) within which research, design, development and deployment must work.

European Fundamental Rights, Principles and Values On the one hand, the recent advances in AI, Data and Robotics technology and applications have fundamentally challenged the ethical values, human rights and safety in the EU and globally. On the other hand, AI, Data and Robotics offer enormous possibilities to raise productivity, address societal and environmental challenges and enhance the quality of life for everyone. The public acceptance of AI, Data and Robotics is a prerequisite for it being trustworthy, ethical and secure, and without public acceptance, its full benefit cannot be realised. The European Commission has already taken action and formulated in its recent communications[4] a vision for an ethical, secure and cutting-edge *AI made in Europe* designed to ensure AI, Data and Robotics operate within an appropriate ethical and legal framework that embeds European values. The Partnership (Zillner et al. 2020) will:

- *Facilitate a multi-stakeholder dialogue* and consensus building around the core issue of trustworthiness by guiding and shaping a common AI, Data and Robotics agenda and fostering research and innovation on trustworthy technologies.

---

[4]Communication Artificial Intelligence on 25 April 2018 (see https://ec.europa.eu/digital-single-market/en/news/communication-artificial-intelligence-europe) and Communication Artificial Intelligence on 7 December 2018 (see https://ec.europa.eu/commission/news/artificial-intelligence-2018-dec-07_en)

- *Seek to promote a common understanding* among stakeholders of the European AI, Data and Robotics ecosystem on the fundamental, rights and values, so that each sector and community are informed and aware of the potential of AI, Data and Robotics as well as the risks and limitations of the current technology and will develop guidance in the responsible implementation of AI, Data and Robotics.
- Establish the basis for identifying and expressing a *European strategic viewpoint* on rights, principles and values by providing clear links to relevant regulation, certification and standardisation.

Capturing Value for Business, Society and People Technical advances in AI, Data and Robotics are now enabling real-world applications. These are leading to improved or new value-added chains being developed and integrated. To capture these new forms of value, AI-based solutions may require innovative business models that redefine the way stakeholders share investments, risk, know-how and data and, consequently, value. This alteration of value flow in existing markets is disruptive and requires stakeholders to alter their business models and revenue streams. These adjustments require new skills, infrastructure and knowledge, and organisations may have to buy in expertise or share data and domain know-how to succeed. This may be incredibly difficult if their underlying digitalisation skills, a prerequisite for AI, Data and Robotics adoption, are weak.

Even incremental improvements or more considerable changes carry risks and may create a reluctance to adopt AI, Data and Robotics. There may be little or no support for change within an organisation or value chain, especially when coupled with a lack of expertise. Successful adoption of AI, Data and Robotics solutions requires a dialogue between the different stakeholders to design a well-balanced and sustainable value network incorporating all stakeholder's interests, roles and assets.

To support the adoption of AI, Data and Robotics applications, the Partnership (Zillner et al. 2020) will stimulate discussions to align supply and demand perspectives of the diverse AI, Data and Robotics value-network partners, with the main focus on application areas and sectors that:

- Are crucial for the European economy
- Relate to critical infrastructure
- Have a social or environmental impact
- Can increase European competitiveness in AI

Policy, Regulation, Certification and Standards (PRCS) The adoption of AI, Data and Robotics depends on a legal framework of approval built on regulation, partly driven by policy, and an array of certification processes and standards driven by industry. As AI, Data and Robotics are deployed successfully in new market areas, regulation and certification can lag behind, thereby creating barriers to adoption.

Similarly, a lack of standards and associated certification and validation methods can hold back the deployment and the creation of supply chains and therefore slow market uptake. In some areas of AI, Data and Robotics, the market will move ahead and wait for regulation to react, but in many application areas existing regulation can present a barrier to adoption and deployment – most notably in applications where there is a close interaction with people, either digitally or physically, or where AI, Data and Robotics are operating in safety or privacy critical environments.

PRCS issues are likely to become a primary area of activity for the AI, Data and Robotics Partnership. Increasingly it is regulation that is the primary lever for the adoption of AI/Data/Robotics systems, particularly when physical interactions are involved or where privacy is a concern. Similarly, the development of standards, particularly around data exchange and interoperability, will be key to the creation of a European AI, Data and Robotics marketplace. Establishing ways that ensure conformity assessments of AI, Data and Robotics will underpin the development of trust that is essential for acceptance and therefore adoption. In addition, the Partnership also has a role to advise on regulation that creates or has the potential to create unnecessary barriers to innovation in AI, Data and Robotics. The Partnership (Zillner et al. 2020) will need to carry out the following activities to progress PRCS issues:

- Identify key stakeholders in each area of PRCS and ensure there is good connectivity between them and to the AI, Data and Robotics Ecosystem.
- Work with stakeholders and the emerging AI, Data and Robotics Ecosystem infrastructure (digital innovation hubs, pilots and data spaces) to identify key issues that impact on adoption and deployment in each major sector.
- Promote best practice in deployment regarding PRCS issues and provide signposts to demonstrators and processes that can accelerate uptake.
- Support and collaborate in standardisation initiatives and the harmonisation of regulation across Europe to create a level AI, Data and Robotics single marketplace and connect with European and global standards and regulatory bodies.
- Foster the responsible testing of AI, Data and Robotics innovation in regulatory sandbox environments.
- Consolidate recommendations towards policy changes and provide support for related impact assessment processes.
- Drive European thinking and needs towards international standardisation bodies.

4.2   Innovation Ecosystem Enablers

The Innovation Ecosystem Enablers are essential ingredients for success in the innovation system. They represent resources that underlie all innovation activities across the sectors and along the innovation chain from research to deployment. Each represents a key area of interest and activity for the Partnership (Zillner et al. 2020), and each presents unique challenges to the rapid development of European AI, Data and Robotics.

Skills and Knowledge As traditional industry sectors undergo an AI, Data and Robotics transformation, so too must their workforces. There is a clear skills gap when it comes to AI, Data and Robotics. However, while there are shortages of people with specific technical skills or domain knowledge, there is also the need to train interdisciplinary experts. AI, Data and Robotics experts need insight into the ethical consequences posed by AI, by machine autonomy and by big data automated processes and services; they need a good understanding of the legal and regulatory landscape, for example, GDPR, and the need to develop and embed trustworthiness, dependability, safety and privacy through the development of appropriate technology.

The Partnership will work through its network to ensure that all stakeholders along the value chain, including citizens and users, have the understanding and skills to work with AI-enabled systems, in the workplace, in the home and online. The Partnership has a critical role to play in bringing together the key stakeholders: academia, industry, professional trainers, formal and informal education networks and policymakers. These collaborations will need to examine regional strengths and needs in terms of skills across the skill spectrum, both technical and non-technical. It is critical to ensure that the skill pipeline is maintained to ensure the AI, Data and Robotics transformation of Europe is not held back. Some concrete actions the Partnership (Zillner et al. 2020) will focus on are as follows:

- Promote equality and diversity within the current and future workforce and ensure diversity and balance in the educational opportunities that drive the skill pipeline.
- Ensure the alignment of curricula and training programmes for AI, Data and Robotics professionals with industry needs.
- Establish AI, Data and Robotics skills, both technical and non-technical, through certification mechanisms for university courses, professional and vocational training, and informal learning.
- Development of complementary short courses related to artificial intelligence aimed at decision makers in industry and public administration and those wishing to upgrade, enhance or acquire AI-based skills.
- Support for secondary education and adult learning to cover STEM skills including the ethical, social and business aspects of AI together with the changing nature of work as well as support for vocational training.

Data for AI In order to further develop AI, Data and Robotics technologies and meet expectations, large volumes of cross-sectoral, unbiased, high-quality and trustworthy data need to be made available. Data spaces, platforms and marketplaces are enablers, the key to unleashing the potential of such data. There are however important business, organisational and legal constraints that can block this scenario such as the lack of motivation to share data due to ownership concerns, loss of control, lack of trust, the lack of foresight in not understanding the value of data or its sharing potential, the lack of data valuation standards in marketplaces, the legal blocks to the free flow of data and the uncertainty around data policies. Additionally, significant technical challenges such as interoperability, data verification and provenance support, quality and accuracy, decentralised data sharing and processing architectures, and maturity and uptake of privacy-preserving technologies for big data have a direct impact on the data made available for sharing. The Partnership (Zillner et al. 2020) will:

- Create the conditions for the development of **trusted European data-sharing frameworks** to enable new data value chain opportunities, building upon existing initiatives and investments (data platforms, i-spaces, big data innovation hubs). Data value chains handling a mix of personal, non-personal, proprietary, closed and open research data need to be supported. The Partnership would promote open datasets and new open benchmarks for AI algorithms, subject to quality validation from both software engineering and functional viewpoints.
- Define specific measures to **incorporate data sharing at the core of the data lifecycle** for greater access to data, encouraging collaboration between data value chain actors in both directions along the chain and across different sectors. Additionally, the Partnership will provide **supportive measures for European businesses** to safely embrace new technologies, practices and policies.
- Facilitate **coordination and harmonisation of member states efforts** and realise the potential of European-wide AI-digital services in the face of global competition. It would guide and influence standards concerning tools for data sharing, privacy preservation, quality verification, collaboration and interaction. Promote standardisation at European level but maintain collaboration with international initiatives for made-in-Europe AI to be adopted worldwide.

Experimentation and Deployment They are central levers for AI/Data/Robotics-based innovation because of the need to deploy in complex physical and digital environments. This includes safe environments for experimentation to explore the data value as well as to test the operation of autonomous actors. AI/Data/Robotics-driven innovations rely on the interplay of different assets, such as data, robotics, algorithms and infrastructure. For that reason, *cooperation* with other partners is central to gaining access to complementary assets. This includes access to the AI, Data and Robotics Ecosystem covering AI platform providers, data scientists, data owners, providers, consumers, specialised consultancy, etc. The Partnership (Zillner et al. 2020) will:

- Stimulate cooperation between all stakeholders in the AI, Data and Robotics value chain around experimentation and deployment.
- Enable access to infrastructure and tools in combination with datasets covering the whole value chain as a basis for doing experiments to support development and deployment.
- Support the creation and linking of DIHs, centres of excellence and all other EC initiatives.
- Support AI/Data/Robotics-based incubators as well as testbed developments as well as promote initiatives that enable SME access to infrastructure and tools at low cost.

- Foster set-ups that bring together industrial users with research excellence and domain experts with data science skills, aiming to fill the gaps between domain/business and technical expertise.

4.3   Cross-Sectorial AI, Data and Robotics Technology Enablers

The last part of the framework is the *technology enablers* for building successful AI products and services. Each embodies the concept that AI, Data and Robotics need to work in unison to achieve optimal function and performance. They represent the fundamental building blocks needed to create AI, Data and Robotics systems of all types.

The sensing and perception and knowledge and learning technology enablers create the data and knowledge on which decisions are made. These are used by the reasoning and decision-making technologies to deliver: edge and cloud based decision making, planning, search and optimisation in systems and the multi-layered decision making necessary for AI, Data and Robotic systems operating in complex environments.

Action and interaction cover the challenges of human interaction, machine to machine interoperation and machine interaction with the human environment. These multiple forms of action and interaction create complex challenges that range from the optimisation of performance to physical safety and social interaction with humans in unstructured and multi-faceted environments.

Systems, hardware, methods and tools provide the technologies that enable the construction and configuring of systems, whether they are built purely on data or on autonomous robots. These tools, methods and processes integrate AI, Data and Robotics technologies into systems and are responsible for ensuring that core system properties and characteristics such as safety, robustness, dependability and trustworthiness can be integrated into the design cycle and tested, validated and ultimately certified for use.

Each technical area overlaps with the other; there are no clear boundaries. Indeed, exciting advances are most often made in the intersections between these five areas and the system-level synergies that emerge from the interconnections between them.

5   A Common European Data Space

For European data economy to develop further and meet expectations, large volumes of cross-sectoral, unbiased, high-quality and trustworthy data need to be made available. The exploration of ethical, secure and trustworthy legal, regulatory and governance frameworks is needed. European values, e.g. democracy, privacy safeguards and equal opportunities, can become the trademark of European data economy technologies, products and practices. Rather than be seen as restrictive, these values enforced by legislation should be considered as a unique competitive advantage in the global data marketplace.

To reflect this new reality, the European data strategy was revised in 2020 to set out a vision for the EU to become a role model for a data-driven society and to create a single market for data to ensure Europe's global competitiveness and data sovereignty. As highlighted by EU Commissioner Thierry Breton[5]: "To be ahead of the curve, we need to develop suitable European infrastructures allowing the storage, the use, and the creation of data-based applications or Artificial Intelligence services. I consider this as a major issue of Europe's digital sovereignty".

Alignment and integration of established data-sharing technologies and solutions, and further developments in architectures and governance models aiming to unlock data silos, would enable data analytics across a European data-sharing ecosystem. This will enable AI-enhanced digital services to make analysis and predictions on European-wide data, thereby combining data and service economies. New business models will help to exploit the value of those data assets through the implementation of AI among participating stakeholders including industry; local, national and European authorities and institutions; research entities; and even private individuals.

As part of the revised data strategy, common European data spaces will ensure that more data becomes available for use in the economy and society while keeping companies and individuals who generate the data in control (*Communication: A European strategy for data* 2020). Platform approaches have proved successful in many areas of technology (Gawer and Cusumano 2014), from supporting transactions among buyers and sellers in marketplaces (e.g. Amazon), innovation platforms that provide a foundation on which to develop complementary products or services (e.g. Windows), to integrated platforms which are a combined transaction and innovation platform (e.g. Android and the Play Store). The idea of large-scale "data" platforms has been touted as a possible next step to support data ecosystems (Curry and Sheth 2018). An ecosystem data platform would have to support continuous, coordinated data flows, seamlessly moving data among systems (Curry and Ojo 2020). Data spaces, platforms and marketplaces are enablers, the key to unleashing the potential of such data. Significant technical challenges such as interoperability, data verification and provenance support, quality and accuracy, decentralised data sharing and processing architectures, and maturity and uptake of privacy-preserving technologies for big data have a direct impact on the data made available for sharing.

The nine initial common European data spaces (Fig. 2) will be the following:

- An **industrial** data space, to support the competitiveness and performance of the EU's industry

---

[5] 15 July 2020: https://ec.europa.eu/commission/presscorner/detail/en/SPEECH_20_1362

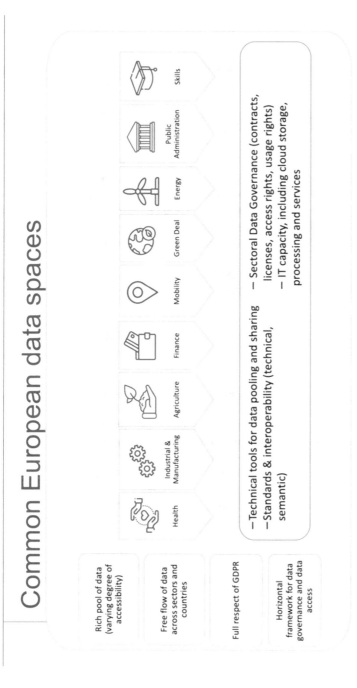

Fig. 2 Common European data spaces (*Communication: A European strategy for data* 2020) (by European Commission licensed under CC BY 4.0)

- A **Green Deal** data space, to use the major potential of data in support of the Green Deal priority actions on issues such as climate change, circular economy, pollution, biodiversity and deforestation
- A **mobility** data space, to position Europe at the forefront of the development of an intelligent transport system
- A **health** data space, essential for advances in preventing, detecting and treating diseases as well as for informed, evidence-based decisions to improve the healthcare systems
- A **financial** data space, to stimulate innovation, market transparency, sustainable finance as well as access to finance for European businesses and a more integrated market
- An **energy** data space, to promote a more substantial availability and cross-sector sharing of data, in a customer-centric, secure and trustworthy manner
- An a**griculture** data space, to enhance the sustainability performance and competitiveness of the agricultural sector through the processing and analysis of data
- Data spaces for **public administrations**, to improve transparency, accountability and efficiency of public spending, fighting corruption, both at EU and national levels
- A **skills** data space, to reduce the skills mismatches between the education and training systems and the labour market needs

6 Summary

AI, Data and Robotics have a tremendous potential to benefit citizens, economy, environment and society. AI, Data and Robotics techniques can extract new value from data to enable data-driven systems with digital capabilities such as perception, reasoning, learning and even autonomous decision making. Data ecosystems are an important driver for data-driven AI to exploit the continued growth of data. We need to establish a solid European AI, Data and Robotics framework as a foundation for deploying AI, Data and Robotics successfully and a common European data space to power this vision. Developing both of these elements together is critical to maximising the future potential of AI and data in Europe.

Acknowledgements Editor and contributors to the BDVA position paper on data-driven AI: Andreas Metzger (paluno, University of Duisburg-Essen), Zoheir Sabeur (University of Southampton), Martin Kaltenböck (Semantic Web Company), Marija Despenic (Philips), Cai Södergard (VTT), Natalie Bertels/Ivo Emanuilov (imec-CiTiP-KU Leuven), Simon Scerri (Fraunhofer), Andrejs Vasiljevs/Tatjana Gornosttaja (Tilde), Axel Ngongo (Technical University of Paderborn), Freek Bomhof (TNO), Yiannis Kompatasiaris and Symeon Papadopoulos (ITI Greece), Nozhae Boujemaa (Inria), Juan-Carlos Perez-Cortes (ITI Valencia), Oscar Lazaro (Innovalia Association).

References

Barocas, S., & Selbst, A. (2016). Big data's disparate impact. *California Law Review*. https://doi.org/10.15779/Z38BG31

Carmichael, L., Stalla-Bourdillon, S., & Staab, S. (2016). Data mining and automated discrimination: A mixed legal/technical perspective. *IEEE Intelligent Systems, 31*(6), 51–55. https://doi.org/10.1109/MIS.2016.96

Communication: A European strategy for data. (2020). Retrieved from https://ec.europa.eu/info/sites/info/files/communication-european-strategy-data-19feb2020_en.pdf

Curry, E. (2016). The big data value chain: Definitions, concepts, and theoretical approaches. In J. M. Cavanillas, E. Curry, & W. Wahlster (Eds.), *New horizons for a data-driven economy: A roadmap for usage and exploitation of big data in Europe.* https://doi.org/10.1007/978-3-319-21569-3_3

Curry, E., & Ojo, A. (2020). Enabling knowledge flows in an intelligent systems data ecosystem. In *Real-time Linked Dataspaces* (pp. 15–43). Berlin: Springer. https://doi.org/10.1007/978-3-030-29665-0_2

Curry, E., & Sheth, A. (2018). Next-generation smart environments: From system of systems to data ecosystems. *IEEE Intelligent Systems, 33*(3), 69–76. https://doi.org/10.1109/MIS.2018.033001418

Gawer, A., & Cusumano, M. A. (2014). Industry platforms and ecosystem innovation. *Journal of Product Innovation Management, 31*(3), 417–433. https://doi.org/10.1111/jpim.12105

OECD. (2014). *Data-driven Innovation for Growth and Well-being.*

Turner, V., Gantz, J. F., Reinsel, D., & Minton, S. (2014). The digital universe of opportunities: Rich data and the increasing value of the internet of things. *Report from IDC for EMC.*

Zillner, S., Curry, E., Metzger, A., Auer, S., & Seidl, R. (Eds.). (2017). *European big data value strategic research & innovation agenda.* Retrieved from Big Data Value Association website: www.bdva.eu

Zillner, S., Gomez, J. A., García Robles, A., & Curry, E. (Eds.). (2018). *Data-driven artificial intelligence for european economic competitiveness and societal progress: BDVA position statement.* Retrieved from www.bdva.eu/sites/default/files/AI-Position-Statement-BDVA-Final-12112018.pdf

Zillner, S, Bisset, D., García Robles, A., Hahn, T., Lafrenz, R., Liepert, B., & Curry, E. (2019). *Strategic research, innovation and deployment agenda for an AI PPP: A focal point for collaboration on artificial intelligence, data and robotics.* Retrieved from http://www.bdva.eu/sites/default/files/AIPPPSRIDA-ConsultationVersion-June2019-Onlineversion.pdf

Zillner, S., Bisset, D., Milano, M., Curry, E., Hahn, T., Lafrenz, R., et al. (2020). *Strategic research, innovation and deployment agenda - AI, data and robotics partnership. Third Release (3rd).* Brussels: BDVA, euRobotics, ELLIS, EurAI and CLAIRE.

Correction to: A Roadmap to Drive Adoption of Data Ecosystems

Sonja Zillner, Laure Le Bars, Nuria de Lama, Simon Scerri,
Ana García Robles, Marie Claire Tonna, Jim Kenneally, Dirk Mayer,
Thomas Hahn, Södergård Caj, Robert Seidl, Davide Dalle Carbonare,
and Edward Curry

Correction to:
Chapter 3 in: E. Curry et al. (eds.),
The Elements of Big Data Value,
https://doi.org/10.1007/978-3-030-68176-0_3

The original version of the chapter was inadvertently published with an error. The affiliation of the author Davide Dalle Carbonare has now been corrected to "Engineering Ingegneria Informatica, Rome, Italy".

The updated online version of this chapter can be found at
https://doi.org/10.1007/978-3-030-68176-0_3